MW00837274

Selected Titles in This Series

38 **Elton P. Hsu,** Stochastic analysis on manifolds, 2002

37 **Hershel M. Farkas and Irwin Kra,** Theta constants, Riemann surfaces and the modular group, 2001

36 **Martin Schechter,** Principles of functional analysis, second edition, 2002

35 **James F. Davis and Paul Kirk,** Lecture notes in algebraic topology, 2001

34 **Sigurdur Helgason,** Differential geometry, Lie groups, and symmetric spaces, 2001

33 **Dmitri Burago, Yuri Burago, and Sergei Ivanov,** A course in metric geometry, 2001

32 **Robert G. Bartle,** A modern theory of integration, 2001

31 **Ralf Korn and Elke Korn,** Option pricing and portfolio optimization: Modern methods of financial mathematics, 2001

30 **J. C. McConnell and J. C. Robson,** Noncommutative Noetherian rings, 2001

29 **Javier Duoandikoetxea,** Fourier analysis, 2001

28 **Liviu I. Nicolaescu,** Notes on Seiberg-Witten theory, 2000

27 **Thierry Aubin,** A course in differential geometry, 2001

26 **Rolf Berndt,** An introduction to symplectic geometry, 2001

25 **Thomas Friedrich,** Dirac operators in Riemannian geometry, 2000

24 **Helmut Koch,** Number theory: Algebraic numbers and functions, 2000

23 **Alberto Candel and Lawrence Conlon,** Foliations I, 2000

22 **Günter R. Krause and Thomas H. Lenagan,** Growth of algebras and Gelfand-Kirillov dimension, 2000

21 **John B. Conway,** A course in operator theory, 2000

20 **Robert E. Gompf and András I. Stipsicz,** 4-manifolds and Kirby calculus, 1999

19 **Lawrence C. Evans,** Partial differential equations, 1998

18 **Winfried Just and Martin Weese,** Discovering modern set theory. II: Set-theoretic tools for every mathematician, 1997

17 **Henryk Iwaniec,** Topics in classical automorphic forms, 1997

16 **Richard V. Kadison and John R. Ringrose,** Fundamentals of the theory of operator algebras. Volume II: Advanced theory, 1997

15 **Richard V. Kadison and John R. Ringrose,** Fundamentals of the theory of operator algebras. Volume I: Elementary theory, 1997

14 **Elliott H. Lieb and Michael Loss,** Analysis, 1997

13 **Paul C. Shields,** The ergodic theory of discrete sample paths, 1996

12 **N. V. Krylov,** Lectures on elliptic and parabolic equations in Hölder spaces, 1996

11 **Jacques Dixmier,** Enveloping algebras, 1996 Printing

10 **Barry Simon,** Representations of finite and compact groups, 1996

9 **Dino Lorenzini,** An invitation to arithmetic geometry, 1996

8 **Winfried Just and Martin Weese,** Discovering modern set theory. I: The basics, 1996

7 **Gerald J. Janusz,** Algebraic number fields, second edition, 1996

6 **Jens Carsten Jantzen,** Lectures on quantum groups, 1996

5 **Rick Miranda,** Algebraic curves and Riemann surfaces, 1995

4 **Russell A. Gordon,** The integrals of Lebesgue, Denjoy, Perron, and Henstock, 1994

3 **William W. Adams and Philippe Loustaunau,** An introduction to Gröbner bases, 1994

2 **Jack Graver, Brigitte Servatius, and Herman Servatius,** Combinatorial rigidity, 1993

1 **Ethan Akin,** The general topology of dynamical systems, 1993

Stochastic Analysis on Manifolds

Stochastic Analysis on Manifolds

Elton P. Hsu

Graduate Studies
in Mathematics

Volume 38

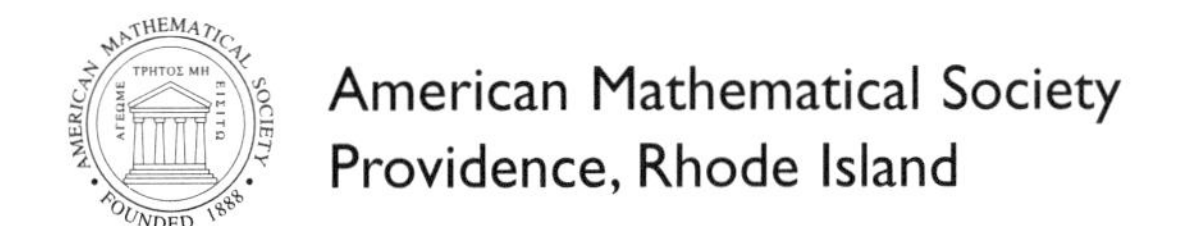

American Mathematical Society
Providence, Rhode Island

2000 *Mathematics Subject Classification.* Primary 58J65, 60J60, 60J65.

ABSTRACT. This book is intended for advanced graduate students and researchers who are familiar with both euclidean stochastic analysis and basic differential geometry. It begins with a brief review of stochastic differential equations on euclidean space. After presenting the basics of stochastic analysis on manifolds, we introduce Brownian motion on a Riemannian manifold and study the effect of curvature on its behavior. We then show how to apply Brownian motion to geometric problems and vice versa by several typical examples, including eigenvalue estimates and probabilistic proofs of the Gauss-Bonnet-Chern theorem and the Atiyah-Singer index theorem for Dirac operators. The book ends with an introduction to stochastic analysis on the path space over a Riemannian manifold, a topic of much current research interest.

Library of Congress Cataloging-in-Publication Data

Hsu, Elton P., 1959–
Stochastic analysis on manifolds / Elton P. Hsu.
p. cm. — (Graduate studies in mathematics, ISSN 1065-7339 ; v. 38)
Includes bibliographical references and index.
ISBN 0-8218-0802-8 (acid-free paper)
1. Stochastic differential equations. 2. Diffusion processes. 3. Differential geometry. I. Title. II. Series.

QA614.9 .H78 2001
514′.74—dc21 2001046052

∞ The paper used in this book is acid-free and falls within the guidelines
established to ensure permanence and durability.
Visit the AMS home page at URL: http://www.ams.org/

10 9 8 7 6 5 4 3 2 1 07 06 05 04 03 02

In Memory of My Beloved Mother
Zhu Pei-Ru (1926-1996)

Quält mich Erinnerung, daß ich verübet,
So manche Tat, die dir das Herz betrübet?
Das schöne Herz, das mich so sehr geliebet?
....
Doch da bist du entgegen mir gekommen,
Und ach! was da in deinem Aug geschwommen,
Das war die süße, langgesuchte Liebe.

Heinrich Heine, *An Meine Mutter.*

Contents

Preface

A large part of modern analysis is centered around the Laplace operator and its generalizations in various settings. On the other hand, under certain technical conditions one can show that the generator of a continuous strong Markov process must be a second order elliptic operator. These are the two facts that make the connection between probability and analysis seem natural. If we adopt the view that Brownian paths are the "characteristic lines" for the Laplace operator, then it is no longer surprising that solutions of many problems associated with the Laplace operator can be explicitly represented by Brownian motion, for it is a well known fact in the theory of partial differential equations that explicit solutions are possible if one knows characteristic lines. While analysts are interested in what happens in average, to a probabilist things usually happen at the path level. For these reasons probability theory, and Brownian motion in particular, has become a convenient language and useful tool in many areas of analysis. The purpose of this book is to explore this connection between Brownian motion and analysis in the area of differential geometry.

Unlike many time-honored areas of mathematics, stochastic analysis has neither a well developed core nor a clearly defined boundary. For this reason the choice of the topics in this book reflects heavily my own interest in the subject; its scope is therefore much narrower than the title indicates. My purpose is to show how stochastic analysis and differential geometry can work together towards their mutual benefit. The book is written mainly from a probabilist's point of view and requires for its understanding a solid background in basic euclidean stochastic analysis. Although necessary geometric facts are reviewed throughout the book, a good knowledge of differential geometry is assumed on the part of the reader. Because of its somewhat unusual dual prerequisites, the book is best suited for highly motivated advanced graduate students in either stochastic analysis or differential geometry and for researchers in these and related areas of mathematics. Notably absent from the book are a collection of exercises commonly associated with books of this kind, but throughout the book there are many proofs which are nothing but an invitation to test the reader's understanding of the topics under discussion.

During the writing of this book, I have greatly benefited from several existant monographs on the subject; these include:

- N. Ikeda and S. Watanabe: *Stochastic Differential Equations and Diffusion Processes*, 2nd edition, North-Holland/Kodansha (1989);
- K. D. Elworthy: *Stochastic Differential Equations on Manifolds*, Cambridge University Press (1982);
- M. Emery: *Stochastic Calculus in Manifolds*, Springer (1989);
- P. Malliavin: *Stochastic Analysis*, Springer (1997).

Overlaps with these works are not significant, and they and the more recent *An Introduction to the Analysis of Paths on a Riemannian Manifold*, American Mathematical Society (2000), by D. W. Stroock, are warmly recommended to the reader.

This book could not have been written without constant support from my wife, who has taken more than her fair share of family duties during its long gestation period. I would like to take this opportunity to thank Elena Kosygina, Tianhong Li, Banghe Li, and Mark A. Pinsky for reading early drafts of various parts of the book and for their valuable suggestions. I would also like to acknowledge many years of financial support through research grants from the National Science Foundation. Most of the book is based on the lectures I have delivered at various places during the late 1990s, notably at Academica Sinica in Beijing (1995), IAS/Park City Mathematics Institute in Princeton (1996), Institut Henri Poincaré in Paris (1998), and Northwestern University in Evanston (1999), and I would like to thank my audiences for their comments and suggestions.

Elton P. Hsu

Hinsdale, IL

October, 2001.

Introduction

Stochastic differential equations on manifolds can be studied in three different but equivalent ways: intrinsic, extrinsic, and local, and we will see examples of all three in this book. We prove the existence and uniqueness of solutions by assuming that the manifold is a closed submanifold of some euclidean space, in order to take advantage of the more familiar theory of such equations in euclidean spaces. Once this is done, the extrinsic point of view is pretty much abandoned. An intrinsic presentation of a result gives it an appealing and compact form, and in many situations shows the relations among the quantites involved more clearly. However, it is often the case that earnest analysis can only be done in a judiciously chosen coordinate system, so local calculations are unavoidable.

A dominating theme of the book is the relation between Brownian motion and curvature, for, aside from topology, it is curvature that distinguishes a manifold from a euclidean space. Except for the first two chapters, the reader will find curvature make its appearance on almost every page; therefore it is important that the reader should have a good grasp on its basic definition and various incarnations. Because this book is mainly intended for probabilists who are interested in geometric applications, basic euclidean stochastic analysis is assumed. Differential geometry is reviewed on the spot, and these brief reviews also familarize the reader with the notations to be used. In order to give some guidance to the reader, in the following I will give a general description of the contents of the book.

CHAPTER 1 introduces basic facts about stochastic differential equations on manifolds. It starts with a review of the euclidean case. Equations on manifolds are solved by embedding them in euclidean spaces (by virtue of

Whitney's embedding theorem). The main result is that every Itô type stochastic differential equation on a manifold can be uniquely solved up to its explosion time. The last section discusses diffusion processes from the viewpoint of martingale problems, and the existence and uniqueness of diffusion processes generated by smooth second order elliptic operators on manifolds are proved by solving stochastic differential equations on manifolds.

CHAPTER 2 studies horizontal lift and stochastic development, two concepts central to the Eells-Elworth-Malliavin construction of Brownian motion on a Riemannian manifold. Necessary background from differentil geometry is explained in the first two sections. The horizontal lift of a general manifold-valued semimartingale is obtained by solving a horizontal stochastic differential equation in the frame bundle. Stochastic line integrals of covariant tensors of orders 1 and 2 along a semimartingale are also introduced here. In the last two sections, we study martingales on manifolds and submanifolds, and prove two local results on convergence and nonconfluence.

CHAPTER 3 starts with a review on the Laplace-Beltrami operator. Brownian motion on a Riemannian manifold is defined as a diffusion generated by half of the operator. This definition is shown to be equivalent to several other descriptions, including the one stating that it is a semimartingale whose anti-development with the Levi-Civita connection is a euclidean Brownian motion. After discussing several typical examples, we investigate the effect of curvature on the radial process of Brownian motion by comparing it with the same process on a radially symmetric manifold. The differential geometry needed for this purpose includes some basic properties of the distance function and the Laplacian comparison theorem. The chapter ends with an estimate on the exit time from a geodesic ball in terms of the lower bound of the Ricci curvature.

CHAPTER 4 explores the connection between heat kernel and Brownian motion, starting from the basic fact that the minimal heat kernel is the transition density function of Brownian motion. Topics included in this chapter are stochastic completeness, the C_0-property (also known as the Feller property) of the heat semigroup, recurrence and transience, and heat kernel comparison theorems. The common feature of these topics is that they can be investigated by examining the radial process of Brownian motion.

CHAPTER 5 studies several problems on the short-time behavior of the heat kernel and Brownian motion. The short-time asymptotic behavior of the heat kernel for points within the cutlocus is the starting point of our discussion. Varadhan's asymptotic relation is then proved for a complete Riemannian manifold. We also describe a general method for computing the short-time asymptotics of the heat kernel at arbitrary two points of a compact Riemannian manifold. for distant points. This method is illustrated by computing the leading term of the heat kernel on a sphere at two antipodal points. In the last section we prove global estimates for the first and second derivatives of the logarithmic heat kernel on a compact Riemannian manifold.

CHAPTER 6 contains two further applications of Brownian motion to geometric problems. The first one is the Dirichlet problem at infinity on a Cartan-Hadamard manifold. This problem is reduced to the problem of angular convergence of Brownian motion. Two typical bounds on the sectional curvature are discussed, namely a constant upper bound and a inverse quadratically vanishing upper bound. The second application is estimating the first nonzero eigenvalue of a compact manifold with nonnegative Ricci curvature using the Kendall-Cranston coupling of Brownian motion.

CHAPTER 7 is devoted to probabilistic proofs of the Atiyah-Singer index theorem. It starts with a discussion on solving heat equations on differential forms using a Feynman-Kac multiplicative functional dictated by the Weitzenböck formula. The easy case of the Gauss-Bonnet-Chern theorem is discussed first, and a probabilistic proof of Patodi's local Gauss-Bonnet-Chern theorem is given. We then explain the algebraic and geometric background for the Atiyah-Singer index theorem for the Dirac operator of a twisted spin bundle on a spin manifold. From a probabilistic viewpoint, the main difference between the two is that the latter requires a more precise asymptotic computation of Brownian holonomy, which is carried out in the last section.

CHAPTER 8 is a basic course on stochastic analysis on the path space over a compact Riemannian manifold. We first prove the quasi-invariance of the Wiener measure on a euclidean path space and use it to derive the integration by parts for the gradient operator. This is followed by a discussion of the spectral properties of the Ornstein-Uhlenbeck operator. The concepts of the gradient operator and the Ornsten-Uhlenbeck operator are then generalized to the path space of a Riemannian manifold. We prove three formulas involving the gradient and the heat kernel due to Bismut. One of them is used to give a proof of Driver's integration by parts formula for the gradient operator in the path space. In preparation for a logarithmic Sobolev inequality on the path space, the Clark-Ocone-Haussman martingale representation theorem is generalized to the setting of the path space over a compact Riemannian manifold. After a general discussion of logarithmic Sobolev inequalities and hypercontractivity, we end the book by proving the logarithmic Sobolev inequality for the path space.

NOTES AND COMMENTS and BIBLIOGRAPHY are limited to the immediate literature I have consulted in preparing this monograph, and are not meant for distributing or adjudicating research credits, for in many cases I have not cited original research publications. They may also serve as a guide for those who are interested in more information or further research on the topics covered in the book.

Chapters 1, 2, 3, and Sections 4.1, 5.1, 5.2, 5.4, 7.1, 7.2 form the core of the book; the remaining part consists mostly of independent topics. Readers who wish to study these topics selectively may find convenient the list of General Notations included at the end of the book.

Chapter 1

Stochastic Differential Equations and Diffusions

We start with with a review of the basic theory of Itô type stochastic differential equations (SDE) on euclidean space driven by general continuous semimartingales. An Itô type stochastic differential equation on a manifold M can be solved by first embedding M into a euclidean space $\mathbb{R}^N$ as a closed submanifold and extending the vector fields of the equation from M to the ambient space. Thus we obtain an equation of the same type on $\mathbb{R}^N$. A solution of the extended equation starting from M will stay in M. Such an extrinsic approach to stochastic differential equations on manifolds has the obvious advantage of avoiding using local coordinates, thus enabling us to streamline the exposition. Since we use an embedding only as an intermediate step, its extrinsic nature should not be a concern once an appropriate uniqueness of the object under consideration is established.

We will pay special attention to the possibility of explosion of a solution. The explosion time is the maximum stopping time up to which a solution of the equation can be defined. If the driving semimartingale runs for all time, then the explosion time coincides with the time at which the solution escapes to infinity.

The last section of this chapter is a brief introduction to the theory of diffusion processes and diffusion measures generated by smooth second order elliptic operators on manifolds. As usual they are defined as solutions of certain martingale problems. An important consequence of the uniqueness of solutions of martingale problems is the strong Markov property, which is used explicitly or implicitly throughout the book.

Let us make a few remarks about the general probabilistic setting in this book. Frequently we work with a probability space $(\Omega, \mathscr{F}, \mathbb{P})$ equipped with a filtration $\mathscr{F}_* = \{\mathscr{F}_t, t \geq 0\}$ of σ-fields contained in $\mathscr{F}$. We always assume that $\mathscr{F} = \lim_{t\uparrow\infty} \mathscr{F}_t$ and call $(\Omega, \mathscr{F}_*, \mathbb{P})$ a filtered probability space. Whenever necessary, we assume that

(1) each $\mathscr{F}_t$ is complete with respect to $\mathbb{P}$, i.e., every null set (a subset of of a set of measure zero) is contained in $\mathscr{F}_t$;

(2) $\mathscr{F}_*$ is right continuous: $\mathscr{F}_t = \bigcap_{s>t} \mathscr{F}_s$.

Since we deal almost exclusively with continuous stochastic processes, we will drop the modifier "continuous" and assume that all stochastic processes are continuous with probability one unless explicitly stated otherwise. In the same spirit, we will also let go the ubiquitous "almost surely," "with probability one," and the like whenever there is no possibility of confusion.

1.1. SDE on euclidean space

In this section we will study Itô type stochastic differential equations on a euclidean space with locally Lipschitz coefficients driven by continuous semimartingales. Later, when we study diffusions generated by a second order elliptic, but not necessarily nondegenerate differential operator, we will have to consider stochastic differential equations with locally Lipschitz coefficients even when the coefficients of the generator are smooth. The main result of this section is the existence and uniqueness of the solution of such an equation up to its explosion time (THEOREM 1.1.8).

Let us first formulate the kind of equations we want to solve on $\mathbb{R}^N$, the euclidean space of dimension N. Such an equation is given by a diffusion coefficient matrix σ and a driving semimartingale Z. We assume that the diffusion coefficient matrix $\sigma = \{\sigma^i_\alpha\} : \mathbb{R}^N \to \mathscr{M}(N, l)$ (the space of $(N \times l)$-matrices) is locally Lipschitz: for each $R > 0$ there is a constant $C(R)$ depending on R such that

$$|\sigma(x) - \sigma(y)| \leq C(R)|x - y|, \qquad \forall x, y \in B(R),$$

where $B(R) = \{x \in \mathbb{R}^N : |x| \leq R\}$. We say that σ is globally Lipschitz if $C(R)$ can be chosen independent of R.

We assume that the driving process $Z = \{Z_t, t \geq 0\}$ is an $\mathbb{R}^l$-valued $\mathscr{F}_*$-semimartingale (adapted to the filtration $\mathscr{F}_*$) defined on a filtered probability space $(\Omega, \mathscr{F}_*, \mathbb{P})$. It is viewed as a column of l real-valued semimartingales: $Z_t = (Z^1_t, \ldots, Z^l_t)^\dagger$. By definition $Z = M + A$, where M is a local

$\mathscr{F}_*$-martingale and A an $\mathscr{F}_*$-adapted process of local bounded variation such that $A_0 = 0$.

Let $X_0 \in \mathscr{F}_0$ be an $\mathbb{R}^N$-valued random variable measurable with respect to $\mathscr{F}_0$. Let τ be an $\mathscr{F}_*$-stopping time and consider the following stochastic differential equation (written in the integral form) for an $\mathbb{R}^N$-valued semimartingale $X = \{X_t, 0 \le t < \tau\}$ defined up to τ:

$$X_t = X_0 + \int_0^t \sigma(X_s)\, dZ_t, \quad 0 \le t < \tau. \tag{1.1.1}$$

Here the stochastic integral is in the Itô sense. We will refer to this equation as $SDE(\sigma, Z, X_0)$. In a typical situation where $Z_t = (W_t, t)$ with an $(l-1)$-dimensional euclidean Brownian motion W and $\sigma = (\sigma_1, b)$ with $\sigma_1 : \mathbb{R}^N \to \mathscr{M}(N, l-1)$ and $b : \mathbb{R}^N \to \mathbb{R}^N$, the equation takes on a more familiar form:

$$dX_t = \sigma_1(X_t)\, dW_t + b(X_t)\, dt.$$

Note that by allowing a solution to run only up to a stopping time, we have incorporated the possibility that a solution may explode in finite time.

The following Itô's formula is a consequence of the usual Itô's formula for semimartingales. For a function $f \in C^2(\mathbb{R}^N)$, we use $f_{x_i}, f_{x_i x_j}$ to denote its first and second partial derivatives with respect to the indicated variables.

Proposition 1.1.1. *Let X be a solution of $SDE(\sigma, Z, X_0)$. Suppose that $f \in C^2(\mathbb{R}^N)$. Then*

$$\begin{aligned}
f(X_t) &= f(X_0) + \int_0^t f_{x_i}(X_s)\, \sigma^i_\alpha(X_s)\, dZ^\alpha_s \\
&\quad + \frac{1}{2}\int_0^t f_{x_i x_j}(X_s)\, \sigma^i_\alpha(X_s)\, \sigma^j_\beta(X_s)\, d\langle Z^\alpha, Z^\beta\rangle_s \\
&= f(X_0) + \int_0^t f_{x_i}(X_s)\sigma^i_\alpha(X_s)\, dM^\alpha_s + \int_0^t f_{x_i}(X_s)\sigma^i_\alpha(X_s)\, dA^\alpha_s \\
&\quad + \frac{1}{2}\int_0^t f_{x_i x_j}(X_s)\, \sigma^i_\alpha(X_s)\, \sigma^j_\beta(X_s)\, d\langle M^\alpha, M^\beta\rangle_s.
\end{aligned} \tag{1.1.2}$$

Proof. Exercise. □

Our goal in this section is to prove the existence and uniqueness of a solution up to its explosion time for the type of stochastic differential equations we just formulated. For this purpose we need an estimate concerning Itô integrals with respect to semimartingales. Let $Z = M + A$ be the canonical decomposition of the semimartingale Z into a local martingale and a process of locally bounded variation. Define

$$Q_t = \sum_{\alpha=1}^{l} \langle M^\alpha, M^\alpha \rangle_t + \sum_{\alpha=1}^{l} \left(|A^\alpha|_t^3 + |A^\alpha|_t \right) + t, \tag{1.1.3}$$

where $|A^\alpha|_t$ stands for the total variation of A^α on $[0, t]$. Adding t in the definition makes Q strictly increasing, so that it has a continuous, strictly increasing inverse $\eta = \{\eta_s, t \geq 0\}$. Since $\{\eta_s \leq t\} = \{s \leq Q_t\}$, each η_s is a finite $\mathscr{F}_*$-stopping time.

Lemma 1.1.2. *There is a constant C depending only on m and l such that for any $\mathscr{F}_*$-adapted, $\mathscr{M}(m, l)$-valued continuous process F and an $\mathscr{F}_*$-stopping time τ,*

$$\mathbb{E} \max_{0 \leq t \leq \tau} \left| \int_0^t F_s \, dZ_s \right|^2 \leq C \, \mathbb{E} \int_0^\tau |F_s|^2 \, dQ_s. \tag{1.1.4}$$

Proof. By considering each component separately, we may assume that $l = m = 1$. We have

$$\max_{0 \leq t \leq \tau} \left| \int_0^t F_s \, dZ_s \right|^2 \leq 2 \max_{0 \leq t \leq \tau} \left| \int_0^t F_s \, dM_s \right|^2 + 2 \max_{0 \leq t \leq \tau} \left| \int_0^t F_s \, dA_s \right|^2.$$

Let $N_t = \displaystyle\int_0^t F_s \, dM_s$. By Doob's inequality we have

$$\mathbb{E} \max_{0 \leq t \leq \tau} |N_t|^2 \leq 4\mathbb{E} \, |N_\tau|^2 = 4\mathbb{E} \langle N \rangle_\tau.$$

Now

$$\langle N \rangle_\tau = \int_0^\tau |F_s|^2 d\langle M \rangle_s \leq \int_0^\tau |F_s|^2 dQ_s,$$

hence

$$\mathbb{E} \max_{0 \leq t \leq \tau} \left| \int_0^t F_s \, dM_s \right|^2 \leq 4\mathbb{E} \int_0^\tau |F_s|^2 dQ_s.$$

On the other hand, by the Cauchy-Schwarz inequality,

$$\begin{aligned} \left| \int_0^t F_s \, dA_s \right|^2 &\leq \left[\int_0^\tau |F_s| \sqrt{|A|_s^2 + 1} \cdot \frac{1}{\sqrt{|A_s|^2 + 1}} d|A|_s \right]^2 \\ &\leq \int_0^\tau |F_s|^2 \left(|A|_s^2 + 1 \right) d|A|_s \cdot \int_0^\tau \frac{d|A|_s}{|A|_s^2 + 1} \\ &\leq \frac{\pi}{6} \int_0^\tau |F_s|^2 dQ_s. \end{aligned}$$

It follows that

$$\mathbb{E} \max_{0 \leq t \leq \tau} \left| \int_0^t F_s \, dZ_s \right|^2 \leq 10 \, \mathbb{E} \int_0^\tau |F_s|^2 \, dQ_s.$$

$\square$

Returning to the $SDE(\sigma, Z, X_0)$, we first consider a case without explosion.

Theorem 1.1.3. *Suppose that σ is globally Lipschitz and X_0 square integrable. Then $SDE(\sigma, Z, X_0)$ has a unique solution $X = \{X_t, t \geq 0\}$.*

Proof. In this proof we use the notation

$$|Y|_{\infty,t} = \max_{0 \leq s \leq t} |Y_s|$$

for a vector-valued process Y. We solve the equation by the usual Picard's iteration. Define a sequence $\{X^n\}$ of semimartingales by $X_t^0 = X_0$ and

$$X_t^n = X_0 + \int_0^t \sigma\left(X_s^{n-1}\right) dZ_s. \tag{1.1.5}$$

We claim that $\{X^n\}$ converges to a continuous semimartingale X which satisfies the equation. Define the increasing process Q as in (1.1.3) and let η be its inverse as before. Each η_T is a stopping time for fixed T. By (1.1.4) and the assumption that σ is globally Lipschitz we have

$$\begin{aligned} \mathbb{E}\left|X^n - X^{n-1}\right|_{\infty,\eta_T}^2 &\leq C_1 \mathbb{E} \int_0^{\eta_T} \left|\sigma\left(X_s^{n-1}\right) - \sigma\left(X_s^{n-2}\right)\right|^2 dQ_s \\ &\leq C_2 \mathbb{E} \int_0^{\eta_T} \left|X_s^{n-1} X_s^{n-2}\right|^2 dQ_s. \end{aligned}$$

Making the change of variable $s = \eta_u$ in the last integral and using $Q_{\eta_u} = u$, we have

$$\mathbb{E}\left|X^n - X^{n-1}\right|_{\infty,\eta_T}^2 \leq C_2 \mathbb{E} \int_0^T \left|X^{n-1} - X^{n-2}\right|_{\infty,\eta_u}^2 du. \tag{1.1.6}$$

For the initial step, from

$$X_t^1 - X_t^0 = \int_0^t \sigma(X_0)\, dZ_s = \sigma(Z_0) Z_t$$

and (1.1.4) again we have

$$\mathbb{E}\left|X^1 - X^0\right|_{\infty,\eta_T}^2 \leq C_3 \mathbb{E}\left\{\left[|X_0|^2 + 1\right] Q_{\eta_T}\right\} = C_3 T \left\{\mathbb{E}|X_0|^2 + 1\right\},$$

which is assumed to be finite. Now we can iterate (1.1.6) to obtain

$$\mathbb{E}\left|X^n - X^{n-1}\right|_{\infty,\eta_T}^2 \leq \frac{(C_4 T)^n}{n!} \left\{\mathbb{E}|X_0|^2 + 1\right\}. \tag{1.1.7}$$

By Chebyshev's inequality, this implies that

$$\mathbb{P}\left\{\left|X^n - X^{n-1}\right|_{\infty,\eta_T} \geq \frac{1}{2^n}\right\} \leq \frac{(4C_4 T)^n}{n!} \left\{\mathbb{E}|X_0|^2 + 1\right\}.$$

By the Borel-Cantelli lemma we have, for fixed $T > 0$,

$$\mathbb{P}\left\{\left|X^n - X^{n-1}\right|_{\infty,\eta_T} \leq \frac{1}{2^n} \text{ for almost all } n\right\} = 1.$$

Since $\eta_s \uparrow \infty$ as $s \uparrow \infty$, it is now clear that $X_t = \lim_{n\to\infty} X_t^n$ exists for each fixed t and that almost surely the convergence is uniform on $[0, \eta_T]$ for each fixed T. This shows that the limiting process X is continuous.

We now justify passing to the limit in (1.1.5). From the inequality

$$|X^m - X^n|_{\infty,\eta_T} \le \sum_{l=n+1}^{m} |X^l - X^{l-1}|_{\infty,\eta_T}$$

we have

$$\sqrt{\mathbb{E}|X^m - X^n|^2_{\infty,\eta_T}} \le \sum_{l=n+1}^{m} \sqrt{\mathbb{E}|X^l - X^{l-1}|^2_{\infty,\eta_T}}.$$

Letting $m \to \infty$ and using Fatou's lemma on the left side and the estimate (1.1.7) on the right side, we obtain

$$\begin{aligned}\sqrt{\mathbb{E}\,|X - X^n|^2_{\infty,\eta_T}} &\le \sum_{l=n+1}^{\infty} \sqrt{\mathbb{E}\,|X^l - X^{l-1}|^2_{\infty,\eta_T}} \\ &\le \sum_{l=n+1}^{\infty} \sqrt{\frac{(C_4T)^l}{l!}} \cdot \sqrt{\mathbb{E}|X_0|^2 + 1},\end{aligned}$$

which implies

$$\lim_{n\to\infty} \mathbb{E}\,|X - X^n|^2_{\infty,\eta_T} = 0. \tag{1.1.8}$$

For the right side of (1.1.5) we have by LEMMA 1.1.2 and (1.1.8)

$$\begin{aligned}&\mathbb{E} \max_{0\le t\le\eta_T} \left| \int_0^t \{\sigma(X_s) - \sigma(X_s^n)\}\, dZ_s \right|^2 \\ &\qquad \le C_5\mathbb{E} \int_0^{\eta_T} |\sigma(X_s) - \sigma(X_s^n)|^2\, dQ_s \\ &\qquad \le C_6\mathbb{E} \int_0^{\eta_T} |X_s - X_s^n|^2\, dQ_s \\ &\qquad \le C_6T\, \mathbb{E}|X - X^n|^2_{\infty,\eta_T} \\ &\qquad \to 0.\end{aligned}$$

Now we can take the limit in (1.1.5) as $n \uparrow \infty$ and obtain, for each fixed $T > 0$,

$$X_{t\wedge\eta_T} = X_0 + \int_0^{t\wedge\eta_T} \sigma(X_s)\, dZ_s.$$

Letting $T \uparrow \infty$, we have

$$X_t = X_0 + \int_0^t \sigma(X_s)\, dZ_s.$$

This shows that X is a semimartingale and satisfies $SDE(\sigma, Z, X_0)$.

For the uniqueness, suppose that Y is another solution. We have

$$X_t - Y_t = \int_0^t \{\sigma(X_s) - \sigma(Y_s)\}\, dZ_s.$$

For a fixed positive integer N let $\tau_N = \inf\{t \geq 0 : |X_t - Y_t| \geq N\}$. We have as before

$$\mathbb{E}|X_t - Y_t|^2_{\infty,\eta_T\wedge\tau_N} \leq C_6 \mathbb{E}\int_0^{\eta_T\wedge\tau_N} |X_s - Y_s|^2\, dQ_s.$$

If we let $V_s = \mathbb{E}|X - Y|^2_{\infty,\eta_s\wedge\tau_N}$, then $V_s \leq N$, and after a change of variables on the right side the above inequality can be written simply as

$$V_T \leq C_6 \int_0^T V_s ds.$$

This implies $V_s = 0$ for all s and hence $X_s = Y_s$ for all $0 \leq s \leq \tau_N$. By the continuity of X and Y we have $\tau_N \uparrow \infty$ as $N \uparrow \infty$. It follows that $X_s = Y_s$ for all $s \geq 0$. □

In the above theorem, the solution X runs for all time because we have assumed that the coefficient matrix σ is globally Lipschitz, which implies that it can grow at most linearly. When the matrix σ is only locally Lipschitz, we have to allow the possibility of explosion. For example, the solution of the equation

$$\frac{dx_t}{dt} = x_t^2, \quad x_0 = 1,$$

is $X_t = 1/(1-t)$, which explodes at time $t = 1$. In view of later applications, we will take a more general viewpoint and define the explosion time of a path in a locally compact metric space M. We use $\widehat{M} = M \cup \{\partial_M\}$ to denote the one-point compactification of M.

Definition 1.1.4. *An M-valued path x with explosion time $e = e(x) > 0$ is a continuous map $x : [0,\infty) \to \widehat{M}$ such that $x_t \in M$ for $0 \leq t < e$ and $x_t = \partial_M$ for all $t \geq e$ if $e < \infty$. The space of M-valued paths with explosion time is called the path space of M and is denoted by $W(M)$.*

We state a few easy but useful facts about explosion times. Recall that an exhaustion of a locally compact metric space M is a sequence of relatively compact open sets $\{O_N\}$ such that $\overline{O}_N \subseteq O_{N+1}$ and $M = \bigcup_{N=1}^\infty O_N$.

Proposition 1.1.5. *(1) Let $\{O_N\}$ be an exhaustion and τ_{O_N} the first exit time from O_N. Then $\tau_{O_N} \uparrow e$ as $N \uparrow \infty$.*

(2) Suppose that $d : M \times M \to \mathbb{R}_+$ is a metric on M with the property that every bounded closed set is compact. Fix a point $o \in M$. Let τ_R be the first exit time from the ball $B(R) = \{x \in M : d(x,o) \leq R\}$. Then $\tau_R \uparrow e$ as $R \uparrow \infty$.

Proof. Exercise. □

We will use Part (2) of the above proposition in two typical situations: (i) M is a complete Riemannian manifold and d is the Riemannian distance function; (ii) M is embedded as a closed submanifold in another complete Riemannian manifold and d is the Riemannian metric of the ambient space.

Let $\mathscr{B}_t = \mathscr{B}_t(W(M))$ be the σ-field generated by the coordinate maps up to time t. Then we have a filtered measurable space $(W(M), \mathscr{B}_*)$, and the lifetime $e : W(M) \to (0, \infty]$ is a $\mathscr{B}_*$-stopping time.

For a locally Lipschitz coefficient matrix σ, a solution of $SDE(\sigma, Z, X_0)$ is naturally a semimartingale defined up to a stopping time. Let us be more precise.

Definition 1.1.6. *Let $(\Omega, \mathscr{F}_*, \mathbb{P})$ be a filtered probability space and τ an $\mathscr{F}_*$-stopping time. A continuous process X defined on the stochastic time interval $[0, \tau)$ is called an $\mathscr{F}_*$-semimartingale up to τ if there exists a sequence of $\mathscr{F}_*$-stopping times $\tau_n \uparrow \tau$ such that for each n the stopped process $X^{\tau_n} = \{X_{t\wedge\tau_n}, t \geq 0\}$ is a semimartingale in the usual sense.*

Definition 1.1.7. *A semimartingale X up to a stopping time τ is a solution of $SDE(\sigma, Z, X_0)$ if there is a sequence of stopping times $\tau_n \uparrow \tau$ such that for each n the stopped process X^{τ_n} is a semimartingale and*

$$X_{t\wedge\tau_n} = X_0 + \int_0^{t\wedge\tau_n} \sigma(X_s)\, dZ_s, \qquad t \geq 0.$$

We are now in a position to show that there is a unique solution X to $SDE(\sigma, Z, X_0)$ up to its explosion time $e(X)$.

Theorem 1.1.8. *Suppose that we are given (i) a locally Lipschitz coefficient matrix $\sigma : \mathbb{R}^N \to \mathscr{M}(N, l)$; (ii) an $\mathbb{R}^l$-valued $\mathscr{F}_*$-semimartingale $Z = \{Z_t, t \geq 0\}$ on a filtered probability space $(\Omega, \mathscr{F}_*, \mathbb{P})$; (iii) an $\mathbb{R}^N$-valued, $\mathscr{F}_0$-measurable random variable X_0. Then there is a unique $W(\mathbb{R}^N)$-valued random variable X which is a solution of $SDE(\sigma, Z, X_0)$ up to its explosion time $e(X)$.*

Proof. We first assume that X_0 is uniformly bounded: $|X_0| \leq K$. For a fixed positive integer $n \geq K$ let $\sigma^n : \mathbb{R}^d \to \mathscr{M}(N, l)$ be a globally Lipschitz function such that $\sigma^n(z) = \sigma(z)$ for $|z| \leq n$. By THEOREM 1.1.3, $SDE(\sigma^n, Z, X_0)$ has a solution X^n. We have $X_t^n = X_t^{n+1}$ when

$$t \leq \tau_n \stackrel{\text{def}}{=} \inf\left\{t \geq 0 : |X_t^n| \text{ or } |X_t^{n+1}| = n\right\}.$$

This follows from the uniqueness part of THEOREM 1.1.3 because $\sigma^n(z) = \sigma^{n+1}(z)$ for $|z| \leq n$ and the two semimartingales X^{n,τ_n} and X^{n+1,τ_n} (X^n and X^{n+1} stopped at τ_n respectively) are solutions of the same equation

$SDE(\sigma^n, Z^{\tau_n}, X_0)$. Now τ_n is the first time the common process reaches the sphere $\partial B(n) = \{z \in \mathbb{R}^N : |z| = n\}$. In particular, we have $\tau_n \le \tau_{n+1}$.

Let $e = \lim_{n\uparrow\infty} \tau_n$, and define a semimartingale X up to time e by $X_t = X_t^n$ for $0 \le t < \tau_n$. Then τ_n is the first time the process X reaches the sphere $\partial B(n)$. From

$$X_t^n = X_0 + \int_0^t \sigma^n\left(X_s^n\right) dZ_s,$$

$X_{t\wedge\tau_n} = X_t^n$, and $\sigma^n(X_s^n) = \sigma(X_s)$ for $s \le \tau_n$, we have

$$X_{t\wedge\tau_n} = X_0 + \int_0^{t\wedge\tau_n} \sigma\left(X_s\right) dZ_s,$$

which means that X is a solution of $SDE(\sigma, Z, X_0)$ up to time e.

We now show that e is the explosion time of X, i.e.,

$$\lim_{t\uparrow\uparrow e} |X_t| = \infty \quad \text{on} \quad \{e < \infty\}.$$

Equivalently, we need to show that for each fixed positive $R \ge K$, there exists a time (not necessarily a stopping time) $t_R < e$ such that $|X_t| \ge R$ for all $t \in [t_R, e)$.

The idea of the proof is as follows. Because the coefficients of the equation are bounded on the ball $B(R+1)$, X needs to spend at least a fixed amount of time (in an appropriate probabilistic sense) when it crosses from $\partial B(R)$ to $\partial B(R+1)$. If $e < \infty$, this can happen only finitely many times. Thus after some time, X either never returns to $B(R)$ or stays inside $B(R+1)$ forever; but the second possibility contradicts the facts that $|X_{\tau_n}| = n$ and $\tau_n \uparrow e$ as $n \uparrow \infty$.

To proceed rigorously, define two sequences $\{\eta_n\}$ and $\{\zeta_n\}$ of stopping times inductively by $\zeta_0 = 0$ and

$$\begin{aligned} \eta_n &= \inf\{t > \zeta_{n-1} : |X_t| = R\}, \\ \zeta_n &= \inf\{t > \eta_n : |X_t| = R+1\}, \end{aligned}$$

with the convention that $\inf\emptyset = e$. If $\zeta_n < e$, the difference $\zeta_n - \eta_n$ is the time X takes to cross from $\partial B(R)$ to $\partial B(R+1)$ for the nth time. By Itô's formula (1.1.2) applied to the function $f(x) = |x|^2$ we have

$$|X_t|^2 = |X_0|^2 + N_t + \int_0^t \Psi_s dQ_s, \quad t < e, \tag{1.1.9}$$

where

$$N_t = 2\int_0^t \sigma_\alpha^i(X_s)X_s^i\, dM_s^\alpha,$$

$$\Psi_s = 2\sigma_\alpha^i(X_s)X_s^i\frac{dA_s^\alpha}{dQ_s} + \sigma_\alpha^i(X_s)\sigma_\beta^i(X_s)\frac{d\langle M^\alpha, M^\beta\rangle_s}{dQ_s}.$$

The definition of Q is given in (1.1.3). It is clear that

$$\langle N\rangle_t = \int_0^t \Phi_s dQ_s,$$

$$\Phi_s = 4\sigma^i_\alpha(X_s)\sigma^j_\beta(X_s)X^i_sX^j_s\,\frac{d\langle M^\alpha, M^\beta\rangle_s}{dQ_s}.$$

By the well known Lévy's criterion, there exists a one-dimensional Brownian motion W such that

$$N_{u+\eta_n} - N_{\eta_n} = W_{\langle N\rangle_{u+\eta_n} - \langle N\rangle_{\eta_n}}.$$

When $\eta_n \le s \le \zeta_n$ we have $|X_s| \le R+1$. Hence there is a constant C depending on R such that $|\Psi_s| \le C$ and $|\Phi_s| \le C$ during the said range of time. From (1.1.9) we have

$$\begin{aligned} 1 &\le |X_{\zeta_n}|^2 - |X_{\eta_n}|^2 \\ &= W_{\langle N\rangle_{\zeta_n} - \langle N\rangle_{\eta_n}} + \int_{\eta_n}^{\zeta_n} \Psi_s dQ_s \\ &\le W^*_{\langle N\rangle_{\zeta_n} - \langle N\rangle_{\eta_n}} + C\left(Q_{\zeta_n} - Q_{\eta_n}\right), \end{aligned}$$

where $W^*_t = \max_{0\le s\le t}|W_s|$. We also have

$$\langle N\rangle_{\zeta_n} - \langle N\rangle_{\eta_n} \le C(Q_{\zeta_n} - Q_{\eta_n}).$$

Now it is clear that the events $\zeta_n < e$ and $Q_{\zeta_n} - Q_{\eta_n} \le (Cn)^{-1}$ together imply that

$$W^*_{1/n} \ge 1 - \frac{1}{n} \ge \frac{1}{2}.$$

Therefore,

$$\begin{aligned} \mathbb{P}\left\{\zeta_n < e, Q_{\zeta_n} - Q_{\eta_n} \le \frac{1}{Cn}\right\} &\le \mathbb{P}\left\{W^*_{1/n} \ge 1/2\right\} \\ &= \sqrt{\frac{2n}{\pi}}\int_{1/2}^\infty e^{-nu^2/2}du \le \sqrt{\frac{8}{\pi n}}\,e^{-n^2/8}. \end{aligned}$$

By the Borel-Cantelli lemma, either $\zeta_n = e$ for some n or $\zeta_n < e$ and $Q_{\zeta_n} - Q_{\eta_n} \ge (Cn)^{-1}$ for all sufficiently large n. The second possibility implies that

$$Q_{\zeta_n} \ge \sum_{m=1}^{n-1}(Q_{\zeta_m} - Q_{\eta_m}) \to \infty,$$

which in turn implies $e > \zeta_n \uparrow \infty$. Thus if $e < \infty$, we must have $\zeta_n = e$ for some n. Let ζ_{n_0} be the last time strictly less than e. Then X never returns to $B(R)$ for $\zeta_{n_0} \le t < e$. This shows that e is indeed the explosion time and X is a solution of $SDE(\sigma, Z, X_0)$ up to its explosion time.

To prove the uniqueness, suppose that Y is another solution up to its explosion time and let ρ_n be the first time either $|X_t|$ or $|Y_t|$ is equal to n.

Then X and Y stopped at time τ_n are solutions of $SDE(\sigma^n, Z^{\rho_n}, X_0)$. By the uniqueness part of PROPOSITION 1.1.3 we have $X_t = Y_t$ for $0 \leq t < \rho_n$, and ρ_n is the time the common process first reaches the sphere $\partial B(n)$. Hence $e(X) = e(Y) = \lim_{n\uparrow\infty} \rho_n$ and $X_t = Y_t$ for all $t \in [0, e(X))$.

Finally we remove the restriction that X_0 is uniformly bounded. For a general X_0 let $\Omega_n = \{|X_0| \leq n\}$, and X^n the solution of $SDE(\sigma, Z, X_0 I_{\Omega_n})$. Define a probability measure $\mathbb{Q}^n$ by

$$\mathbb{Q}^n(C) = \frac{\mathbb{P}(C \cap \Omega_n)}{\mathbb{P}(\Omega_n)}.$$

Since $\Omega_n \in \mathscr{F}_0$, both X^n and X^{n+1} satisfy the same $SDE(\sigma, Z, X_0 I_{\Omega_n})$ under $\mathbb{Q}^n$. Hence by the uniquenss we have just proved they must coincide, i.e., $X^n = X^{n+1}$ and $e(X^n) = e(X^{n+1})$ on Ω_n. In view of the fact that $\mathbb{P}(\Omega_n) \uparrow 1$ as $n \uparrow \infty$, we can define a new process by $X = X^n$ and $e = e(X^n)$ on Ω_n. It is clear that X is a semimartingale and satisfies the equation up to its explosion time. The uniqueness follows from the observation that if Y is another solution, then it must be a solution to the same equation with the initial condition $X_0 I_{\Omega_n}$ under $\mathbb{Q}^n$; therefore it must coincide with X on the set Ω_n for all n. □

For future reference we need the following slightly more general form of uniqueness.

Proposition 1.1.9. *Suppose that σ is locally Lipschitz. Let X and Y be two solutions of $SDE(\sigma, Z, X_0)$ up to stopping times τ and η respectively. Then $X_t = Y_t$ for $0 \leq t < \tau \wedge \eta$. In particular, if X is a solution up to its explosion time $e(X)$, then $\eta \leq e(X)$ and $X_t = Y_t$ for $0 \leq t < \eta$.*

Proof. Exercise. □

In contrast to the pathwise uniqueness discussed above, the weak uniqueness (also called uniqueness in law) asserts roughly that if (Z, X_0) and $(\hat{Z}, \hat{X}_0)$ (possibly defined on different filtered probability spaces) have the same law, then the solutions X and $\hat{X}$ of $SDE(\sigma, Z, X_0)$ and $SDE(\sigma, \hat{Z}, \hat{X}_0)$ also have the same law. For simplicity, we restrict ourselves to the typical situation where the driving semimartingale has the special form $Z_t = (W_t, t)$ with an l-dimensional euclidean $\mathscr{F}_*$-Brownian motion W, and correspondingly the coefficient matrix has the form (σ, b). In this case the equation becomes

$$X_t = X_0 + \int_0^t \sigma(X_s)\, dW_s + \int_0^t b(X_s)\, ds. \tag{1.1.10}$$

We will need the weak uniqueness for this type of equations in SECTION 1.3 when we discuss the uniqueness of diffusion measures generated by second order elliptic operators. Note that W is an $\mathscr{F}_*$-Brownian motion if it is a Brownian motion adapted to $\mathscr{F}_*$ such that $\mathscr{F}_t$ is independent of $\{W_{s+t} - W_t, t \geq 0\}$; or equivalently, W has the Markov property with respect to the filtration $\mathscr{F}_*$.

Theorem 1.1.10. *Suppose that σ and b are locally Lipschitz. Then the weak uniqueness holds for the equation (1.1.10). More precisely, suppose that $\widehat{X}$ is the solution of*

$$\widehat{X}_t = \widehat{X}_0 + \int_0^t \sigma(\widehat{X}_s)\, d\widehat{W}_s + \int_0^t b(\widehat{X}_s)\, ds, \tag{1.1.11}$$

where $\widehat{W}$ is a Brownian motion defined on another filtered probability space $(\widehat{\Omega}, \widehat{\mathscr{F}_*}, \widehat{\mathbb{P}})$ and $\widehat{X}_0 \in \widehat{\mathscr{F}_0}$ has the same law as X_0. Then $\widehat{X}$ and X have the same law.

Proof. We pass to the product probability space

$$(\mathbb{R}^N \times W(\mathbb{R}^l), \mathscr{B}(\mathbb{R}^N) \times \mathscr{B}(W(\mathbb{R}^l))_*, \mu_0 \times \mu^W),$$

where μ_0 is the common distribution of X_0 and $\hat{X}_0$, and μ^W is the law of an l-dimensional Brownian motion. A point in this space is denoted by (x_0, w). The equation on this space

$$J_t = x_0 + \int_0^t \sigma(J_s)\, dw_s + \int_0^t b\,(J_s)\, ds \tag{1.1.12}$$

has a unique solution up to its explosion time. Now the map $J : \mathbb{R}^N \times W(\mathbb{R}^l) \to W(\mathbb{R}^N)$ is a measurable map well defined $\mu_0 \times \mu^W$-almost everywhere. Since the law of (X_0, B) is $\mu_0 \times \mu^W$, the composition $J(X_0, W) : \Omega \to W(\mathbb{R}^d)$ is well defined. By (1.1.12), $J(X_0, W)$ satisfies (1.1.10). Now by the pathwise uniqueness for (1.1.10) we have $X = J(X_0, W)$. Likewise we have $\widehat{X} = J(\widehat{X}_0, \widehat{W})$. Because (X_0, W) and $(\widehat{X}_0, \widehat{W})$ have the same law, we conclude that X and $\hat{X}$ also have the same law. □

The following nonexplosion criterion is well known.

Proposition 1.1.11. *Suppose that Z is defined on $[0, \infty)$. If σ is locally Lipschitz and there is a constant C such that $|\sigma(x)| \leq C(1 + |x|)$, then the solution of $SDE(\sigma, Z, X_0)$ does not explode.*

Proof. We may assume that X_0 is uniformly bounded (see the last part of the proof of THEOREM 1.1.8). Let η be the inverse of Q as before (see (1.1.3)) and $\tau_N = \inf\{t > 0 : |X_t| = N\}$. From

$$X_t = X_0 + \int_0^t \sigma(X_s)\, dZ_s$$

and LEMMA 1.1.2 we have

$$\mathbb{E}|X|^2_{\infty,\eta_T\wedge\tau_N} \le 2\mathbb{E}|X_0|^2 + C\mathbb{E}\int_0^{\eta_T\wedge\tau_N}\left\{|X_s|^2+1\right\}dQ_s$$
$$\le 2\mathbb{E}|X_0|^2 + CT + C\int_0^T \mathbb{E}|X|^2_{\infty,\eta_s\wedge\tau_N}ds.$$

Hence by Gronwall's lemma

$$\mathbb{E}|X|^2_{\infty,\eta_T\wedge\tau_N} \le \left\{2\mathbb{E}|X_0|^2 + CT\right\}e^{CT}.$$

Letting $N\uparrow\infty$ and using the fact that $\tau_N\uparrow e(X)$, we see that, by defining $|X_t|=\infty$ for $t\ge e$,

$$\mathbb{E}|X|^2_{\infty,\eta_T\wedge e(X)}<\infty, \quad \text{hence} \quad \mathbb{P}\left\{|X|_{\infty,\eta_T\wedge e(X)}<\infty\right\}=1.$$

This implies that $\eta_T < e(X)$. Now $e(X)=\infty$ follows from the fact that $\eta_T\uparrow\infty$ as $T\uparrow\infty$. □

Example 1.1.12. (Exponential martingale) Suppose that N be a semimartingale on $\mathbb{R}^1$ and $N_0=0$. Consider the equation

$$X_t = 1 + \int_0^t X_s\,dN_s.$$

The solution is

$$X_t = \exp\left[N_t - \frac{1}{2}\langle N, N\rangle_t\right].$$

If N is a local martingale. Then X is called an exponential martingale. We leave the proof of the following properties of exponential martingales as an exercise:

(1) X is a nonnegative supermartingale; hence $\mathbb{E}X_t\le 1$ for all $t\ge 0$;

(2) X is a martingale if and only if $\mathbb{E}X_t = 1$ for all $t\ge 0$;

(3) If $\mathbb{E}\exp\{\alpha\langle N\rangle_t\}$ is finite for some $\alpha>1/2$, then $\mathbb{E}X_t=1$.

In fact $\mathbb{E}X_t = 1$ holds if $\mathbb{E}\exp\{\langle N\rangle_t/2\}$ is finite, or even $\mathbb{E}\exp\{N_t/2\}$ is finite, but the proof is considerably more difficult.

Example 1.1.13. (Ornstein-Uhlenbeck process) The equation on $\mathbb{R}^1$

$$dX_t = dZ_t - X_, dt$$

can be solved explicitly:

$$X_t = e^{-t}X_0 + \int_0^t e^{-(t-s)}dZ_s.$$

When Z is a 1-dimensional euclidean Brownian motion, X is a diffusion generated by

$$L = \frac{1}{2}\left(\frac{d}{dx}\right)^2 - x\frac{d}{dx}$$

and is called an Ornstein-Uhlenbeck process. Let X^x be the solution starting from x. Then

$$\mathbb{E}f(X_t^x) = \int_{\mathbb{R}^1} f\left(e^{-t}x + \sqrt{1-e^{-2t}}y\right)\mu(dy),$$

where

$$\mu(dx) = \frac{1}{\sqrt{2\pi}}e^{-x^2/2}dx$$

is the standard Gaussian measure on $\mathbb{R}^1$. The Gaussian measure μ is invariant for the process, i.e.,

$$\int_{\mathbb{R}} \mathbb{E}\left\{f(X_t^x)\right\}\mu(dx) = \int_{\mathbb{R}} f(x)\mu(dx).$$

Example 1.1.14. (Brownian bridge) Let B be a d-dimensional euclidean Brownian motion. Then the process $t \mapsto X_t = B_t - tB_1$ is called a Brownian bridge. Let $\mathscr{G}_t = \sigma\left\{B_s, s \le t; B_1\right\}$. Prove the following facts as an exercise:

(1) X is a semimartingale with respect to $\mathscr{G}_*$, and there is a $\mathscr{G}_*$-adapted Brownian motion W such that X is the solution of

$$X_t = W_t - \int_0^t \frac{X_s ds}{1-s}.$$

This equation can be solved explicitly:

$$X_t = (1-t)\int_0^t \frac{dW_s}{1-s};$$

(2) There is a Brownian motion U such that

$$X_t = (1-t)\, U_{t/(1-t)}, \quad 0 \le t < 1;$$

(3) The reversed process $t \mapsto X_t^* = X_{1-t}$ is also a Brownian bridge, i.e., it has the same law as X. □

If the driving semimartingale Z itself is defined only up to a stopping time τ, we can consider $SDE(\sigma, Z^{\tau_n}, X_0)$ for a sequence of stopping times $\tau_n \uparrow\uparrow \tau$.

Theorem 1.1.15. *Let Z be a semimartingale defined up to a stopping time τ. Then there is a unique solution X to $SDE(\sigma, Z, X_0)$ up to the stopping time $e(X) \wedge \tau$. If Y is another solution up to a stopping time $\eta \le \tau$, then $\eta \le e(X) \wedge \tau$ and $X_t = Y_t$ for $0 \le t < e(X) \wedge \eta$.*

Proof. Exercise. □

We now turn to the Stratonovich formulation of stochastic differential equations. The advantage of this formulation is that Itô's formula appears in the same form as the fundamental theorem of calculus; therefore stochastic calculus in this formulation takes a more familiar form (compare (1.1.2) with (1.1.14) below). This is a very convenient feature when we study stochastic differential equations on manifolds. However, it often happens that useful probabilistic and geometric information reveals itself only after we separate martingale and bounded variation components.

Suppose that V_α, $\alpha = 1, \ldots, l$, are smooth vector fields on $\mathbb{R}^N$. Each V_α can be regarded as a function $V_\alpha : \mathbb{R}^N \to \mathbb{R}^N$ so that $V = (V_1, \ldots, V_l)$ is an $\mathscr{M}(N, l)$-valued function on $\mathbb{R}^d$. Let Z and X_0 be as before and consider the Stratonovich stochastic differential equation

$$X_t = X_0 + \int_0^t V(X_s) \circ dZ_s,$$

where the stochastic integral is in the Stratonovich sense. To emphasize the fact that V is a set of l vector fields, we rewrite the equation as

$$X_t = X_0 + \int_0^t V_\alpha(X_s) \circ dZ_s^\alpha. \tag{1.1.13}$$

Converting the Stratonovich integral into the equivalent Itô integral, we obtain the equivalent Itô formulation of the equation:

$$X_t = X_0 + \int_0^t V_\alpha(X_s)\, dZ_s^\alpha + \frac{1}{2}\int_0^t \nabla_{V_\beta} V_\alpha(X_s)\, d\langle Z^\alpha, Z^\beta\rangle_s.$$

Here $\nabla_{V_\beta} V_\alpha$ is the derivative of V_α along V_β. For future reference we record here Itô's formula in this setting in the following proposition.

Proposition 1.1.16. *Let X be a solution to the equation (1.1.13) and $f \in C^2(\mathbb{R}^d)$. Then*

$$f(X_t) = f(X_0) + \int_0^t V_\alpha f(X_s) \circ dZ_s, \qquad 0 \le s < e(X). \tag{1.1.14}$$

Proof. Exercise. □

1.2. SDE on manifolds

The discussion in the preceding section makes it clear that solutions of stochastic differential equations on manifolds should be sought in the space of manifold-valued semimartingales.

Definition 1.2.1. *Let M be a differentiable manifold and $(\Omega, \mathscr{F}_*, \mathbb{P})$ a filtered probability space. Let τ be an $\mathscr{F}_*$-stopping time. A continuous, M-valued process X defined on $[0, \tau)$ is called an M-valued semimartingale if $f(X)$ is a real-valued semimartingale on $[0, \tau)$ for all $f \in C^\infty(M)$.*

By Itô's formula it is easy to see that when $M = \mathbb{R}^N$ this definition gives the usual $\mathbb{R}^N$-valued semimartingales.

Remark 1.2.2. It can be shown that if $f(X)$ is a real-valued martingale for all $f \in C_K^\infty(M)$ (smooth functions on M with compact support), then X is an M-valued semimartingale; in other words, the set of test functions can be reduced to $C_K^\infty(M)$. On the other hand, if X is an M-valued semimartingale, then by Itô's formula, $f(X)$ is a real-valued semimartingale for all $f \in C^2(M)$; see PROPOSITION 1.2.7

A stochastic differential equation on a manifold M is defined by l vector fields $V_1, \ldots, V_l$ on M, an $\mathbb{R}^l$-valued driving semimartingale Z, and an M-valued random variable $X_0 \in \mathscr{F}_0$, serving as the initial value of the solution. We write the equation symbolically as

$$dX_t = V_\alpha(X_t) \circ dZ_s^\alpha$$

and refer to it as $SDE(V_1, \ldots, V_l; Z, X_0)$. In view of PROPOSITION 1.1.16 we make the following definition.

Definition 1.2.3. *An M-valued semimartingale X defined up to a stopping time τ is a solution of $SDE(V_1, \ldots, V_l, Z, X_0)$ up to τ if for all $f \in C^\infty(M)$,*

$$f(X_t) = f(X_0) + \int_0^t V_\alpha f(X_s) \circ dZ_s^\alpha, \quad 0 \le t < \tau. \tag{1.2.1}$$

It is an easy consequence of Itô's formula (1.1.14) that if (1.2.1) holds for $f_1, \ldots, f_k$, then it automatically holds for any smooth function of them. We see later that if M is embedded in a euclidean space $\mathbb{R}^N$, then it is enough to verify (1.2.1) for the coordinate functions.

The advantage of the Stratonovich formulation is that stochastic differential equations on manifolds in this formulation transform consistently under diffeomorphisms between manifolds. Let $\Gamma(TM)$ denote the space of smooth vector fields on a manifold M (the space of sections of the tangent bundle TM). A diffeomorphism $\Phi : M \to N$ between two manifolds induces a map $\Phi_* : \Gamma(TM) \to \Gamma(TN)$ between the vector fields on the respective manifolds by the prescription

$$(\Phi_* V)f(y) = V(f \circ \Phi)(x), \quad y = \Phi(x), \quad f \in C^\infty(N).$$

Equivalently, if V is the tangent vector of a curve C on M, then $\Phi_* V$ is the tangent vector of the curve $\Phi \circ C$ on N.

Proposition 1.2.4. *Suppose that $\Phi : M \to N$ is a diffeomorphism and X a solution of $SDE(V_1, \ldots, V_l; Z, X_0)$. Then $\Phi(X)$ is a solution of $SDE(\Phi_* V_1, \ldots, \Phi_* V_l; Z, \Phi(X_0))$ on N.*

Proof. Let $Y = \Phi(X)$ and $f \in C^\infty(N)$. Applying (1.2.1) to $f \circ \Phi \in C^\infty(M)$ and using $V_\alpha(f \circ \Phi)(X_s) = (\Phi_* V_\alpha) f(Y_s)$, we obtain

$$\begin{aligned} f(Y_t) &= f(Y_0) + \int_0^t V_\alpha(f \circ \Phi)(X_s) \circ dZ_s^\alpha \\ &= f(Y_0) + \int_0^t (\Phi_* V_\alpha) f(Y_s) \circ dZ_s^\alpha. \end{aligned}$$

Hence $Y = \Phi(X)$ is a solution of $SDE(\Phi_* V_1, \ldots, \Phi_* V_l; Z, \Phi(X_0))$. □

We will prove that $SDE(V_1, \ldots, V_l; Z, X_0)$ has a unique solution up to its explosion time. Our strategy is to reduce the equation to an equation on a euclidean space of the same type by the well known Whitney's embedding theorem (see de Rham [**63**]).

Theorem 1.2.5. (Whitney's embedding theorem) *Suppose that M is a differentiable manifold. Then there exists an embedding $i : M \to \mathbb{R}^N$ for some N such that the image $i(M)$ is a closed subset of $\mathbb{R}^N$.*

It is known from differential topology that $N = 2\dim M + 1$ will do. We often identify M with the image $i(M)$ and assume that M itself is a closed submanifold of $\mathbb{R}^N$. Note that we assume that M does not have a boundary. The fact that M is a *closed* submanifold of $\mathbb{R}^N$ (i.e., M is a closed subset of $\mathbb{R}^N$) is very important to us, for it allows us to identify the point at infinity of M with that of $\mathbb{R}^N$.

Proposition 1.2.6. *Let M be a (noncompact) closed submanifold of $\mathbb{R}^N$ and $\widehat{M} = M \cup \{\partial_M\}$ its one-point compactification. A sequence of points $\{x^n\}$ in M converges to ∂_M in $\widehat{M}$ if and only if $|x^n|_{\mathbb{R}^N} \to \infty$.*

Proof. Exercise. □

Suppose that M is a closed submanifold of $\mathbb{R}^N$. A point $x \in M$ has N coordinates $\{x^1, \ldots, x^N\}$ as a point in $\mathbb{R}^N$. The following proposition shows that the N coordinate functions $f^i(x) = x^i$ can serve as a natural set of test functions for Itô's formula on M.

Proposition 1.2.7. *Suppose that M is a closed submanifold of $\mathbb{R}^N$. Let $f^1, \ldots, f^N$ be the coordinate functions. Let X be an M-valued continuous process.*

(i) X is a semimartingale on M if and only if it is an $\mathbb{R}^N$-valued semimartingale, or equivalently, if and only if $f^i(X)$ is a real-valued semimartingale for each $i = 1, \ldots, N$.

(ii) X is a solution of $SDE(V_1, \ldots, V_l; Z, X_0)$ up to a stopping time σ if and only if for each $i = 1, \ldots, N$,

$$f^i(X_t) = f^i(X_0) + \int_0^t V_\alpha f^i(X_s) \circ dZ_s^\alpha, \quad 0 \le t < \sigma. \tag{1.2.2}$$

Proof. (i) Suppose that X is an M-valued semimartingale. Each f^i is a smooth function on M, thus by definition $f^i(X) = X^i$ is a real-valued semimartingale. This means that X is a $\mathbb{R}^N$-valued semimartingale. Conversely, suppose that X lives on M and is an $\mathbb{R}^N$-valued semimartingale. Since M is closed in $\mathbb{R}^N$, a function $f \in C^\infty(M)$ can be extended to a function $\tilde{f} \in C^\infty(\mathbb{R}^N)$. Hence $f(X) = \tilde{f}(X)$ is a real-valued semimartingale, and by definition X is an M-valued semimartingale.

(ii) If X is a solution, then (1.2.2) holds because each $f^i \in C^\infty(M)$. Now suppose (1.2.2) holds and $f \in C^\infty(M)$. Take an extension $\tilde{f} \in C^\infty(\mathbb{R}^N)$ of f. Then $f(X_t) = \tilde{f}(f^1(X_t), \cdots, f^N(X_t))$. By Itô's formula,

$$\begin{aligned} d\{f(X_t)\} &= f_{x_i}(f^1(X_t), \cdots, f^N(X_t)) \circ d\{f^i(X_t)\} \\ &= f_{x_i}(f^1(X_t), \cdots, f^N(X_t)) \circ V_\alpha f^i(X_t) \circ dZ_t^\alpha \\ &= \{(f_{x_i}(X_t^1, \cdots, X_t^N) V_\alpha f^i(X_t)\} \circ dZ_t^\alpha \\ &= V_\alpha f(X_t) \circ dZ^\alpha. \end{aligned}$$

In the last step we have used the chain rule for differentiating composite functions. □

Returning to the $SDE(V_1, \ldots, V_l; Z, X_0)$, we fix an embedding of M into $\mathbb{R}^N$ and regard M as a closed submanifold of $\mathbb{R}^N$. Each vector field V_α is at the same time a smooth, $\mathbb{R}^N$-valued function on M and can be extended to a vector field $\tilde{V}_\alpha$ on $\mathbb{R}^N$. From the discussion in the last section the equation

$$X_t = X_0 + \int_0^t \tilde{V}_\alpha(X_s) \circ dZ_s^\alpha \tag{1.2.3}$$

on $\mathbb{R}^N$ has a unique solution X up to its explosion time $e(X)$. Since X starts from M and the vector fields $\tilde{V}_\alpha$ are tangent to M on M, it is expected, as in ordinary differential equations, that X never leaves M. Once this fact is established, $e(X)$ is also the explosion time of X as a semimartingale on M by PROPOSITION 1.2.6.

Proposition 1.2.8. *Let X be the solution of the extended equation (1.2.3) up to its explosion time $e(X)$ and $X_0 \in M$. Then $X_t \in M$ for $0 \le t < e(X)$.*

Proof. Without loss of generality we may assume that X_0 is uniformly bounded (see the end of the proof of THEOREM 1.1.8). Let $d_{\mathbb{R}^N}(x, M)$ be the euclidean distance from x to M. Because M is closed and without boundary, the function $f(x) = d_{\mathbb{R}^N}(x, M)^2$ is smooth in a neighborhood of M. Multiplying by a suitable cut-off function, we may assume that $f \in C^\infty(\mathbb{R}^N)$. Since the vector fields $\tilde{V}_\alpha$ are tangent to M along the submanifold M, a local calculation shows that the functions $\tilde{V}_\alpha f$ and $\tilde{V}_\alpha \tilde{V}_\beta f$ vanish along M at the rate of the square of the distance $d_{\mathbb{R}^N}(x, M)$. Hence there is a neighborhood U of M such that (i) $f(x) = 0$ for an $x \in U$ if and only if $x \in M$; (ii) for each $R > 0$, there exists C depending on R such that

$$|\tilde{V}_\alpha f(x)| \le C f(x), \quad |\tilde{V}_\alpha \tilde{V}_\beta f(x)| \le C f(x) \tag{1.2.4}$$

for all $x \in U \cap B(0; R)$. Define the stopping times:

$$\begin{aligned} \tau_R &= \inf \{t > 0 : X_t \notin B(R)\}, \\ \tau_U &= \inf \{t > 0 : X_t \notin U\}, \\ \tau &= \tau_U \wedge \tau_R. \end{aligned}$$

Consider the process X before τ, the first exit time from $U \cap B(R)$. Applying Itô's formula, we have

$$f(X_t) = \int_0^t V_\alpha f(X_s)\, dZ_s^\alpha + \frac{1}{2} \int_0^t V_\alpha V_\beta f(X_s)\, d\langle Z^\alpha, Z^\beta \rangle_s.$$

Using LEMMA 1.1.2 to the right side of the above equation, we have

$$\begin{aligned} &\mathbb{E}\, |f(X)|^2_{\infty, \eta_t \wedge \tau} \le \\ &\qquad C \sum_{\alpha,\beta} \mathbb{E} \int_0^{\eta_t \wedge \tau} \left\{ |V_\alpha f(X_s)|^2 + |V_\alpha V_\beta f(X_s)|_s^2 \right\} dQ_s. \end{aligned}$$

We now make the time change $s \mapsto \eta_s$ and use (1.2.4), which is permissible because $X_s \in U \cap B(R)$ if $s \le \tau$. This yields the inequality

$$\mathbb{E}\, |f(X)|^2_{\infty, \eta_t \wedge \tau} \le C_1 \int_0^t \mathbb{E} |f(X)|^2_{\infty, \eta_s \wedge \tau} ds.$$

It follows that $f(X_s) = 0$ for $s \le \eta_t \wedge \tau$, or

$$X_s \in M \quad \text{for} \quad 0 \le s < \eta_t \wedge \tau_U \wedge \tau_R.$$

Letting t and then R go to infinity, we see that $X_s \in M$ for $0 \le s < e(X) \wedge \tau_U$, from which we conclude easily that X stays on M up to $e(X)$. □

We are now in a position to prove the main result of this section.

Theorem 1.2.9. *There is a unique solution of $SDE(V_1, \ldots, V_l; Z, X_0)$ up to its explosion time.*

Proof. By PROPOSITION 1.2.8 the solution X of $SDE(\tilde{V}_1, \ldots, \tilde{V}_l; Z, X_0)$ stays in M up to its explosion time and satisfies (1.2.3). But (1.2.3) is nothing but a rewriting of (1.2.2); hence X is a solution of $SDE(V_1, \ldots, V_l; Z, X_0)$ by PROPOSITION 1.2.7 (ii).

If Y is another solution up to a stopping time τ, then, regarded as a semimartingale on $\mathbb{R}^N$, it is also a solution of $SDE(\tilde{V}_1, \ldots, \tilde{V}_l; Z, X_0)$ up to τ. By the uniqueness statement in PROPOSITION 1.1.9, Y must coincide with X on $[0, \tau)$. □

Remark 1.2.10. The case where the driving semimartingale runs up to a stopping time can be discussed as in the euclidean case; see THEOREM 1.1.15.

1.3. Diffusion processes

The analytic significance of diffusion processes derives from their relation with second order elliptic operators. Diffusion theory of smooth second order elliptic operators can be adequately treated by stochastic differential equations. In this section we discuss diffusion processes generated by such operators from the viewpoint of martingale problems. The main results of this section are the uniqueness in law and the strong Markov property of such processes. For a complete exposition of martingale problems, the reader is referred to Stroock and Varadhan [**68**], which is also our main reference for this section.

Throughout this section L denotes a smooth second order elliptic, but not necessarily nondegenerate, differential operator on a differentiable manifold M. If $f \in C^2(M)$ and $\omega \in W(M)$, we let

$$M^f(\omega)_t = f(\omega_t) - f(\omega_0) - \int_0^t Lf(\omega_s)\, ds, \qquad 0 \le t < e(\omega). \tag{1.3.1}$$

Definition 1.3.1. *(i) An $\mathscr{F}_*$-adapted stochastic process $X : \Omega \to W(M)$ defined on a filtered probability space $(\Omega, \mathscr{F}_*, \mathbb{P})$ is called a diffusion process generated by L (or simply an L-diffusion) if X is an M-valued $\mathscr{F}_*$-semimartingale up to $e(X)$ and*

$$M^f(X)_t = f(X_t) - f(X_0) - \int_0^t Lf(X_s)\, ds, \quad 0 \le t < e(X),$$

is a local $\mathscr{F}_*$-martingale for all $f \in C^\infty(M)$.

(ii) A probability measure μ on the standard filtered path space $(W(M), \mathscr{B}(W(M))_*)$ is called a diffusion measure generated by L (or simply an L-diffusion measure) if

$$M^f(\omega)_t = f(\omega_t) - f(\omega_0) - \int_0^t Lf(\omega_s)\, ds, \quad 0 \le t < e(\omega),$$

is a local $\mathscr{B}(W(M))_*$-martingale for all $f \in C^\infty(M)$.

For a given L, the relation between L-diffusion measures and L-diffusion processes is as follows. If X is an L-diffusion, then its law $\mu^X = \mathbb{P} \circ X^{-1}$ is an L-diffusion measure; conversely, if μ is an L-diffusion measure on $W(M)$, then the coordinate process $X(\omega)_t = \omega_t$ on $(W(M), \mathscr{B}(W(M))_*, \mu)$ is an L-diffusion process.

We show that, given a smooth second order elliptic operator and a probability distribution μ_0 on M, there exists a unique L-diffusion measure whose initial distribution is μ_0. The approach is similar to the one we have used before for stochastic differential equations on manifolds, namely, we embed M in $\mathbb{R}^N$ as a closed submanifold and extend L to an operator $\tilde{L}$ of the same type on the ambient space. The $\tilde{L}$-diffusion X is then constructed by solving a stochastic differential equation on $\mathbb{R}^N$, and we verify that it in fact lives on M and is an L-diffusion.

In the following we will use $\mathscr{S}_+(l)$ to denote the space of $(l \times l)$-symmetric, positive definite (more precisely, nonnegative definite) matrices. Let $d = \dim M$. In a local coordinate system $x = \{x^i\}$ covering a neighborhood U, the operator L has the form

$$L = \frac{1}{2}\sum_{i,j=1}^{d} a^{ij}(x)\frac{\partial}{\partial x^i}\frac{\partial}{\partial x^j} + \sum_{i=1}^{d} b^i(x)\frac{\partial}{\partial x^i}, \tag{1.3.2}$$

where both $a = \{a^{ij}\} : U \to \mathscr{S}_+(d)$ and $b = \{b^i\} : U \to \mathbb{R}^d$ are smooth functions. Define

$$\Gamma(f,g) = L(fg) - fLg - gLf, \qquad f, g \in C^\infty(M).$$

A direct computation shows that

$$\Gamma(f,g) = a^{ij}\frac{\partial f}{\partial x^i}\frac{\partial g}{\partial x^j}. \tag{1.3.3}$$

It is easy to verify that if $f^1, \ldots, f^N$, are N smooth functions on M, then the square matrix $\{\Gamma(f^\alpha, f^\beta)\}$ is positive definite everywhere on M.

We assume that M is a closed submanifold of $\mathbb{R}^N$ and denote the coordinates of the ambient space by $z = \{z^1, \ldots, z^N\}$. The coordinate functions on M are $f^\alpha(z) = z^\alpha$, the αth coordinate of the point $z \in M$. Let

$$\tilde{a}^{\alpha\beta} = \Gamma(f^\alpha, f^\beta), \quad \tilde{b}^\alpha = Lf^\alpha.$$

They are smooth functions on M and $\{\tilde{a}^{\alpha\beta}\}$ is positive definite. We extend $\tilde{a}$ and $\tilde{b}$ smoothly to the ambient space so that they take values in $\mathscr{S}_+(N)$ and $\mathbb{R}^N$ respectively. The operator

$$\tilde{L} = \frac{1}{2}\sum_{\alpha,\beta=1}^{N} \tilde{a}^{\alpha\beta}\frac{\partial^2}{\partial z^\alpha \partial z^\beta} + \sum_{\alpha=1}^{N} \tilde{b}^\alpha \frac{\partial}{\partial z^\alpha} \tag{1.3.4}$$

is an extension of L to the ambient space in the following sense.

Lemma 1.3.2. *Suppose that $f \in C^\infty(M)$. Then for any $\tilde{f} \in C^\infty(\mathbb{R}^N)$ which extends f from M to $\mathbb{R}^N$, we have $\tilde{L}\tilde{f} = Lf$ on M.*

Proof. Let $x = \{x^i\}$ be a local coordinate system on M. Then on M we have $f(x) = \tilde{f}(f^1(x), \cdots, f^N(x))$. Using (1.3.2), we compute Lf by the chain rule and see that it agrees with $\tilde{L}\tilde{f}$. □

The next step is to construct an $\tilde{L}$-diffusion on $\mathbb{R}^N$ by solving a stochastic differential equation of the form

$$X_t = X_0 + \int_0^t \tilde{\sigma}(X_s)\, dW_s + \int_0^t \tilde{b}(X_s) ds, \tag{1.3.5}$$

where W is an N-dimensional euclidean Brownian motion and X_0 is an $\mathbb{R}^N$-valued random variable independent of W. Using Itô's formula we verify easily that a process X satifying (1.3.5) is an $\tilde{L}$ diffusion with $\tilde{L}$ given in (1.3.4) if $\tilde{a} = \tilde{\sigma}\tilde{\sigma}^\dagger$. The obvious choice is to take $\tilde{\sigma} = \tilde{a}^{1/2}$, the unique (symmetric) positive definite matrix square root of a. In order that our previous theory apply to the equation (1.3.5), we need the following lemma.

Lemma 1.3.3. *Let $S : \mathbb{R}^N \to \mathscr{S}_+(l)$ be twice continuously differentiable. Then its positive definite matrix square root $S^{1/2} : \mathbb{R}^N \to \mathscr{S}_+(l)$ is locally Lipschitz.*

Proof. Since the problem is local we may assume that all derivatives of S up to the second order are uniformly bounded. By considering $S + \lambda I$ and then letting $\lambda \downarrow 0$, we may assume that S is strictly positive definite and prove the following bound for the first derivatives of $T = S^{1/2}$ independent of the lower bound of the eigenvalues of S:

$$\sup_{x\in\mathbb{R}^N} |\nabla T(x)_{ij}| \le \sqrt{2C}l,$$

where

$$C = \sup_{x\in\mathbb{R}^N,\, 1\le k,m\le l} \left\| \left\{ \frac{\partial^2 S_{km}(x)}{\partial x^i \partial x^j} \right\} \right\|_{2,2}.$$

We first prove the special case of a positive function, i.e., $l = 1$. Let $x, y \in \mathbb{R}^N$. There is a point z on the line segment joining x and y such that

$$0 \le f(y) = f(x) + \nabla f(x) \cdot (y - x) + \frac{1}{2}(y - x)^\dagger \nabla^2 f(z)(y - x).$$

Hence for all $y \in \mathbb{R}^N$,

$$f(x) + \nabla f(x) \cdot (y - x) + \frac{1}{2}C|y - x|^2 \ge 0.$$

Letting $y = x - \nabla f(x)/C$, we obtain $|\nabla f(x)| \le \sqrt{2Cf(x)}$.

For the general case, fix an $x \in \mathbb{R}^d$ and assume for the time being that $S(x)$ is diagonal. We have

$$S(y) = S(x) + \nabla S(x) \cdot (y - x) + O(|x - y|^2),$$
$$T(y) = T(x) + \nabla T(x) \cdot (y - x) + O(|x - y|^2).$$

Comparing the coefficients of the first order terms in $T(y)T(y) = S(y)$, we obtain

$$\nabla T(x)T(x) + T(x)\nabla T(x) = \nabla S(x).$$

Comparing the (i, j)-entries, we have

$$\nabla T(x)_{ij} = \frac{\nabla S(x)_{ij}}{T(x)_{ii} + T(x)_{jj}}. \tag{1.3.6}$$

Now the functions $f_\pm(x) = (e_i \pm e_j)^\dagger S(x)(e_i \pm e_j)$ (e_i and e_j are the ith and jth coordinate unit vectors in $\mathbb{R}^l$) are nonnegative on $\mathbb{R}^d$. Applying the special case proved above to these two functions, we have $|\nabla f_\pm(x)| \le \sqrt{8Cf_\pm(x)}$. Hence,

$$\begin{aligned}
4|\nabla S(x)_{ij}| &= |\nabla f_+(x) - \nabla f_-(x)| \\
&\le \sqrt{8C}\left\{\sqrt{f_+(x)} + \sqrt{f_-(x)}\right\} \\
&\le 4\sqrt{C}\sqrt{f_+(x) + f_-(x)} \\
&= 4\sqrt{2C}\sqrt{S(x)_{ii} + S(x)_{jj}} \\
&\le 4\sqrt{2C}\left\{T(x)_{ii} + T(x)_{jj}\right\}.
\end{aligned}$$

From this and (1.3.6) we have $|\nabla T(x)_{ij}| \le \sqrt{2C}$.

If $S(x)$ is not diagonal at x, let O be an orthogonal matrix such that $OS(x)O^\dagger$ is diagonal. Then $(OSO^\dagger)^{1/2} = OTO^\dagger$. Applying the above argument to $OSO^\dagger$, we see that every entry of the matrix $O\nabla T(x)O^\dagger$ is bounded by $\sqrt{2C}$. It follows by simple linear algebra that every entry of $\nabla T(x)$ is bounded by $\sqrt{2C}l$. □

Now it is not difficult to prove the existence of an L-diffusion measure with a given initial distribution.

Theorem 1.3.4. *Let L be a smooth second order elliptic operator on a differentiable manifold M and μ_0 a probability measure on M. Then there exists an L-diffusion measure with initial distribution μ_0.*

Proof. Let $\tilde{a}, \tilde{b}$ be as before and $\tilde{\sigma} = \tilde{a}^{1/2}$ the unique positive matrix square root of $\tilde{a}$. By the preceding lemma, $\tilde{\sigma}$ is locally Lipschitz on $\mathbb{R}^N$. Let W be an N-dimensional euclidean Brownian motion and X_0 an M-valued random variable independent of W whose distribution is μ_0. The solution X of

(1.3.5) is an $\tilde{L}$-diffusion. We claim that the law $\mu^X = \mathbb{P} \circ X^{-1}$ of X in the path space $W(\mathbb{R}^N)$ is concentrated on $W(M)$ and is an L-diffusion measure with initial distribution μ_0.

The proof that X lives on M is similar to that of PROPOSITION 1.2.8. We need only to observe that $f(x) = d_{\mathbb{R}^N}(x, M)^2$ has the property that for each $R > 0$ there exists C depending on R such that $|\tilde{L}f(x)| \le Cf(x)$ in a neighborhood of $M \cap B(R)$, and this fact can be verified easily from the definition of $\tilde{L}$ in (1.3.4) by a local computation.

We next show that X is an L-diffusion. Since X is an $\tilde{L}$-diffusion, for every $\tilde{f} \in C^\infty(\mathbb{R}^N)$,

$$\tilde{f}(X_t) = \tilde{f}(X_0) + \text{local martingale} + \int_0^t \tilde{L}\tilde{f}(X_s)\, ds, \quad 0 \le t < e(X).$$

Suppose that $f \in C^\infty(M)$ and let $\tilde{f}$ be a smooth extension of f to $\mathbb{R}^N$. In view of LEMMA 1.3.2 and the fact that X lives on M the above equality becomes

$$f(X_t) = f(X_0) + \text{local martingale} + \int_0^t Lf(X_s)\, ds, \quad 0 \le t < e(X).$$

This shows that X is an L-diffusion and its law μ^X is an L-diffusion measure with the initial distribution $\mu_0 = \mathbb{P} \circ X_0^{-1}$. □

In order to prove the uniqueness in law of an L-diffusion measure we need the following extension of Lévy's criterion for Brownian motion.

Lemma 1.3.5. *Let M be an $\mathbb{R}^N$-valued local martingale on a probability space $(\Omega, \mathscr{F}_*, \mathbb{P})$ such that*

$$\langle M, M^\dagger \rangle_t = \int_0^t a(s)\, ds$$

for an $\mathscr{F}_*$-adapted, $\mathscr{S}_+(N)$-valued process $\{a(s), s \ge 0\}$. Let $\sigma(s)$ be the (symmetric) positive definite matrix square root of $a(s)$. Let W be an $\mathbb{R}^N$-valued Brownian motion on another probability space $(\Pi, \mathscr{G}_*, \mathbb{Q})$. On the product probability space $(\Omega \times \Pi, \mathscr{F}_* \times \mathscr{G}_*, \mathbb{P} \times \mathbb{Q})$ define $M(\omega, \pi) = M(\omega)$ and $\sigma(\omega, \pi) = \sigma(\omega)$. Then there exists an N-dimensional euclidean Brownian motion B defined on the product probability space such that

$$M_t = \int_0^t \sigma(s)\, dB_s.$$

Proof. If σ is nondegenerate, then

$$B_t = \int_0^t \sigma(s)^{-1} dM_s$$

is a Brownian motion on $\mathbb{R}^N$ by the usual Lévy's criterion. If $\sigma(s)$ is degenerate we need to use the extra Brownian motion W to "fill in the gaps."

We regard a positive definite symmetric matrix σ as a linear transform on $\mathbb{R}^N$ and denote by P_σ the orthogonal projection onto its range $\mathrm{Ran}(\sigma)$. Since σ is symmetric we have $\mathbb{R}^N = \mathrm{Ker}(\sigma) \oplus \mathrm{Ran}(\sigma)$, and the restriction $\sigma : \mathrm{Ran}(\sigma) \to \mathrm{Ran}(\sigma)$ is an isomorphism. The generalized inverse σ^{-1} is a linear transform defined as follows:

$$\sigma^{-1}z = \begin{cases} 0, & \text{if } z \in \mathrm{Ker}(\sigma); \\ y, & \text{if } z \in \mathrm{Ran}(\sigma); \end{cases}$$

where y is the unique element in $\mathrm{Ran}(\sigma)$ such that $\sigma y = z$. It is easy to verify that σ^{-1} is again symmetric and

$$\sigma\sigma^{-1} = \sigma^{-1}\sigma = P_\sigma.$$

After these algebraic preparations we define, on the product probability space,

$$B_t = \int_0^t \sigma(s)^{-1} dM_s + \int_0^t \left(I - P_{\sigma(s)}\right) dW_s.$$

Its quadratic variation matrix is

$$\begin{aligned} d\langle B, B^\dagger\rangle_t &= \left\{\sigma(t)^{-1} dM_t + \left(I - P_{\sigma(t)}\right) dW_t\right\} \\ &\qquad \cdot \left\{\sigma(t)^{-1} dM_t + \left(I - P_{\sigma(t)}\right) dW_t\right\}^\dagger \\ &= \sigma(t)^{-1} d\langle M, M^\dagger\rangle_t \sigma(t)^{-1} \\ &\qquad + \left(I - P_{\sigma(t)}\right) d\langle W, W^\dagger\rangle_t \left(I - P_{\sigma(t)}\right) \\ &= \sigma(t)^{-1} a(t) \sigma(t)^{-1} dt + \left(I - P_{\sigma(t)}\right) dt \\ &= P_{\sigma(t)} dt + \left(I - P_{\sigma(t)}\right) dt \\ &= I dt. \end{aligned}$$

Note that since W and M are independent martingales with respect to the filtration $\mathscr{F}_* \times \mathscr{G}_*$, we have

$$d\langle W, M^\dagger\rangle_t = d\langle M, W^\dagger\rangle_t = 0.$$

Now, by the usual Lévy's criterion B is an N-dimensional euclidean Brownian motion. To verify the representation for M, we first note that

$$\left(I - P_{\sigma(t)}\right) dM_t = 0, \qquad \text{or} \qquad P_{\sigma(t)}\, dM_t = dM_t,$$

because its quadratic variation is

$$\begin{aligned} &\left(I - P_{\sigma(t)}\right) d\langle M, M^\dagger\rangle_t \left(I - P_{\sigma(t)}\right)^\dagger \\ &\qquad = \left(I - P_{\sigma(t)}\right) a(t) \left(I - P_{\sigma(t)}\right)^\dagger dt = 0. \end{aligned}$$

Finally we have

$$\sigma(t)\, dB_t = \sigma(t)\sigma(t)^{-1} dM_t - \Gamma_{\sigma(t)} dM_t = dM_t,$$

and this completes the proof. □

Theorem 1.3.6. *An L-diffusion measure with a given initial distribution is unique.*

Proof. Suppose that μ is an L-diffusion measure on $W(M)$. Let $\{X_t\}$ be the coordinate process on $W(M)$, i.e., $X_t(\omega) = \omega_t$ for $\omega \in W(M)$. Let $f^\alpha(z) = z^\alpha$ as before and $f^{\alpha\beta} = f^\alpha f^\beta$. The components of the coordinate process X_t in $\mathbb{R}^N$ are $X_t^\alpha = f^\alpha(X_t)$. Under the measure μ, we have

$$X_t^\alpha = X_0^\alpha + M_t^{f^\alpha} + \int_0^t \tilde{b}^\alpha(X_s)\, ds, \tag{1.3.7}$$

where $\tilde{b}^\alpha = Lf^\alpha$. We construct a stochastic differential equation for the coordinate process $\{X_t^\alpha\}$ and then use the uniqueness in law on $\mathbb{R}^N$ in THEOREM 1.1.10.

The Doob-Meyer decomposition of $X_t^\alpha X_t^\beta = f^{\alpha\beta}(X_t)$ can be computed in two ways, namely by the martingale property of μ applied to the function $f^{\alpha\beta}$,

$$X_t^\alpha X_t^\beta = X_0^\alpha X_0^\beta + M_t^{f^{\alpha\beta}} + \int_0^t Lf^{\alpha\beta}(X_s)\, ds,$$

and by Itô's formula applied to the product of two semimartingales X_t^α and X_t^β (see (1.3.7)),

$$\begin{aligned} X_t^\alpha X_t^\beta = X_0^\alpha X_0^\beta &+ \int_0^t Lf^\alpha(X_s)\, dM_s^{f^\beta} + \int_0^t Lf^\beta(X_s)\, dM_s^{f^\alpha} \\ &+ \int_0^t \left\{ f^\alpha(X_s) Lf^\beta(X_s) + f^\beta(X_s) Lf^\alpha(X_s) \right\} ds \\ &+ \langle M^{f^\alpha}, M^{f^\beta} \rangle_t. \end{aligned}$$

Comparing the bounded variation parts of the two expressions for $X_t^\alpha X_t^\beta$, we obtain

$$\langle X^\alpha, X^\beta \rangle_t = \langle M^{f^\alpha}, M^{f^\beta} \rangle_t = \int_0^t \Gamma(f^\alpha, f^\beta)(X_s)\, ds, \tag{1.3.8}$$

where Γ is defined in (1.3.3). The matrix $\left\{ \Gamma(f^\alpha, f^\beta) \right\}$ is symmetric and positive definite. Denote by $\tilde{\sigma}$ its symmetric positive definite square root. By LEMMA 1.3.5, on an extended probability space there exists an N-dimensional Brownian motion B such that

$$M_t^{f^\alpha} = \int_0^t \tilde{\sigma}^{\alpha\beta}(X_s)\, dB_s^\beta$$

and (1.3.7) becomes

$$X_t = X_0 + \int_0^t \tilde{\sigma}(X_s)\, dB_s + \int_0^t \tilde{b}(X_s)\, ds. \tag{1.3.9}$$

This is the equation for X we are looking for. Now μ remains the law of X in the extended probability space; it is therefore the law of the solution of (1.3.9). Because the coefficients in the equation are determined by L (and a fixed embedding of M into $\mathbb{R}^N$), every L-diffusion measure is the law of the solution of the same equation (1.3.9). Therefore by the uniqueness in law of solutions of stochastic differential equations with local Lipschitz coefficients (THEOREM 1.1.10) an L-diffusion measure with a given initial distribution must be unique □

When the generator L is understood, we use $\mathbb{P}_{\mu_0}$ to denote the L-diffusion measure with initial distribution μ_0. If μ_0 is concentrated at a point x, we simply denote it by $\mathbb{P}_x$. Now consider the filtered probability space $(W(M), \mathscr{B}_*, \mathbb{P}_{\mu_0})$, where $\mathscr{B}_* = \mathscr{B}(W(M))_*$ is the standard filtration on $W(M)$. Let $\{\mathbb{P}^{\tilde{\omega}}, \tilde{\omega} \in W(M)\}$ be the regular conditional probabilities of $\mathbb{P}_{\mu_0}$ with respect to $\mathscr{B}_0$. This means that $\mathbb{P}^{\tilde{\omega}}$ is a probability measure on $(W(M), \mathscr{B}_*)$ for each $\tilde{\omega} \in W(M)$ such that

(i) $\tilde{\omega} \mapsto \mathbb{P}^{\tilde{\omega}}(C)$ is $\mathscr{B}_0$-measurable for each $C \in \mathscr{B}_\infty$;

(ii) $\mathbb{P}_{\mu_0}(C \cap B) = \int_B \mathbb{P}^{\tilde{\omega}}(C)\mathbb{P}_{\mu_0}(d\tilde{\omega})$ for $B \in \mathscr{B}_0$ and $C \in \mathscr{B}_\infty$;

(iii) $\mathbb{P}^{\tilde{\omega}}\{\omega \in \Omega : \omega(0) = \tilde{\omega}(0)\} = 1$.

See Stroock and Varadhan [**68**], 12-17, for a detailed discussion on regular conditional probabilities. Since M^f (defined in (1.3.1)) is a local $\mathscr{B}_*$-martingale under $\mathbb{P}_{\mu_0}$, Property (ii) implies by a simple argument that it is also a local $\mathscr{B}_*$-martingale under $\mathbb{P}^{\tilde{\omega}}$ for $\mathbb{P}_{\mu_0}$-almost all $\tilde{\omega}$. Thus $\mathbb{P}^{\tilde{\omega}}$ is an L-diffusion measure. Property (iii) says that the initial distribution of $\mathbb{P}^{\tilde{\omega}}$ is concentrated on $\tilde{\omega}(0)$. It follows from the uniqueness of L-diffusion measures that $\mathbb{P}^{\tilde{\omega}} = \mathbb{P}_{\tilde{\omega}(0)}$, the unique L-diffusion measure starting from $\tilde{\omega}(0)$. Using Property (ii) again, we now have

$$\mathbb{P}_{\mu_0}(C) = \int_M \mathbb{P}_x(C)\mu_0(dx). \tag{1.3.10}$$

An important consequence of the uniqueness of L-diffusion measures is the strong Markov property for such measures. In the following we will use $X_{\tau+*} = \{X_{\tau+t}, t \geq 0\}$ to denote the process X shifted in time by τ.

Theorem 1.3.7. *Suppose that X is an L-diffusion process on a probability space $(\Omega, \mathscr{F}_*, \mathbb{P})$ and τ an $\mathscr{F}_*$-stopping time. Then for any $C \in \mathscr{B}_\infty$,*

$$\mathbb{P}\{X_{\tau+*} \in C | \mathscr{F}_\tau\} = \mathbb{P}_{X_\tau}(C), \tag{1.3.11}$$

$\mathbb{P}$-almost surely on $\{\tau < e(X)\}$. In particular, the shifted process $X_{\tau+*}$ is an L-diffusion with respect to the shifted filtration $\mathscr{F}_{\tau+*}$.

Proof. It is enough to show that for any $A \in \mathscr{F}_\tau$ contained in $\{\tau < e(X)\}$,

$$\mathbb{P}\{A \cap [X_{\tau+*} \in C]\} = \mathbb{E}\{\mathbb{P}_{X_\tau}(C); A\}. \tag{1.3.12}$$

Consider the following measure on the path space $W(M)$:

$$\mathbb{Q}(C) = \mathbb{P}\{A \cap [X_{\tau+*} \in C]\}, \qquad C \in \mathscr{B}_\infty.$$

In general $\mathbb{Q}$ is not a probability measure, but we can normalize it by dividing by $\mathbb{P}(A)$. Because X is an L-diffusion, $M^f(X)$ is a local $\mathscr{F}_*$-martingale under $\mathbb{P}$; hence

$$M^f(X_{\tau+*})_t = M^f(X)_{t+\tau} - M^f(X)_\tau$$

is a local $\mathscr{F}_{\tau+*}$-martingale under $\mathbb{P}$. Without loss of generality (see REMARK 1.2.2) we assume that f has compact support on M so that the above process is in fact an integrable martingale. This means that for $t \geq s$,

$$\mathbb{E}\left\{M^f(X_{\tau+*})_t \middle| \mathscr{F}_{\tau+s}\right\} = M^f(X_{\tau+*})_s.$$

Let $D \in \mathscr{B}_s$. Then

$$X_{\tau+*}^{-1}(D) = \{\omega \in \Omega : X_{\tau+*}(\omega) \in D\} \in \mathscr{F}_{\tau+s},$$

and we have

$$\mathbb{E}\left\{M^f(X_{\tau+*})_t; A \cap X_{\tau+*}^{-1} D\right\} = \mathbb{E}\left\{M^f(X_{\tau+*})_s; A \cap X_{\tau+*}^{-1} D\right\}.$$

By the definition of $\mathbb{Q}$, the above relation is equivalent to

$$\mathbb{E}^{\mathbb{Q}}\left\{M_t^f; D\right\} = \mathbb{E}^{\mathbb{Q}}\left\{M_s^f; D\right\}.$$

This being true for all $D \in \mathscr{B}_s$, we have

$$\mathbb{E}^{\mathbb{Q}}\left\{M^f \mid \mathscr{B}_s\right\} = M_s^f,$$

which shows that M^f is a martingale under $\mathbb{Q}$ for every $f \in C^\infty(M)$ with compact support. By definition, $\mathbb{Q}$ is an L-diffusion measure. Its initial distrubtion is given by

$$\mathbb{Q}\{\omega_0 \in G\} = \mathbb{P}\{A \cap [X_\tau \in G]\}, \qquad G \in \mathscr{B}(M).$$

By the uniqueness of L-diffusion measures and (1.3.10), we must have

$$\mathbb{Q}(C) = \int_{W(M)} \mathbb{P}_x(C)\,\mathbb{P}\{X_\tau \in dx; A\}.$$

This is equivalent to the identity (1.3.12) we wanted to prove. $\square$

We can also state the strong Markov property exclusively in terms of the family of L-diffusion measures $\{\mathbb{P}_x, x \in M\}$. Let $\theta_t : W(M) \to W(M)$ be the shift operator:

$$(\theta_t \omega)_s = \omega_{t+s}, \qquad \omega \in W(M).$$

Corollary 1.3.8. *Suppose that τ is a $\mathscr{B}_*$-stopping time. Then for any $C \in \mathscr{B}_\infty$,*

$$\mathbb{P}_x\left\{\theta_\tau^{-1} C | \mathscr{B}_\tau\right\} = \mathbb{P}_{X_\tau}(C)$$

$\mathbb{P}_x$-almost surely on $\{\tau < e\}$

Proof. Exercise. □

Chapter 2

Basic Stochastic Differential Geometry

In this chapter we introduce horizontal lift and stochastic development of a manifold-valued semimartingales, two concepts central to the Eells-Elworthy-Malliavin construction of Brownian motion on a Riemannian manifold. In differential geometry, for a manifold is equipped with a connection, it is possible to lift a smooth curve on M to a horizontal curve on the frame bundle $\mathscr{F}(M)$ by solving an ordinary differential equation, and this horizontal curve corresponds uniquely to a smooth curve (its anti-development) in the euclidean space of the same dimension. Up to an action by the general linear group there is a one-to-one correspondence between the set of smooth curves on the manifold starting from a fixed point and their anti-developments in the euclidean space. We show that an analagous construction can be carried out for semimartingales on a manifold equipped with a connection. This construction is realized by solving an appropriate horizontal stochastic differential equation on the frame bundle.

It is possible to approach stochastic horizontal lift and stochastic development by smooth approximation, but we choose the approach by stochastic differential equations because it is technically simpler and is more consistent with the overall theme of the book. It also allows us to give a relatively smooth treatment of stochastic line integrals, which appear on various occasions during the discussion of manifold-valued martingales in the later part of the chapter.

For a manifold equipped with a connection, a special class of semimartingales, namely, that of manifold-valued martingales can be defined. The most important example of manifold-valued martingales is Brownian motion on a Riemannian manifold, which will be the topic of the next chapter. In this chapter we restrict ourselves to some general properties of manifold-valued martingales and two local results on convergence and nonconfluence. A more detailed exposition of manifold-valued martingales can be found in *Stochastic Calculus in Manifolds* by Emery [**24**].

We devote the first two sections of the chapter to a review of necessary background materials in differential geometry, mainly some basic facts about frame bundle, connection, and tensor fields. Other less frequently used geometric concepts are introduced and reviewed as they are needed. Basic differential geometry is only discussed to the extent needed for the understanding of this chapter. The discussion is therefore rather brief; it also serves as an opportunity for setting up the notations to be used throughout the book. More systematic treatment of the topics covered here can be found in *Geometry of Manifolds* by Bishop and Crittenden [**4**], and *Foundations of Differential Geometry*, Volume 1, by Kobayashi and Nomizu [**53**]. SECTIONS 3 and 4 introduce the concepts of horizontal lift, development, anti-development in the context of semimartingales, and stochastic line integrals. SECTION 5 contains a general discussion of manifold-valued martingales and a proof of the local convergence theorem. In the last SECTION 6 we study martingales on submanifolds and prove the local nonconfluence property of manifold-valued martingales.

2.1. Frame bundle and connection

Let M be a differentiable manifold of dimension d. The tangent space at x is denoted by T_xM and the tangent bundle by TM. The space $\Gamma(TM)$ of smooth sections of the tangent bundle is just the set of vector fields on M. A connection on M is a convention of differentiating a vector field along another vector field. It is therefore given by a map

$$\nabla : \Gamma(TM) \times \Gamma(TM) \to \Gamma(TM)$$

with the following properties: for $X, Y, Z \in \Gamma(TM)$ and $f, g \in C^\infty(M)$:

1) $\nabla_{fX+gY} Z = f\nabla_X Z + g\nabla_Y Z$,
2) $\nabla_X(Y + Z) = \nabla_X W + \nabla_X Z$,
3) $\nabla_X(fY) = f\nabla_X Y + X(f)Y$.

$\nabla_X Y$ is called covariant differentiation of Y along X. In local coordinates a connection is expressed in terms of its Christoffel symbols. Let $x = \{x^1, \ldots, x^d\}$ be a local chart on an open subset O of M. Then the vector fields $X_i = \dfrac{\partial}{\partial x^i}$, $i = 1, \ldots, d$, span the tangent space T_xM at each point $x \in O$, and the Christoffel symbols Γ_{ij}^k are functions on O defined uniquely by the relation

$$\nabla_{X_i} X_j = \Gamma_{ij}^k X_k.$$

To calculate $\nabla_X Y$ at a point x_0, it is enough to know the value of Y along a curve $\{x_t\}$ from x_0 whose tangent at $t = 0$ is $\dot{x}_0 = X_{x_0}$.

Suppose that M is a manifold equipped with a connection. A vector field V along a curve $\{x_t\}$ on M is said to be parallel along the curve if $\nabla_{\dot{x}} V = 0$ at every point of the curve. In this case the vector V_{x_t} at x_t is the parallel transport of V_{x_0} along the curve. In local coordinates, if $x_t = \{x_t^i\}$ and $V_{x_t} = v^i(t) X_i$, then V is parallel if and only if its components $v^i(t)$ satisfy the system of first order ordinary differential equations

$$\dot{v}^k(t) + \Gamma_{jl}^k(x_t)\dot{x}_t^j v^l(t) = 0. \tag{2.1.1}$$

Hence locally a parallel vector field V along a curve $\{x_t\}$ is uniquely determined by its initial value V_{x_0}.

A curve $\{x_t\}$ on M is called a geodesic if $\nabla_{\dot{x}} \dot{x} = 0$ along $\{x_t\}$, i.e., if the tangent vector field itself is parallel along the curve. From (2.1.1) and the fact that $\dot{x}_t = \dot{x}_t^i X_i$ we obtain the ordinary differential equation for geodesics:

$$\ddot{x}_t^k + \Gamma_{jl}^k(x_t)\dot{x}_t^j \dot{x}_t^l = 0. \tag{2.1.2}$$

Thus a geodesic is uniquely determined by its initial position x_0 and initial direction $\dot{x}_0$.

Let us now see how the connection ∇ manifests itself on the frame bundle $\mathscr{F}(M)$ of M. A frame at x is an $\mathbb{R}$-linear isomorphism $u : \mathbb{R}^d \to T_x M$. Let $e_1, \ldots, e_d$ be the coordinate unit vectors of $\mathbb{R}^d$. The the tangent vectors $ue_1, \ldots, ue_d$ make up a basis (or equivalently, a frame) for the tangent space T_x. We use $\mathscr{F}(M)_x$ to denote the space of all frames at x. The general linear group $GL(d, \mathbb{R})$ acts on $\mathscr{F}(M)_x$ by $u \mapsto ug$, where ug denotes the composition

$$\mathbb{R}^d \xrightarrow{g} \mathbb{R}^d \xrightarrow{u} T_x M. \tag{2.1.3}$$

The frame bundle

$$\mathscr{F}(M) = \bigcup_{x \in M} \mathscr{F}(M)_x$$

can be made into a differentiable manifold of dimension $d + d^2$, and the canonical projection $\pi : \mathscr{F}(M) \to M$ is a smooth map. The group $GL(d, \mathbb{R})$ acts on $\mathscr{F}(M)$ fibre-wise; each fibre $\mathscr{F}(M)_x$ is diffeomorphic to $GL(d, \mathbb{R})$, and $M = \mathscr{F}(M)/GL(d, \mathbb{R})$. In differential geometry terminology, these facts make $(\mathscr{F}(M), M, GL(d, \mathbb{R}))$ into a principal bundle with structure group $GL(d, \mathbb{R})$. With the standard action of $GL(d, \mathbb{R})$ on $\mathbb{R}^d$, the tangent bundle is simply the associated bundle

$$TM = \mathscr{F}(M) \times_{GL(d,\mathbb{R})} \mathbb{R}^d, \qquad (u, e) \mapsto ue.$$

The tangent space $T_u\mathscr{F}(M)$ of the frame bundle is a vector space of dimension $d + d^2$. A tangent vector $X \in T_u\mathscr{F}(M)$ is called vertical if it is tangent to the fibre $\mathscr{F}(M)_{\pi u}$. The space of vertical vectors at u is denoted by $V_u\mathscr{F}(M)$; it is a subspace of $T_u\mathscr{F}(M)$ of dimension d^2.

Now assume that M is equipped with a connection ∇. A curve $\{u_t\}$ in $\mathscr{F}(M)$ is just a smooth choice of frames at each point of the curve $\{\pi u_t\}$ on M. The curve $\{u_t\}$ is called horizontal if for each $e \in \mathbb{R}^d$ the vector field $\{u_t e\}$ is parallel along $\{\pi u_t\}$. A tangent vector $X \in T_u\mathscr{F}(M)$ is called horizontal if it is the tangent vector of a horizontal curve from u. The space of horizontal vectors at u is denoted by $H_u\mathscr{F}(M)$; it is a subspace of dimension d, and we have the decomposition

$$T_u\mathscr{F}(M) = V_u\mathscr{F}(M) \oplus H_u\mathscr{F}(M).$$

It follows that the canonical projection $\pi : \mathscr{F}(M) \to M$ induces an isomorphism $\pi_* : H_u\mathscr{F}(M) \to T_{\pi u}M$, and for each $X \in T_xM$ and a frame u at x, there is a unique horizontal vector X^*, the horizontal lift of X to u, such that $\pi_* X^* = X$. Thus if X is a vector field on M, then X^* is a horizontal vector field on $\mathscr{F}(M)$.

The above discussion shows that a connection ∇ on M gives rise to a choice of linear complement of the vertical subspace $V_u\mathscr{F}(M)$ in $T_u\mathscr{F}(M)$ at each point u. The converse is also true, namely a smooth assignment $u \mapsto H_u\mathscr{F}(M)$ of a d-dimensional subspace of $T_u\mathscr{F}(M)$ linearly complement to $V_u\mathscr{F}(M)$ at each point $u \in \mathscr{F}(M)$ corresponds uniquely to a connection ∇ such that $H_u\mathscr{F}(M)$ is its horizontal vector space.

Given a curve $\{x_t\}$ and a frame u_0 at x_0, there is a unique horizontal curve $\{u_t\}$ such that $\pi u_t = x_t$. It is called the horizontal lift of x_t from u_0. The linear map

$$\tau_{t_0 t_1} = u_{t_1} u_{t_0}^{-1} : T_{x_{t_0}}M \to T_{x_{t_1}}M$$

is independent of the choice of the initial frame u_0 and is called the parallel translation (or parallel transport) along $\{x_t\}$.

Remark 2.1.1. Although the horizontal lift $\{u_t\}$ of $\{x_t\}$ is obtained by solving an ordinary differential equation along the curve, the solution of the equation does not blow up. A detailed proof can be found in Kobayashi and Nomizu [**53**], 69–70. See also LEMMA 2.3.7 below. □

For each $e \in \mathbb{R}^d$, the vector field H_e on $\mathscr{F}(M)$ defined at $u \in \mathscr{F}(M)$ by the relation

$$H_e(u) = (ue)^* = \text{ the horizontal lift of } ue \in T_{\pi u}M \text{ to } u$$

is a horizontal vector field on $\mathscr{F}(M)$. Let $e_1, \ldots, e_d$ be the coordinate unit vectors of $\mathbb{R}^d$. Then $H_i = H_{e_i}$, $i = 1, \ldots, d$, are the fundamental horizontal fields of $\mathscr{F}(M)$; they span $H_u\mathscr{F}(M)$ at each $u \in \mathscr{F}(M)$.

The action of $GL(d, \mathbb{R})$ on $\mathscr{F}(M)$ preserves the the fundamental horizontal fields in the sense described in the following proposition.

Proposition 2.1.2. *Let $e \in \mathbb{R}^d$ and $g \in GL(d, \mathbb{R})$. Then*

$$g_* H_e(u) = H_{ge}(gu), \qquad u \in \mathscr{F}(M),$$

where $g_* : T_u\mathscr{F}(M) \to T_{ug}\mathscr{F}(M)$ is the action of g on the tangent bundle $T\mathscr{F}(M)$ induced by the canonical action $g : \mathscr{F}(M) \to \mathscr{F}(M)$ defined in (2.1.3).

Proof. Exercise. □

A local chart $x = \{x^i\}$ on a neighborhood $O \subseteq M$ induces a local chart on $\widetilde{O} = \pi^{-1}(O)$ in $\mathscr{F}(M)$ as follows. Let $X_i = \partial/\partial x^i$, $1 \leq i \leq d$, be the moving frame defined by the local chart. For a frame $u \in \widetilde{O}$ we have $ue_i = e_i^j X_j$ for some matrix $e = (e_j^i) \in GL(d, \mathbb{R})$. Then $(x, e) = (x^i, e_j^i) \in \mathbb{R}^{d+d^2}$ is a local chart for $\widetilde{O}$. In terms of this chart, the vertical subspace $V_u\mathscr{F}(M)$ is spanned by $X_{kj} = \partial/\partial e_j^k$, $1 \leq j, k \leq d$, and the vector fields $\{X_i, X_{ij}, 1 \leq i, j \leq d\}$ span the tangent space $T_u\mathscr{F}(M)$ for every $u \in \widetilde{O}$. We will need the local expression for the fundamental horizontal vector field H_i.

Proposition 2.1.3. *In terms of the local chart on $\mathscr{F}(M)$ described above, at $u = (x, e) = (x^i, e_j^k) \in \mathscr{F}(M)$ we have*

$$H_i(u) = e_i^j X_j - e_i^j e_m^l \Gamma_{jl}^k(x) X_{km}, \tag{2.1.4}$$

where

$$X_i = \frac{\partial}{\partial x^i}, \qquad X_{km} = \frac{\partial}{\partial e_m^k}.$$

Proof. Recall that $H_i(u) = (ue_i)^*$ is the horizontal lift of ue_i, where e_i is the ith coordinate unit vector of $\mathbb{R}^d$. Let $t \mapsto u_t = (x_t, e(t))$ be a horizontal curve starting from $u_0 = u$ such that $\pi_* \dot{u}_0 = ue_i$. By definition

$$H_i(u) = \dot{u}_0 = \dot{x}_0^j X_j + \dot{e}_m^k(0) X_{km}. \tag{2.1.5}$$

The vector field $t \mapsto u_t e_m = e_m^k(t) X_k$ is parallel along $t \mapsto x_t = \{x_t^i\}$ for each m. By the ordinary differential equation for parallel vector fields (2.1.1) we have

$$\dot{e}_m^k(t) + \Gamma_{jl}^k(x_t) \dot{x}_t^j e_m^l(t) = 0,$$

from which we obtain, at $t = 0$,

$$\dot{e}_m^k(0) = -e_i^j e_m^l \Gamma_{jl}^k(x). \tag{2.1.6}$$

On the other hand, $\pi_* \dot{u}_0 = ue_i$ is equivalent to $\dot{x}_0^j = e_i^j$. Using this and (2.1.6) in (2.1.5), we obtain the desired formula for $H_i(u)$. □

Let $\{u_t\}$ be a horizontal lift of a differentiable curve $\{x_t\}$ on M. Since $\dot{x}_t \in T_{x_t}M$, we have $u_t^{-1}\dot{x}_t \in \mathbb{R}^d$. The anti-development of the curve $\{x_t\}$ (or of the horizontal curve $\{u_t\}$) is a curve $\{w_t\}$ in $\mathbb{R}^d$ defined by

$$w_t = \int_0^t u_s^{-1}\dot{x}_s\, ds. \tag{2.1.7}$$

Note that w depends on the choice of the initial frame u_0 at x_0 but in a simple way: if $\{v_t\}$ is another horizontal lift of $\{x_t\}$ and $u_0 = v_0 g$ for a $g \in GL(d, \mathbb{R})$, then the anti-development of $\{v_t\}$ is $\{gw_t\}$. From $u_t \dot{w}_t = \dot{x}_t$ we have

$$H_{\dot{w}_t}(u_t) = \widetilde{u_t \dot{w}_t} = \widetilde{\dot{x}}_t = \dot{u}_t.$$

Hence the anti-development $\{w_t\}$ and the horizontal lift $\{u_t\}$ of a curve $\{x_t\}$ on M are connected by an ordinary differential equation on $\mathscr{F}(M)$:

$$\dot{u}_t = H_i(u_t)\, \dot{w}_t^i. \tag{2.1.8}$$

We can also start from a curve $\{w_t\}$ in $\mathbb{R}^d$ and a frame u_0 at a point x_0. The unique solution of the above equation is a horizontal curve $\{u_t\}$ in $\mathscr{F}(M)$ and is called the development of $\{w_t\}$ in $\mathscr{F}(M)$, and its projection $t \mapsto x_t = \pi u_t$ is called the development of $\{w_t\}$ in M. Note again that $\{x_t\}$ depends on the choice of the initial frame. The procedure of passing from $\{w_t\}$ to $\{x_t\}$ is referred to as "rolling without slipping."

If M is a Riemannian manifold and $\langle \cdot, \cdot \rangle$ its Riemannian metric, then we can restrict ourselves to a smaller set of frames, namely the orthonormal frames. Let $\mathscr{O}(M)$ be the orthonormal frame bundle. By definition an element in $\mathscr{O}(M)$ is a euclidean isometry $u : \mathbb{R}^d \to T_xM$. The action group is correspondingly reduced from $GL(d, \mathbb{R})$ to the orthogonal group $O(d)$, and $\mathscr{O}(M)$ is a principal fibre bundle with the structure group $O(d)$.

The parallel translation associated with a general connection ∇ may not preserve the orthogonality of a frame. If it does, the connection is said to be compatible with the Riemannian metric. This happens if and only if for every triple of vector fields X, Y, Z on M,

$$\nabla_X \langle Y, Z \rangle = \langle \nabla_X Y, Z \rangle + \langle X, \nabla_Y Z \rangle.$$

If a Riemannian manifold is equipped with such a connection, everything we have said so far in this section about the general linear frame bundle $\mathscr{F}(M)$ carries over, *mutatis mutandis*, to the orthonormal frame bundle $\mathscr{O}(M)$. In particular, the formula (2.1.4) for the fundamental horizontal vector fields in PROPOSITION 2.1.3 is still valid with the caveat that now $\left\{x^i, e_j^i, 1 \le i \le j\right\}$ is a set of local coordinates for $\mathscr{O}(M)$. Although each individual X_{ij} may not be tangent to the fibre $\mathscr{O}(M)_x$, we leave it an exercise for the reader to verify that the linear combinations $e_m^l \Gamma_{jl}^k X_{km}$ are in fact tangent to $\mathscr{O}(M)$, which makes (2.1.4) a vector field on $\mathscr{O}(M)$.

2.2. Tensor fields

For each $x \in M$, let $T_x^*M = (T_xM)^*$ be the cotangent space at x, namely the dual space of the tangent space T_xM (the space of linear functions on T_xM). The cotangent bundle $T^*M = \bigcup_{x \in M} T_x^*M$ is a differentiable manifold, and a section $\theta \in \Gamma(T^*M)$ of the cotangent bundle is called a 1-form on M. As such, it is a smooth assignment of a linear functional θ_x on each T_xM at each point $x \in M$. For a vector field $X \in \Gamma(TM)$, the map $\theta(X) : x \mapsto \theta_x(X_x)$ is the contraction of the vector X and the 1-form θ (or the evaluation of θ_x on X_x for each x). In general, the bundle of (r, s)-tensors is

$$T^{r,s}M = \bigcup_{x \in M} T_xM^{\otimes r} \otimes_{\mathbb{R}} T_x^*M^{\otimes s}.$$

An (r, s)-tensor θ on M is a section of the vector bundle $T^{r,s}M$. For each $x \in M$, the value of the tensor field $\theta_x \in \mathrm{Hom}_{\mathbb{R}}(T_xM^{\otimes s}, T_xM^{\otimes r})$, the linear space of $\mathbb{R}$-multilinear maps from T_xM to $T_xM^{\otimes i}$ with s arguments. In local coordinates $x = \{x^r\}$ with $X_i = \partial//\partial x^i$, it is customary to denote the frame on T^*M dual to $\{X_i\}$ by $\{dx^i\}$, i.e., $dx^i(X_j) = \delta_j^i$, the Kronecker symbols. In terms of this basis for T^*M, an (r, s)-tensor locally can be written in the form

$$\theta = \theta_{j_1 \cdots j_s}^{i_1 \cdots i_r} X_{i_1} \otimes \cdots \otimes X_{i_r} \otimes dx^{j_1} \otimes \cdots \otimes dx^{j_s}.$$

The covariant differentiation ∇, which is originally defined for vector fields, is now extended to tensor fields by assuming that it is a derivation commuting with contractions. Thus for two tensor fields θ and ψ, we have

$$\nabla_X(\theta \otimes \psi) = \nabla_X\theta \otimes \psi + \theta \otimes \nabla_X\psi.$$

In particular if θ is a 1-form and X a vector field, then $\nabla_X\theta$ is uniquely determined by the relation

$$(\nabla_X\theta)(Y) = X\theta(Y) - \theta(\nabla_XY), \quad \forall Y \in \Gamma(TM).$$

The reader can verify easily that the assignment $(X, Y) \mapsto (\nabla_X\theta)(Y)$ makes $\nabla\theta$ into a (0,2)-tensor field. In general if θ is an (r, s)-tensor field, then its covariant differentiation $\nabla\theta$ is an $(r, s + 1)$-tensor field.

We can realize covariant differentiation in the frame bundle $\mathscr{F}(M)$. At each frame u, the vectors $X_i = ue_i, i = 1, \ldots, d$, form a basis at T_xM. Let $\{X^i\}$ be the dual frame on T_x^*M. Then an (r, s)-tensor θ can be expressed uniquely as

$$\theta = \theta_{j_1 \cdots j_s}^{i_1 \cdots i_r} X_{i_1} \otimes \cdots \otimes X_{i_r} \otimes X^{j_1} \otimes \cdots \otimes X^{j_s}.$$

The scalarization of θ at u is defined by

$$\widetilde{\theta}(u) = \theta^{i_1 \cdots i_r}_{j_1 \cdots j_s} e_{i_1} \otimes \cdots \otimes e_{i_r} \otimes e^{j_1} \otimes \cdots \otimes e^{j_s},$$

where again $\{e_i\}$ is the canonical basis for $\mathbb{R}^d$ and $\{e^i\}$ the corresponding dual basis. Thus if θ is an (r,s)-tensor field on M, then its scalarization

$$\tilde{\theta} : \mathscr{F}(M) \to \mathbb{R}^{\otimes r} \otimes \mathbb{R}^{*\otimes s}$$

is a vector space-valued function on $\mathscr{F}(M)$, a fact we will often take advantage of. This function is $O(d)$-equivariant in the sense that $\widetilde{\theta}(ug) = g\widetilde{\theta}(u)$, where the g on the right side means the usual extension of the action of $O(d)$ from $\mathbb{R}$ to the tensor space $\mathbb{R}^{\otimes r} \otimes \mathbb{R}^{*\otimes s}$. Conversely, every $O(d)$-invariant function on $\mathscr{F}(M)$ is the scalarization of a tensor field on M.

The following proposition shows how covariant differentiatiation on a manifold is realized on its frame bundle.

Proposition 2.2.1. *Let $X \in \Gamma(TM)$ and $\theta \in \Gamma(T^{r,s}M)$. Then the scalarization of the covariant derivative $\nabla_X \theta$ is given by*

$$\widetilde{\nabla_X \theta} = X^* \widetilde{\theta},$$

where X^* is the horizontal lift of X.

Proof. We prove the typical case where $\theta = Y$ is a vector field, and leave the general case as an exercise.

Let $\{x_t\}$ be a smooth curve on M such that $\dot{x}_0 = X$ and $\{u_t\}$ a horizontal lift of $\{x_t\}$. Let $\tau_t = u_t u_0^{-1}$ be the parallel transport along the curve. We claim that

$$\nabla_X Y = \frac{d}{dt} \tau_t^{-1} Y(x_t), \tag{2.2.1}$$

where the derivative is evaluated at $t = 0$. To see this, let $e_i(t) = u_t e_i$, where e_i is the ith coordinate unit vector of $\mathbb{R}^d$. Then $\{e_i(t)\}$ is horizontal along $\{x_t\}$. Now let

$$Y(x_t) = a^i(t) e_i(t),$$

i.e., $\{a^i(t)\}$ are the coordinates of $Y(x_t)$ in the basis $\{e_i(t)\}$. Taking the covariant differentiation of both sides along X and using the fact that $\{e_i(t)\}$ are parallel along the curve, we have

$$\nabla_X Y = \dot{a}_i(0) e_i(0).$$

On the other hand, we have

$$\tau_t^{-1} Y(x_t) = a^i(t) e_i(0).$$

Differentiating this equation with respect to t, we obtain

$$\frac{d}{dt} \tau_t^{-1} Y(x_t) = \dot{a}^i(0) e_i(0).$$

The desired relation (2.2.1) follows immediately.

Now by the definition of the horizontal lift of X we have $X^* = \dot{u}_0$. Hence from $\widetilde{Y}(u_t) = u_t^{-1}Y(x_t)$ we have

$$X^*\widetilde{Y} = \frac{d}{dt}\widetilde{Y}(u_t) = u_0^{-1}\frac{d}{dt}\tau_t^{-1}Y(x_t) = u_0^{-1}\nabla_X Y(x_0) = \widetilde{\nabla_X Y},$$

which is what we wanted. □

We mentioned above that for an (r, s)-tensor field θ, the covariant differentiation $\nabla\theta$ is an $(r, s+1)$-tensor field. Of special importance to us is the Hessian$\nabla^2 f$ of a smooth function f on M. It is the covariant differentiation of the 1-form ∇f (denoted also by df) and is therefore a $(0, 2)$-tensor. We verify easily from definition that for two vector fields X and Y,

$$\nabla^2 f(X, Y) = X(Yf) - (\nabla_X Y)f, \tag{2.2.2}$$

a relation which can also serve as the definition of $\nabla^2 f$. Passing to the frame bundle $\mathscr{F}(M)$, we have

$$\nabla^2 f(ue_i, ue_j) = H_i H_j \widetilde{f}(u), \qquad u \in \mathscr{F}(M), \tag{2.2.3}$$

a relation which the reader is invited to verify. Here, as always, H_i are the fundamental horizontal vector fields and $\widetilde{f} = f \circ \pi$ is the lift of f to $\mathscr{F}(M)$. In local coordinates the Hessian can be expressed in terms of the Christoffel symbols as

$$\nabla^2 f(X_i, X_j) = f_{ij} - \Gamma^k_{ij} f_k, \qquad X_i = \frac{\partial}{\partial x^i}, \tag{2.2.4}$$

where $f_i = \partial f/\partial x^i$ and similarly for f_{ij}. This follows from (2.2.2) and the definition of Christoffel symbols $\nabla_{X_i} X_j = \Gamma^k_{ij} X_k$.

There are two important tensor fields derived from a connection ∇. The torsion of ∇ is a $(1, 2)$-tensor defined by

$$T(X, Y) = \nabla_X Y - \nabla_Y X - [X, Y], \quad X, Y \in \Gamma(TM),$$

where $[X, Y]$ is the Lie bracket of the vector fields X and Y. The connection ∇ is called torsion-free if $T = 0$ on M. For a torsion-free connection, the Hessian and the Christoffel symbols are symmetric:

$$\nabla^2 f(X, Y) = \nabla^2 f(Y, X), \qquad \Gamma^i_{jk} = \Gamma^i_{kj}.$$

On a Riemannian manifold M, the Levi-Civita connection is the unique torsion-free connection which is compatible with the Riemannian metric. Starting from Chapter 3, we will use exclusively the Levi-Civita connection, so we will not see the torsion tensor at all.

The curvature of ∇ is a (1,3)-tensor defined by

$$R(X, Y, Z) = \nabla_X \nabla_Y Z - \nabla_Y \nabla_X Z - \nabla_{[X,Y]} Z, \quad X, Y, Z \in \Gamma(TM).$$

A considerable portion of this book is occupied with the interaction between Brownian motion and curvature, and this is very natural in view of the following fundamental fact from differential geometry: a Riemannian manifold with vanishing torsion and curvature tensors is locally isometric to a euclidean space.

2.3. Horizontal lift and stochastic development

Stochastic differential equations on a manifold are a convenient and useful tool for generating semimartingales on a manifold M from ones on $\mathbb{R}^N$. If a manifold M is equipped with a connection, then there are invariantly defined fundamental horizontal vector fields H_i on the frame bundle $\mathscr{F}(M)$, and many things we have said about smooth curves on M in the last two sections can be generalized to semimartingales on M. In particular, if the equation (2.1.8) interpreted properly—this usually means replacing the usual integral by the corresponding Stratonovich stochastic integral—we can develop a semimartingale W on $\mathbb{R}^d$ into a horizontal semimartingale U on $\mathscr{F}(F)$, and then project it down to a semimartingale on M ("rolling without slipping"). Conversely, we can lift a semimartingale X on M to a horizontal semimartingale U on $\mathscr{F}(M)$ and then to a semimartingale W on $\mathbb{R}^d$. Once a horizontal lift U_0 of the initial value X_0 is fixed (i.e., $\pi U_0 = X_0$), the correspondence $W \longleftrightarrow X$ is one-to-one. Because euclidean semimartingales are easier to handle than manifold-valued semimartingales, we can use this geometrically defined correspondence to our advantage. Later we will see that a connection also gives rise to the notion of manifold-valued martingales. As expected, for semimartingales stochastic development and horizontal lift are obtained by solving stochastic differential equations driven by either $\mathbb{R}^d$-valued or M-valued semimartingales. But unlike the case of smooth curves, these equations are not local at a fixed time.

Consider the following SDE on the frame bundle $\mathscr{F}(M)$:

$$dU_t = H(U_t) \circ dW_t, \tag{2.3.1}$$

where W is an $\mathbb{R}^d$-valued semimartingale. Whenever necessary, we will use the more precise notation

$$dU_t = \sum_{i=1}^{d} H_i(U_t) \circ dW_t^i.$$

In writing the above equation, we have implicitly assumed that M has been equipped with a connection, and $\{H_i\}$ are the corresponding fundamental horizontal vector fields on $\mathscr{F}(M)$. We now give a few definitions. All processes are defined on a fixed filtered probability space $(\Omega, \mathscr{F}_*, \mathbb{P})$ and are $\mathscr{F}_*$-adapted.

Definition 2.3.1. *(i) An $\mathscr{F}(M)$-valued semimartingale U is said to be horizontal if there exists an $\mathbb{R}^d$-valued semimartingale W such that (2.3.1) holds. The unique W is called the anti-development of U (or of its projection $X = \pi U$).*

(ii) Let W be an $\mathbb{R}^d$-valued semimartingale and U_0 an $\mathscr{F}(M)$-valued, $\mathscr{F}_0$-measurable random variable. The solution U of the SDE (2.3.1) is called a (stochastic) development W in $\mathscr{F}(M)$. Its projection $X = \pi U$ is called a (stochastic) development of W in M.

(iii) Let X be an M-valued semimartingale. An $\mathscr{F}(M)$-valued horizontal semimartingale U such that its projection $\pi U = X$ is called a (stochastic) horizontal lift of X.

In the correspondences $W \longleftrightarrow U \longleftrightarrow X$ the only transitions which need explanation are $X \mapsto U$ and $U \mapsto W$. We will prove the existence of a horizontal lift by deriving a stochastic differential equation for it on the frame bundle $\mathscr{F}(M)$ driven by X. For this purpose we assume that M is a closed submanifold of $\mathbb{R}^N$ and regard $X = \{X^\alpha\}$ as an $\mathbb{R}^N$-valued semimartingale. For each $x \in M$, let $P(x) : \mathbb{R}^N \to T_xM$ be the orthogonal projection from $\mathbb{R}^N$ onto the subspace $T_xM \subseteq \mathbb{R}^N$. Then intuitively we have, on $\mathbb{R}^N$,

$$X_t = X_0 + \int_0^t P(X_s) \circ dX_s.$$

This identity will be proved in LEMMA 2.3.3 below. Rewriting this more explicitly, we have

$$dX_t = P_\alpha(X_t) \circ dX_t^\alpha.$$

Once we are convinced that this holds, the obvious candidate for the horizontal lift U of X is the solution of the following equation on $\mathscr{F}(M)$:

$$dU_t = \sum_{\alpha=1}^N P_\alpha^*(U_t) \circ dX_t^\alpha, \tag{2.3.2}$$

where $P*_\alpha(u)$ is the horizontal lift of $P_\alpha(\pi u)$ We now prove that the solution of (2.3.2) is indeed a horizontal lift of X. We start with two simple geometric facts. First, let $f = \{f^\alpha\} : M \to \mathbb{R}^N$ be the coordinate function. Its lift $\widetilde{f} : \mathscr{F}(M) \to \mathbb{R}^N$ defined by

$$\widetilde{f}(u) = f(\pi u) = \pi u \in M \subseteq \mathbb{R}^N$$

is nothing but the projection $\pi : \mathscr{F}(M) \to M$ written as an $\mathbb{R}^N$-valued function on $\mathscr{F}(M)$. The change of notation from π to $\widetilde{f}$ emphasizes that it is regarded as a vector-valued function rather than a map. Again let e_i be the ith coordinate unit vector in $\mathbb{R}^d$.

Lemma 2.3.2. *Let $\widetilde{f} : \mathscr{F}(M) \to M \subseteq \mathbb{R}^N$ be the projection function. The following two identities hold on $\mathscr{F}(M)$:*

$$P_\alpha^* \widetilde{f}(u) = P_\alpha(\pi u), \tag{2.3.3}$$

$$\sum_{\alpha=1}^N P_\alpha(\pi u) H_i \widetilde{f}^\alpha(u) = ue_i. \tag{2.3.4}$$

Proof. If $\{u_t\}$ is the horizontal lift from $u_0 = u$ of a curve $\{x_t\}$ with $\dot{x}_0 = P_\alpha(\pi u)$, then $P_\alpha^*(u) = \dot{u}_0$. Hence

$$P_\alpha^* f(u) = \frac{d\left\{\widetilde{f}(u_t)\right\}}{dt} = \frac{d\,(\pi u_t)}{dt} = \dot{x}_0 = P_\alpha(\pi u).$$

This proves the first identity. The proof of the second inequality is similar. If $\{v_t\}$ is the horizontal lift from $v_0 = u$ of a curve $\{y_t\}$ on M with $\dot{y}_0 = ue_i$, then

$$H_i \widetilde{f}(u) = \frac{d\left\{\widetilde{f}(v_t)\right\}}{dt} = \frac{d\,(\pi v_t)}{dt} = \dot{y}_0 = ue_i. \tag{2.3.5}$$

This shows that $H_i \widetilde{f}(u) \in T_{\pi u}M$. Hence $P(\pi u) H_i \widetilde{f}(u) = H_i \widetilde{f}(u)$ and we have

$$\sum_{\alpha=1}^N P_\alpha(\pi u) H_i \widetilde{f}^\alpha(u) = P(\pi u) H_i \widetilde{f}(u) = H_i \widetilde{f}(u) = ue_i,$$

which proves the second identity. □

Next we prove a useful fact about semimartingales on an embedded submanifold. It implies in particular that every submartingale on a manifold is a solution of a Stratonovich type stochastic differential equation on the manifold.

Lemma 2.3.3. *Suppose that M is a closed submanifold of $\mathbb{R}^N$. For each $x \in M$, let $P(x) : \mathbb{R}^N \to T_xM$ be the orthogonal projection from $\mathbb{R}^N$ to the tangent space T_xM. If X is an M-valued semimartingale, then*

$$X_t = X_0 + \int_0^t P(X_s) \circ dX_s. \tag{2.3.6}$$

Proof. Let $\{\xi_\alpha\}$ be the canonical basis for $\mathbb{R}^N$. Define

$$P_\alpha(x) = P(x)\,\xi_\alpha, \quad Q_\alpha(x) = \xi_\alpha - P(x)\,\xi_\alpha.$$

Then $P_\alpha(x)$ is tangent to M, $Q_\alpha(x)$ is normal to M, and $P_\alpha + Q_\alpha = \xi_\alpha$. Let

$$Y_t = X_0 + \int_0^t P_\alpha(X_s) \circ dX_t^\alpha. \tag{2.3.7}$$

We first verify that Y lives on M. Let f be a smooth nonnegative function on $\mathbb{R}^N$ which vanishes only on M. By Itô's formula,

$$f(Y_t) = f(X_0) + \int_0^t P_\alpha f(X_t) \circ dX_t^\alpha.$$

But if $x \in M$, then $P_\alpha(x) \in T_x M$ and $P_\alpha f(x) = 0$. Hence $P_\alpha f(X_t) = 0$ and $f(Y_t) = 0$, which shows that $Y_t \in M$.

For each $x \in \mathbb{R}^N$ let $h(x)$ be the point on M closest to x. Since M is a closed submanifold, $h : \mathbb{R}^N \to M \subseteq \mathbb{R}^N$ is a well defined smooth function in a neighborhood of M and is constant on each line segment perpendicular to M. This means that $Q_\alpha h(x) = 0$ for $x \in M$, since Q_α is normal to the manifold. As a consequence, regarding ξ_α as a vector field on $\mathbb{R}^N$, we have

(2.3.8) $$P_\alpha h(x) = P_\alpha h(x) + Q_\alpha h(x) = \xi_\alpha h(x), \quad x \in M.$$

Now we have

$$\begin{aligned}
Y_t &= h(Y_t) && Y_t \in M \\
&= X_0 + \int_0^t P_\alpha h(X_s) \circ dX_t^\alpha && \text{(2.3.7) and Itô's formula} \\
&= X_0 + \int_0^t \xi_\alpha h(X_s) \circ dX_s^\alpha && \text{(2.3.8)} \\
&= h(X_t) && \text{Itô's formula on } \mathbb{R}^N \\
&= X_t && X_t \in M.
\end{aligned}$$

This completes the proof. □

Theorem 2.3.4. *A horizontal semimartingale U on the frame bundle $\mathscr{F}(M)$ has a unique anti-development W. In fact,*

(2.3.9) $$W_t = \int_0^t U_s^{-1} P_\alpha(X_s) \circ dX_s^\alpha,$$

where $X_t = \pi U_t$.

Proof. W should be the $\mathbb{R}^d$-valued semimartingale defined by

$$dU_t = H_i(U_t) \circ dW_t^i.$$

Let $\widetilde{f}$ be the projection function defined just before Lemma 2.3.2. From $\widetilde{f}(U_t) = \pi U_t = X_t$ we have

$$dX_t = H_i \widetilde{f}(U_t) \circ dW_t^i,$$

or equivalently,

$$dX_t^\alpha = H_i \widetilde{f}^\alpha(U_t) \circ dW_t^i.$$

Now multiplying both sides by $U^{-1}P_\alpha(X_t) \in \mathbb{R}^d$ and using (2.3.4), we have

$$U_t^{-1}P_\alpha(X_t) \circ dX_t^\alpha = U_t^{-1}P_\alpha(X_t) H_i \widetilde{f}^\alpha(U_t) \circ dW_t^i = e_i \, dW_t^i,$$

which is equivalent to (2.3.9). □

Now we can prove the main theorem.

Theorem 2.3.5. *Suppose that $X = \{X_t, 0 \le t < \tau\}$ is a semimartingale on M up to a stopping time τ, and U_0 an $\mathscr{F}(M)$-valued $\mathscr{F}_0$-random variable such that $\pi U_0 = X_0$. Then there is a unique horizontal lift $\{U_t, 0 \le t < \tau\}$ of X starting from U_0.*

Proof. Let U be the unique solution of (2.3.2). By LEMMA 2.3.7 below it is well defined up to the stopping time τ. We verify that it is a horizontal lift of X. Since it is obviously horizontal, all we need to show is $\pi U = X$. As before, let $f : \mathscr{F}(M) \to M \subseteq \mathbb{R}^N$ be the projection function $\pi : \mathscr{F}(M) \to M$, regarded as an $\mathbb{R}^N$-valued function. Let $Y_t = f(U_t) = \pi U_t$. We need to show that $X = Y$. From (2.3.2) for U, we have by (2.3.3)

$$dY_t = P_\alpha^* f(U_t) \circ dX_t^\alpha = P_\alpha(Y_t) \circ dX^\alpha.$$

This is an equation for the semimartingale Y on M driven by the $\mathbb{R}^N$-valued semimartingale X with the initial condition $Y_0 = \pi U_0 = X_0$. On the other hand, by LEMMA 2.3.3 X is a solution of the same equation. Hence by uniqueness we must have $X = Y$.

We now show that a horizontal lift with a given initial starting frame is unique by verifying that any other horizontal lift Π of X will satisfy the same equation (2.3.2) as U. Since Π is horizontal, there is an $\mathbb{R}^d$-valued semimartingale W such that

$$d\Pi_t = H_i(\Pi_t) \circ dW_t^i \tag{2.3.10}$$

By (2.3.9) in THEOREM 2.3.4 we have

$$dW_t = \Pi_t^{-1} P_\alpha(X_t) \circ dX_t^\alpha.$$

Substituting this into (2.3.10) and using the fact that the horizontal lift $P_\alpha^*(\Pi_t)$ of $P_\alpha(X_t)$ is given by

$$P_\alpha^*(\Pi_t) = \sum_{i=1}^d \left\{\Pi^{-1} P_\alpha(X_t)\right\}^i H_i(\Pi_t),$$

we find that

$$d\Pi_t = P_\alpha^*(\Pi_t) \circ dX_t^\alpha,$$

i.e., Π satisfies the same equation as U. □

Example 2.3.6. Let $M = \mathbb{R}^1$ equipped with a general connection given by $\nabla_e e = \Gamma e$, where e is the usual unit vector field on $\mathbb{R}^1$: $e(f) = f'$, and $\Gamma \in C^\infty(\mathbb{R}^1)$. Define

$$G(x) = \int_0^x \Gamma(y)dy, \qquad \phi(x) = \int_0^x e^{-G(y)}dy.$$

Let X be a semimartingale.

(i) With the canonical embedding $GL(\mathbb{R}^1) = \mathbb{R}^1 \times \mathbb{R}^1$, the horizontal lift of X is given by

$$U_t = \left(X_t, e^{-G(X_t)} \right).$$

(ii) The anti-development of X is $W = \phi(X)$.

(iii) We call X a ∇-martingale if its anti-development W is a local martingale. Then X is a ∇-martingale if and only if there is a local martingale M (in the usual sense) such that

$$X_t = X_0 + M_t - \frac{1}{2} \int_0^t \Gamma(X_s) \, d\langle M, M \rangle_s.$$

In this case,

$$W_t = \int_0^t e^{G(X_s)} \, dM_s. \qquad \square$$

We now show that the horizontal lift of a semimartingale can be defined on the maximal time interval on which the semimartingale is defined, namely, there is no explosion in the vertical direction.

Lemma 2.3.7. *Let X be a semimartingale on M defined up to a stopping time τ. Then a horizontal lift U of X is also defined up to τ.*

Proof. We need to use some special features of the equation (2.3.2) for the horizontal lift U, because normally a solution is defined only up to its own explosion time, which may be strictly smaller than the stopping time up to which the driving semimartingale is defined. By a typical stopping time argument, we can assume without loss of generality that $\tau = \infty$, i.e., the semimartingale X is defined on all of $[0, \infty)$. We can also assume that there is a relatively compact neighborhood O covered by a local chart $x = \left\{ x^i \right\}$ such that $X_t \in O$ for all $t \geq 0$, and $u = \left\{ x^i, e^i_j \right\}$ be the corresponding local chart on $\mathscr{F}(M)$ defined just after PROPOSITION 2.1.2. As before we use the notations

$$X_i = \frac{\partial}{\partial x^i}, \quad X_{ij} = \frac{\partial}{\partial e^i_j}.$$

These are vector fields on $\mathscr{F}(M)$. Define the function

$$h(u) = \sum_{1 \leq i,j \leq d} |e^i_j|^2.$$

It is enough to show that $h(U_t)$ does not explode. For this purpose we first need to write the horizontal lift P^*_α of P_α in the local coordinates. Since $ue_i = e^j_i X_j$ by definition, we have $X_q = f^i_q ue_i$, where $\left\{ f^j_i \right\}$ is the inverse matrix of $\left\{e_i^j\right\}$. The horizontal lift H_i of ue_i is given by LEMMA 2.1.3; hence the horizontal lift of X_q is

$$X_q^* = X_q - e_m^l \Gamma_{ql}^k X_{km}.$$

If $P_\alpha(x) = p_\alpha^q(x) X_q$, then the horizontal lift of $P_\alpha(x)$ is

$$P_\alpha^*(u) = p_\alpha^q X_q - p_\alpha^q e_m^l \Gamma_{qj}^k(x) X_{km}.$$

This is the local expression for $\widetilde{P}_\alpha$ we are looking for. The point is that the coefficients Γ_{ij}^k and p_α^i are uniformly bounded on the relatively compact neighborhood O. It is therefore clear from the definition of the function h that there is a constant C such that

$$|P_\alpha^* h| \le Ch, \quad |P_\alpha^* P_\beta^* h| \le Ch. \tag{2.3.11}$$

Now from $dU_t = P_\alpha^*(U_t) \circ dX^\alpha$ and Itô's formula,

$$h(U_t) = h(U_0) + \int_0^t P_\alpha^* h(U_s)\, dX^\alpha + \frac{1}{2}\int_0^t P_\alpha^* P_\beta^* h(U_s)\, d\langle X^\alpha, X^\beta\rangle_s.$$

According to (2.3.11), the integrands grows at most linearly in $h(U_t)$. We can now follow the proof of PROPOSITION 1.1.11 and conclude that $h(U_t)$ does not explode. □

Let X be a semimartingale on M. Two horizontal lifts U and V of the semimartingale X are related in a simple way: $U_t = V_t V_0^{-1} U_0$. From this fact we see that

$$\tau_{t_1 t_2}^X = U_{t_2} U_{t_1}^{-1} : T_{X_{t_1}} M \to T_{X_{t_2}} M, \qquad 0 \le t_1 < t_2 < e(X)$$

is independent of the choice of the initial frame in the definition of U. It is called the (stochastic) parallel transport from X_{t_1} to X_{t_2} along X.

We have shown in LEMMA 2.3.3 that a semimartingale X on a manifold M is always a solution of a stochastic differential equation on M. The following result is often convenient in applications.

Proposition 2.3.8. *Let a semimartingale X on a manifold M be the solution of $SDE(V_1, \ldots, V_N; Z, X_0)$ and let V_α^* be the horizontal lift of V_α to the frame bundle $\mathscr{F}(M)$. Then the horizontal lift U of X is the solution of $SDE(V_1^*, \ldots, V_N^*; Z, U_0)$, and the anti-development of X is given by*

$$W_t = \int_0^t U_s^{-1} V_\alpha(X_s) \circ dZ_s^\alpha.$$

Proof. We assume that M is a submanifold of $\mathbb{R}^N$ and use the notations introduced before. The proof of the first assertion is similar to the first part of the proof of THEOREM 2.3.5. If $f : \mathscr{F}(M) \to M \subseteq \mathbb{R}^N$ is the projection function as before, then it is easy to verify that $V_\alpha^* f(u) = V_\alpha(x)$.

Let U be the solution of the horizontal equation $SDE(V_1^*, \ldots, V_N^*; Z, U_0)$ and $Y_t = f(U_t) = \pi U_t$. Differentiating, we have

$$dY_t = V_\alpha^* f(U_t) \circ d\, Z_t^\alpha = V_\alpha(Y_t) \circ d\, Z_t^\alpha.$$

This is the equation for X. Hence $Y = X$, i.e., U is the horizontal lift of X.

For the second assertion, we have from THEOREM 2.3.4

$$dW_t = U_t^{-1} P_\beta(X_t) \circ dX_t^\beta.$$

In our case $dX_t = V_\alpha(X_t) \circ d\, Z_t^\alpha$, or in components

$$dX_t^\beta = V_\alpha^\beta(X_t) \circ d\, Z_t^\alpha,$$

where $\left\{V_\alpha^\beta\right\}$ are the component of the vector V_α in $\mathbb{R}^N$. Since V_α is tangent to M, it is easy to verify that

$$V_\alpha = \sum_{\beta=1}^N V_\alpha^\beta P_\beta.$$

It follows that

$$dW_t = U_t^{-1} \left[\sum_{\beta=1}^N V_\alpha^\beta(X_t) P_\beta(X_t)\right] \circ d\, Z_t^\alpha = U_t^{-1} V_\alpha(X_t) \circ d\, Z_t^\alpha,$$

as desired. □

2.4. Stochastic line integrals

Let θ be a 1-form on a manifold M and $\{x_t\}$ a smooth curve on M. Then the line integral of θ along $\{x_t\}$ is defined as

$$\int_{x[0,t]} \theta = \int_0^t \theta(\dot{x}_s)\, ds. \tag{2.4.1}$$

Let us find an expression for the line integral which can be directly extended to semimartingales. Fix a frame u_0 at x_0 and let the curve $\{u_t\}$ be the horizontal lift of $\{x_t\}$ from u_0. By definition the anti-development of $\{x_t\}$ is a $\mathbb{R}^d$-valued curve $\{w_t\}$ defined by $\dot{w}_t = u_t^{-1}\dot{x}_t$. Hence

$$\theta(\dot{x}_t) = \theta(u_t \dot{w}_t) = \theta(ue_i)\, \dot{w}_t^i$$

and

$$\int_{x[0,t]} \theta = \int_0^t \theta(\dot{x}_s) ds = \int_0^t \theta(u_s e_i)\, \dot{w}_s^i ds.$$

Clearly the right side is independent of the choice of the connection and the initial frame u_0. We now make the following definition.

Definition 2.4.1. *Let θ be a 1-form on M and X an M-valued semimartingale. Let U be a horizontal lift of X and W its anti-development (with respect to any connection). Then the (stochastic) line integral of η along X is defined by*

$$\int_{X[0,t]} \theta = \int_0^t \theta(U_s e_i) \circ d\,W_s^i.$$

Let $\widetilde{\theta} : \mathscr{F}(M) \to \mathbb{R}^d$ be the scalarization of θ, namely $\widetilde{\theta}(u) = \{\theta(ue_i)\}$. Then we can also write

$$\int_{X[0,t]} \theta = \int_0^t \widetilde{\theta}(U_s) \circ \, d\,W_s^\dagger.$$

We have defined the line integral along X in terms of its horizontal lift and anti-development as a matter of convenience. The following result gives a more direct expression, and it shows again that the definition is independent of the connection chosen for the manifold. Note that we have shown that every semimartingale on M is a solution of an Itô type stochastic differential equation on M (LEMMA 2.3.3).

Proposition 2.4.2. *Let θ be a 1-form on M and X the solution of the equation $dX_t = V_\alpha(X_t) \circ d\,Z_t^\alpha$. Then*

$$\int_{X[0,t]} \theta = \int_0^t \theta(V_\alpha)(X_s) \circ d\,Z_s^\alpha.$$

Proof. From LEMMA 2.3.8 we have $d\,W_t = U_t^{-1} V_\alpha(X_t) \circ d\,Z_t^\alpha$. Hence the differential of the line integral is

$$\widetilde{\theta}(U_t) \circ d\,W_t^\dagger = \left\langle \widetilde{\theta}(U_t), U_t^{-1} V_\alpha(X_t) \right\rangle \circ d\,Z_t^\alpha = \theta(V_\alpha)(X_t) \circ d\,Z_t^\alpha.$$

□

In view of the above result, it is reasonable to write the differential of the stochastic line integral symbolically as $\theta \circ dX_t$.

Example 2.4.3. If θ is an exact 1-form, i.e., $\theta = df$ for a smooth function f, then

$$\int_{X[0,t]} df = f(X_t) - f(X_0).$$

□

Example 2.4.4. Let $x = \{x^i\}$ be a local chart on M and $\theta(x) = \theta_i(x)\,dx^i$. Let $X = \{X^i\}$ be a semimartingale on M. We have

$$dX_t = V_i(X_t) \circ dX_t^i, \qquad \text{where} \quad V_i = \frac{\partial}{\partial x^i}.$$

Then $\theta(V_i) = \theta_i$, and we have from PROPOSITION 2.4.2

$$\int_{X[0,t]} \theta = \int_0^t \theta_i(X_s) \circ dX_s^i.$$

This is the local expression for the line integral. □

The anti-development W of a horizontal semimartingale U on $\mathscr{F}(M)$ is the line integral of the so-called solder form θ along U. This is an $\mathbb{R}^d$-valued 1-form θ on $\mathscr{F}(M)$ defined by

$$\theta(Z)(u) = u^{-1}(\pi_* Z)$$

for a vector field Z on $\mathscr{F}(M)$. In particular, $\theta(H_i) = e_i$.

Proposition 2.4.5. *Let U be a horizontal semimartingale on $\mathscr{F}(M)$. Then its the corresponding anti-development is given by*

$$W_t = \int_{U[0,t]} \theta,$$

where θ is the solder form on $\mathscr{F}(M)$.

Proof. Exercise. □

We now define the quadratic variation of a semimartingale with respect to a $(0,2)$-tensor. By definition a $(0,2)$-tensor $h \in \Gamma(T^*M \otimes T^*M)$ is a section of the vector bundle $T^*M \otimes T^*M$, namely, a smooth assignment of a tensor $h_x \in T_x^*M \otimes T_x^*M = \mathrm{Hom}(T_xM \otimes T_xM, \mathbb{R})$ (the space of linear functionals on the tensor product $T_xM \otimes T_xM$) at each point $x \in M$. Its scalarization $\widetilde{h}$ is an $\mathbb{R}^{d*} \otimes_{\mathbb{R}} \mathbb{R}^{d*}$-valued function on the frame bundle $\mathscr{F}(M)$ and we have

$$\widetilde{h}(u)(e,f) = h(ue, uf), \qquad e, f \in \mathbb{R}^d \text{ and } u \in \mathscr{F}(M).$$

Note that we write $h(a \otimes b)$ as $h(a,b)$ for a (0,2)-tensor on a vector space V and $a, b \in V$. We now make the following definition (cf. DEFINITION 2.4.1).

Definition 2.4.6. *Let h be a (0,2)-tensor on M and X an M-valued semimartingale. Let U be a horizontal lift of X and W its anti-development. Then the h-quadratic variation of X is*

$$\int_0^t h(dX_s, dX_s) = \int_0^t \widetilde{h}(dW_t, dW_t),$$

or more precisely,

$$\int_0^t h(dX_s, dX_s) = \int_0^t h(U_s e_i, U_s e_j)\, d\langle W^i, W^j\rangle_s. \tag{2.4.2}$$

A (0,2)-tensor h is called symmetric if $h(A,B)=h(B,A)$, and antisymmetric if $h(A,B)=-h(B,A)$, where A,B are vector fields on M. For an arbitrary (0,2)-tensor h, its symmetric part is defined by

$$h^{\rm sym}(A,B)=\frac{h(A,B)+h(B,A)}{2}.$$

Proposition 2.4.7. *Let $h^{\rm sym}$ be the symmetric part of h. Then*

$$\int_0^t h(dX_s,dX_s)=\int_0^t h^{\rm sym}(dX_s,dX_s).$$

In particular, if h is antisymmetric, then

$$\int_0^t h(dX_s,dX_s)=0.$$

Proof. Exercise. □

We have the following analogue of PROPOSITION 2.4.2.

Proposition 2.4.8. *Let h be a (0,2)-tensor on M and X the solution of $SDE(V_1,\ldots,V_N,Z;X_0)$. Then*

$$\int_0^t h(dX_s,dX_s)=\int_0^t h(V_\alpha,V_\beta)(X_s)\,d\langle Z^\alpha,Z^\beta\rangle_s.$$

Proof. Exercise. □

Example 2.4.9. Let $x=\{x^i\}$ be a global coordinate system on M. A (0,2)-tensor can be written as

$$h(x)=h_{ij}(x)\,dx^i\otimes dx^j,\qquad h_{ij}=h\left(\frac{\partial}{\partial x^i},\frac{\partial}{\partial x^j}\right).$$

If $X=\{X^i\}$ is a semimartingale on M, then

$$\int_0^t h(dX_s,dX_s)=\int_0^t h_{ij}(X_s)\,d\langle X^i,X^j\rangle_s.$$

Example 2.4.10. For $f\in C^\infty(M)$, the Hessian $\nabla^2 f$ is a $(0,2)$-tensor. Using (2.2.3), we have by DEFINITION 2.4.6

$$\int_0^t \nabla^2 f(dX_s,dX_s)=\int_0^t H_iH_j\widetilde{f}(U_s)\,d\langle W^i,W^j\rangle_s.$$ □

Example 2.4.11. If $f,g\in C^\infty(M)$, then

$$\int_0^t (d\,f\otimes d\,g)(dX_s,dX_s)=\langle f(X),g(X)\rangle_t.$$ □

A symmetric (0,2)-tensor θ is said to be positive definite if $\theta(X,X)\geq 0$ for every vector X.

Proposition 2.4.12. *If θ is a positive definite (0,2)-tensor, then the θ-quadratic variation of a semimartingale X is nondecreasing.*

Proof. We have

$$\int_0^t h(dX_s, dX_s) = \int_0^t h(U_s e_i, U_s e_j)\, d\langle W^i, W^j\rangle_s.$$

By the assumption the matrix $\{h(U_s e_i, U_s e_j)\}$ is symmetric and positive definite. Let $\{m_i^k(s)\}$ be its positive definite matrix square root and

$$J_t^k = \int_0^t m_i^k(s)\, dW_s^i.$$

Then we have

$$\int_0^t h(dX_s, dX_s) = \int_0^t m_i^k(s) m_j^k(s)\, d\langle W^i, W^j\rangle_s = \langle J, J^\dagger\rangle_t.$$

The result follows immediately. □

Remark 2.4.13. The above result can be strengthened as follows. Let

$$D = \{x \in M : h \text{ is positive definite at } x\}.$$

Then $t \mapsto \displaystyle\int_0^t h(dX_s, dX_s)$ is nondecreasing whenever $X_t \in D$. We leave the proof to the reader. □

2.5. Martingales on manifolds

The concept of (local) martingales on a euclidean space can be extended to a differentiable manifold equipped with a connection. The definition is straightforward.

Definition 2.5.1. *Suppose that M is a differentiable manifold equipped with a connection ∇. An M-valued semimartingale X is called a ∇-martingale if its anti-development W with respect to the connection ∇ is an $\mathbb{R}^d$-valued local martingale.*

A ∇-martingale is also called a Γ-martingale in the literature, with the Γ presumably referring to the Christoffel symbols of the connection ∇. Here is a minor point of confusion: for $M = \mathbb{R}^d$ with the usual euclidean connection, an M-valued martingale is a *local* martingale on $\mathbb{R}^d$.

We give an alternative definition without referring to anti-development. Recall that $\nabla^2 f$ is the Hessian of $f \in C^\infty(M)$.

Proposition 2.5.2. *An M-valued semimartingale X is a ∇-martingale if and only if*

$$N^f(X)_t \stackrel{\text{def}}{=} f(X_t) - f(X_0) - \frac{1}{2}\int_0^t \nabla^2 f(dX_s, dX_s)$$

is a $\mathbb{R}$-valued local martingale for every $f \in C^\infty(M)$.

Proof. The horizontal lift U of X satisfies $dU_t = H_i(U_t) \circ dW_t^i$, where W is a semimartingale. Let $\widetilde{f} = f \circ \pi$ be the lift of f to $\mathscr{F}(M)$. Applying Itô's formula to $\widetilde{f}(U)$, we have

$$\begin{aligned} & f(X_t) - f(X_0) \\ = & \int_0^t H_i\widetilde{f}(U_s)\,dW_s^i + \frac{1}{2}\int_0^t H_iH_j\widetilde{f}(U_s)\,d\langle W^i, W^j\rangle \\ = & \int_0^t H_i\widetilde{f}(U_s)\,dW_s^i + \frac{1}{2}\int_0^t \nabla^2 f(dX_s, dX_s). \end{aligned}$$

Here in the last step we have used EXAMPLE 2.4.10. Therefore for any $f \in C^\infty(M)$,

$$N^f(X)_t = \int_0^t H_i\widetilde{f}(U_s)\,dW_s^i. \tag{2.5.1}$$

Now, if X is a ∇-martingale, then W is a local martingale, and so is $N^f(X)$. For the converse, we assume that M is embedded in some euclidean space $\mathbb{R}^N$. Let $f : M \to \mathbb{R}^N$ be the coordinate function $f(x) = x$ and $\widetilde{f}$ its lift to $\mathscr{F}(M)$. In LEMMA 2.3.2) (see (2.3.5)), we have shown that $H_i\widetilde{f}(u) = ue_i$; hence by (2.5.1) we have, as processes in $\mathbb{R}^N$,

$$N^f(X)_t = \int_0^t U_s e_i\,dW_s^i = \int_0^t U_s\,dW_s.$$

Note that $U_s \in \mathscr{M}(d, N)$ is an isomorphism from $\mathbb{R}^d$ onto $T_{X_s}M$. To solve for W from the above equation we define $V_s \in \mathscr{M}(N, d)$ by

$$V_s\xi = \begin{cases} U_s^{-1}\xi, & \text{if } \xi \in T_{X_s}M; \\ 0, & \text{if } \xi \perp T_{X_s}M. \end{cases}$$

Then $V_sU_se = e$ for $e \in \mathbb{R}^d$, and we obtain

$$\int_0^t V_s\,dN^f(X)_s = \int_0^t V_sU_s\,dW_s = W_t.$$

If $N^f(X)$ is a local martingale, then W is also a local martingale. This completes the proof. □

Remark 2.5.3. The last part of the above proof makes it clear that when M is a submanifold of $\mathbb{R}^N$, X is a martingale on M if and only if $N^{f^\alpha}, \alpha = 1, \ldots, N$, are local martingales, i.e., we can reduce the set of test functions in the above proposition to the coordinate functions $\{f^1, \ldots, f^N\}$. □

We now give a local version of the above proposition. This chacterization is sometimes taken as the definition of ∇-martingales. PROPOSITION 2.5.2 can be proved from this local version by a stopping time argument. We will prove this local characterization using DEFINITION 2.5.1.

Proposition 2.5.4. *Suppose that $x = \{x^i\}$ is a local chart on M and $X = \{X^i\}$ a semimartingale on M. Then X is a ∇-martingale if and only if*

$$X_t^i = X_0^i + \textit{local martingale} - \frac{1}{2}\int_0^t \Gamma_{jk}^i(X_s)\, d\langle X^j, X^k\rangle_s.$$

Proof. The proof is an exericise in writing the anti-development W of X in terms of N and vice versa, where N is the $\mathbb{R}^d$-valued semimartingale defined by

$$N_t^i = X_t^i - X_0^i + \frac{1}{2}\int_0^t \Gamma_{jk}^i(X_s)\, d\langle X^j, X^k\rangle_s.$$

To start with, we apply the equation $dU_t = H_i(U_t) \circ dW_t^i$ to local coordinate functions $u \mapsto \left\{x^i(u), e_j^i(u)\right\}$ and use the local formulas for the horizontal vector fields in PROPOSITION 2.1.3. This gives the equations

$$\begin{cases} dX_t^i = e_j^i(t) \circ dW_t^j, \\ de_j^i(t) = -\Gamma_{kl}^i(X_t) e_j^l(t) \circ dX_t^k. \end{cases}$$

Here we have written $e_j^i(t) = e_j^i(U_t)$ to simplify the notation. If $\left\{f_j^i\right\}$ is the matrix inverse of $\left\{e_j^i\right\}$, then we have from the first equation

$$dW_t^j = f_k^j(t) \circ dX_t^k.$$

Now we have

$$\begin{aligned} dX_t^i &= e_j^i(t) dW_t^i + \frac{1}{2} d\langle e_j^i, W^j\rangle_t \\ &= e_j^i(t)\, dW_t^i - \frac{1}{2}\Gamma_{kl}^i(X_t) e_j^l(t) f_m^j(t) d\langle X^k, X^m\rangle_t \\ &= e_j^i(t)\, dW_t^i - \frac{1}{2}\Gamma_{kl}^i(X_t)\, d\langle X^k, X^l\rangle. \end{aligned}$$

It follows that

$$dN_t^i = e_j^i(t)\, dW_t^i \quad \text{and} \quad dW_t^i = f_j^i(t)\, dN_t^i.$$

Therefore N is a local martingale if and only if W is a local martingale, and the desired result follows immediately. □

We now discuss the convergence problem for manifold-valued martingales. It is well known that a real-valued local martingale $\{X_t, t \geq 0\}$ converges as $t \to \infty$ if it is uniformly bounded. When we consider manifold-valued martingales, this does not hold in general. For example, if W is a Brownian motion on $\mathbb{R}^1$, then it is easy to check that $X = (\sin W, \cos W)$ is a martingale on the unit circle $\mathbb{S}^1$ with the usual connection because its development is just W. It is clear that X does not converge, although it is uniformly bounded. However, we will show that the convergence does hold locally.

The first thing we need to do is to choose a good local chart to work with. When a manifold is equipped with a connection there is a natural local coordinate system obtained by the exponential map based at a point. Let us review some geometric facts about exponential maps.

Let M be a manifold with a connection ∇ and $o \in M$. The exponential map $\exp_o : T_oM \to M$ is defined in a neighborhood of o as follows. Let $X \in T_oM$, and let C_X be the unique geodesic from o such that $\dot{C}_X(0) = X$. If X is sufficiently small, $C_X(t)$ is well defined up to time $t = 1$ and we define

$$\exp X = C_X(1), \qquad X \in T_oM.$$

Locally C_X is obtained by solving second order ordinary differential equations (see (2.1.2)). With the help of these equations, it is an easy exercise to show that the exponential map is a diffeomorphism from T_oM to M in a neighborhood of the origin. A local coordinate system is then obtained by identifying T_oM with $\mathbb{R}^d$ via a fixed frame $u_o : \mathbb{R}^d \to T_oM$, namely,

$$\mathbb{R}^d \xrightarrow{u_o} T_oM \xrightarrow{\exp_o} M.$$

If M is a Riemannian manifold, we can take an orthonormal basis on T_oM and set up the usual cartesian coordinates on T_o. Through the exponential map, these coordinates define a system of of local coordinates in a neighborhood of o on the manifold M. These are called the normal coordinates at o.

We will need the following property of this local chart.

Lemma 2.5.5. *Let $x = \{x^i\}$ be a local coordinate system obtained from the exponential map at o. Then its Christoffel symbols are anti-symmetric at o, i.e., $\Gamma^i_{jk} + \Gamma^i_{kj} = 0$. In particular, if the connection ∇ is torsion-free, then its Christoffel symbols vanish at o.*

Proof. By the definition of the exponential map, the curve

$$t \mapsto \left\{ x^i = t;\ x^k = 0, k \neq i \right\}$$

is a geodesic whose tangent field is $X_i = \partial/\partial x^i$. Hence $\nabla_{X_i} X_i = 0$ along the curve and, in particular, at o. This shows that $\Gamma^k_{ii} = 0$ at o. For two distinct indices i and j, the curve

$$t \mapsto \left\{ x^i = x^j = t;\ x^k = 0, k \neq i, j \right\}$$

is also a geodesic whose tangent field is $X = X_i + X_j$. Hence $\nabla_X X = 0$ along the curve and, at o,

$$\Gamma^k_{ii} + \Gamma^k_{ij} + \Gamma^k_{ji} + \Gamma^k_{jj} = 0.$$

Since the first and the last terms vanish at o, we have $\Gamma^k_{ij} + \Gamma^k_{ji} = 0$. If ∇ is torsion-free, then $\Gamma^k_{ij} = \Gamma^k_{ji}$, and we have $\Gamma^k_{ij} = 0$. □

Now we come to the local convergence theorem for manifold-valued martingales. Let M be a manifold and O an open set on M. We say that a path $x : \mathbb{R}_+ \to M$ lies eventually in O if there exists a random time T such that $X_t \in O$ for all $t \geq T$. Note that T may not be a stopping time.

Theorem 2.5.6. *Suppose that M is a manifold equipped with a connection. Every point of M has a neighborhood O with the following property: if X is a ∇-martingale, then*

$$\{X \text{ lies eventually in } O\} \subseteq \{\lim_{t\uparrow\infty} X_t \text{ exists}\}. \tag{2.5.2}$$

Proof. Let $o \in M$ and choose a local coordinate system $x = \left\{x^i\right\}$ in a neighborhood of o defined by the exponential map at o. Let O be a relatively compact neighborhood of o such that $\overline{O}$ is covered by the local coordinate system. Let f^i be a smooth function on M with compact support such that on O

$$f^i(x) = x^i + \sum_{j=1}^d (x^j)^2.$$

Shrinking O if necessary, we may assume that $f = \left\{f^i\right\} : O \to \mathbb{R}^d$ is a diffeomorphism onto its image. Using (2.2.4) we see that the Hessian of f^i is given by

$$\nabla^2 f^i = \left[2\delta_{jk} - \Gamma^i_{jk} - 2x^l \Gamma^l_{jk} \right] dx^j \otimes dx^k.$$

Let h^i be the symmetric part of $\nabla^2 f^i$:

$$h^i = \left[2\delta_{jk} - \frac{\Gamma^i_{jk} + \Gamma^i_{kj}}{2} - x^l \left(\Gamma^l_{jk} + \Gamma^l_{kj} \right) \right] dx^j \otimes dx^k.$$

By LEMMA 2.5.5 the coefficients of h^i are δ_{jk} at o; hence by continuity h is strictly positive definite in a neighborhood of o. Replacing O by a smaller one if necessary, we may assume that each h^i is strictly positive definite everywhere on O. We claim that O defined this way has the desired property.

Suppose that X is a martingale on M defined on a filtered probability space $(\Omega, \mathscr{F}_*, \mathbb{P})$. Let

$$\Omega_O = \{X \text{ lies eventually in } O\}.$$

Because $f : \overline{O} \to f(\overline{O})$ is a diffeomorphism, it is enough to show that on Ω_O each $f^i(X_t)$ converges as $t \uparrow \infty$. We have

$$f^i(X_t) = f^i(X_0) + N_t^i + \int_0^t \nabla^2 f^i(dX_s, dX_s),$$

where N^i is a real-valued local martingale. By PROPOSITION 2.4.7, $\nabla^2 f^i$ may be replaced by its symmetric part; hence

$$f^i(X_t) = f^i(X_0) + N_t^i + \int_0^t h^i(dX_s, dX_s), \tag{2.5.3}$$

Now we use PROPOSITION 2.4.12 and the remark after it. If $\omega \in \Omega_O$, the h^i-quadratic variation of X is eventually nondecreasing because X stays O, on which h^i is positive definite. This fact implies that the third term on the right side of (2.5.3) is bounded from below. On the other hand, because f^i is uniformly bounded, the left side is uniformly bounded. Hence N^i is bounded from above. But N^i is a local martingale, hence a time-changed Brownian motion $N_t^i = B_{\langle N^i \rangle_t}$. Since a Brownian motion path is bounded from neither side, the only possibility is that $\lim_{t\uparrow\infty} \langle N^i \rangle_t = \langle N^i \rangle_\infty$ exists and is finite, and $\lim_{t\uparrow\infty} N_t^i = B_{\langle N^i \rangle_\infty}$. It follows from (2.5.3) that $\lim_{t\uparrow\infty} f^i(X_t)$ exists, and the proof is completed. □

2.6. Martingales on submanifolds

Let N and M be two differentiable manifolds with connections ∇^N and ∇^M respecitvely. Suppose that M is a submanifold of N. In the first part of this section we study two questions:

(1) When is a ∇^N-martingale on N which lives on the submanifold M a martingale on M?

(2) When is a ∇^M-martingale on the submanifold M a martingale on N?

The results we obtain in the course of answering these two questions will be used to prove the local nonconfluence property of manifold-valued martingales.

Naturally the answers to the questions posed above depend on how the two connections ∇^M and ∇^N are related. Geometrically, the relation of a connection on a manifold to a connection on an ambient manifold is described by the second fundamental form. It is a map of the following type:

$$\Pi : \Gamma(TM) \times \Gamma(TM) \to \Gamma(TN|_M).$$

Since M is a submanifold of N, we have $T_xM \subseteq T_xN$ for $x \in M$. Both covariant derivatives $\nabla^N_X Y$ and $\nabla^M_X Y$ make sense and belong to T_xN, and we define

$$\Pi(X,Y) = \nabla^N_X Y - \nabla^M_X Y, \quad X, Y \in \Gamma(TM). \tag{2.6.1}$$

The reader can verify easily that

$$\Pi(fX, gY) = fg\Pi(X,Y), \qquad f, g \in C^\infty(M).$$

This implies by a well known argument in tensor analysis that the value of $\Pi(X,Y)$ at a point depends only on the values of the vector fields at the same point. Thus Π is a tensor field on M taking values in the vector bundle $TN|_M$, or $\Pi_x \in \mathrm{Hom}(T_xM \otimes T_xM, T_xN)$ for $x \in M$. The submanifold M is called totally geodesic if the second fundamental form Π vanishes everywhere on M.

Recall that the torsion of a connection ∇ is a $(1,2)$-tensor defined by

$$T(X,Y) = \nabla_X Y - \nabla_Y X - [X,Y],$$

where $[X,Y]$ is the Lie bracket of X and Y. By the definition of Π we have

$$\Pi(X,Y) - \Pi(Y,X) = T^N(X,Y) - T^M(X,Y).$$

This shows that Π is symmetric if both ∇^N and ∇^M are torsion-free.

We have the following characterization of totally geodesic submanifolds.

Proposition 2.6.1. *Suppose that M and N are two differentiable manifolds equipped with connections ∇^M and ∇^N resepectively and M is a submanifold of N. If M is totally geodesic in N, then every ∇^M-geodesic in M is also a ∇^N-geodesic in N. The converse also holds if both connections are torsion-free.*

Proof. Suppose that C is a ∇^M-geodesic in M. Then $\nabla^M_{\dot C}\dot C = 0$. By the definition of the second fundamental form we have

$$\nabla^N_{\dot C}\dot C = \nabla^M_{\dot C}\dot C + \Pi(\dot C, \dot C) = 0.$$

Hence C is a ∇^N-geodesic in N.

Conversely, let $X \in T_xM$ and let C be the ∇^M-geodesic on M such that $C(0) = 0$ and $\dot C(0) = X$. Because C is also ∇^N-geodesic, we have both $\nabla^M_{\dot C}\dot C = 0$ and $\nabla^N_{\dot C}\dot C = 0$ along C. This means $\Pi(\dot C, \dot C) = 0$. In particular, $\Pi(X,X) = 0$. But the connections are torsion-free, which implies that Π

is symmetric; hence by symmetrization $\mathit{\Pi}(X, Y) = 0$ for all vector fields X and Y, and M is totally geodesic in N. □

We will need the following result.

Lemma 2.6.2. *Let M be a submanifold of $\mathbb{R}^l$ and $f : M \to \mathbb{R}^l$ the coordinate function on M. Let $\widetilde{f} = f \circ \pi$ be its lift to the frame bundle $\mathscr{F}(M)$. Then*

$$H_i H_j \widetilde{f}(u) = \mathit{\Pi}(ue_i, ue_j).$$

Proof. Let $\{u_t\}$ be a horizontal curve such that $u_0 = u$ and $\dot{u}_0 = H_i(u)$. The vector field $X(x_t) = u_t e_j$ is ∇^M-parallel along $x_t = \pi u_t$, namely, $\nabla^M_{ue_i} X = 0$; hence by definition $\mathit{\Pi}(ue_i, ue_j) = \nabla^{\mathbb{R}^l}_{ue_i} X$. Now, from (2.4.2) we have $H_j \widetilde{f}(u_t) = u_t e_j = X(x_t)$. It follows that

$$H_i H_j \widetilde{f}(u) = \frac{d\{X(x_t)\}}{dt} = \nabla^{\mathbb{R}^l}_{ue_i} X = \mathit{\Pi}(ue_i, ue_j).$$

□

Example 2.6.3. Let N be a Riemannian manifold with its Levi-Civita connection and M a submanifold of N with the induced metric. Then the Levi-Civita connection of M is given by

$$\nabla^M_X Y = \text{ the orthogonal projection of } \nabla^N_X Y \text{ to } T_x M.$$

In this case the second fundamental form $\mathit{\Pi}(X, Y)_x \in T_x^\perp M$, the orthogonal complement of $T_x M$ in $T_x N$. In particular $\operatorname{Ran}(\mathit{\Pi}_x) \cap T_x M = \{0\}$ at each point $x \in M$.

The mean curvature of M at x is defined to be the trace of the second fundamental form:

$$H = \sum_{i=1}^{d} \mathit{\Pi}(e_i, e_i),$$

where $\{e_i\}$ is an orthonormal basis for $T_x M$. The submanifold M is called minimal if $H = 0$ on M. If M is a hypersurface of N (codimension 1) and both M and N are oriented, we let n be the unit normal vector field on M such that the basis of $T_x N$ obtained by appending n to an oriented basis of $T_x M$ is positively oriented. In this case the second fundamental form $\mathit{\Pi}$ is usually identified with the (real-valued) symmetric (0,2)-tensor $(X, Y) \mapsto \langle \mathit{\Pi}(X, Y), n \rangle$, and the mean curvature is given by $H = \operatorname{trace} \mathit{\Pi}$.

A hypersurface is said to be convex (strictly convex) if $\mathit{\Pi}$ is positive (strictly positive) definite at every point of M. □

In the following, we will use E and W to denote the anti-developments of a semimartingale X on M (hence also a semimartingale on N) with respect to the connections ∇^N and ∇^M respectively. Since a martingale on a manifold is by definition a semimartingale whose anti-development is a local martingale, the natural starting point is to find a relation between the anti-developments E and W.

Theorem 2.6.4. *Let N be a manifold of dimension l equipped with a connection ∇^N and M a submanifold of N of dimension d equipped with a connection ∇^M. Suppose that:*

(1) *X is a semimartingale on M;*
(2) *U is a horizontal lift of X in $\mathscr{F}(M)$ with respect to ∇^M, and W the corresponding $\mathbb{R}^d$-valued anti-development;*
(3) *V is a horizontal lift of X in $\mathscr{F}(N)$ with respect to ∇^N, and E the corresponding $\mathbb{R}^l$-valued anti-development.*

Then

$$E_t = \int_0^t V_s^{-1} U_s dW_s + \frac{1}{2}\int_0^t V_s^{-1} \Pi(dX_s, dX_s), \tag{2.6.2}$$

where Π is the second fundamental form of M.

Proof. According to (2.4.2),

$$\Pi(dX_s, dX_s) = \Pi(U_t e_i, U_t e_j)\, d\left\langle W^i, W^j\right\rangle_t.$$

We have

$$\Pi(U_t e_i, U_t e_j) \in T_{X_t}M \subseteq T_{X_t}N \quad \text{and} \quad V_s^{-1}\Pi(U_t e_i, U_t e_j) \in \mathbb{R}^l.$$

Thus a more precise but admittedly somewhat clumsy rewriting of (2.6.2) is

$$E_t = \int_0^t V_s^{-1} U_s e_i\, dW_s^i + \int_0^t V_s^{-1} \Pi(U_s e_i, U_s e_i)\, d\langle W^i, W^j\rangle_s.$$

We first show the special case when $N = \mathbb{R}^l$ equipped with the euclidean connection. In this case $E = X$ and $V = I_{\mathbb{R}^l}$, the identity map on $\mathbb{R}^l$. Let $f(x) = x \in \mathbb{R}^l$ on M and let $\widetilde{f} = f \circ \pi$ be its lift to $\mathscr{F}(M)$. By Itô's formula,

$$X_t = X_0 + \int_0^t H_i\widetilde{f}(U_s)\, dW_s^i + \frac{1}{2}\int_0^t H_i H_j \widetilde{f}(U_s)\langle dW^i, dW^j\rangle_s. \tag{2.6.3}$$

From (2.3.5) and Lemma 2.6.2,

$$H_i\widetilde{f}(u) = ue_i, \qquad H_i H_j \widetilde{f}(u) = \Pi(ue_i, ue_j).$$

Substituting these relations into (2.6.3), we obtain

$$X_t = X_0 + \int_0^t U_s dW_s + \frac{1}{2}\int_0^t \Pi(dX_s, dX_s), \tag{2.6.4}$$

which is (2.6.2) for the special case $N = \mathbb{R}^l$.

For a general manifold N, we assume that it is a submanifold of $\mathbb{R}^m$ for some m. Thus M is also a submanifold of $\mathbb{R}^m$. Let $\Pi^{M,\mathbb{R}^m}$ and $\Pi^{N,\mathbb{R}^m}$ be the second fundamental forms of M and N in $\mathbb{R}^m$ respectively. From the definition (2.6.1) we have, for vector fields X, Y on M,

$$\Pi(X,Y) = \Pi^{M,\mathbb{R}^m}(X,Y) - \Pi^{N,\mathbb{R}^m}(X,Y).$$

Note that this equation should be understood as a vector equation in $\mathbb{R}^m$. By (2.6.4) applied to the two embeddings $M \subseteq \mathbb{R}^m$ and $N \subseteq \mathbb{R}^m$,

$$X_t = X_0 + \int_0^t U_s dW_s + \frac{1}{2}\int_0^t \Pi^{M,\mathbb{R}^m}(dX_s, dX_s),$$

$$X_t = X_0 + \int_0^t V_s dE_s + \frac{1}{2}\int_0^t \Pi^{N,\mathbb{R}^m}(dX_s, dX_s).$$

It follows that

$$\begin{aligned}
E_t &= \int_0^t V_s^{-1} dX_s - \frac{1}{2}\int_0^t V_s^{-1}\Pi^{N,\mathbb{R}^m}(dX_s, dX_s) \\
&= \int_0^t V_s^{-1} U_s dW_s + \frac{1}{2}\int_0^t V_s^{-1}\Pi^{M,\mathbb{R}^m}(dX_s, dX_s) \\
&\quad - \frac{1}{2}\int_0^t V_s^{-1}\Pi^{N,\mathbb{R}^m}(dX_s, dX_s) \\
&= \int_0^t V_s^{-1} U_s dW_s + \frac{1}{2}\int_0^t V_s^{-1}\Pi(dX_s, dX_s).
\end{aligned}$$

□

Let us draw a few corollaries from the above theorem.

Corollary 2.6.5. *Suppose that M is a totally geodesic submanifold of N. Then every ∇^M-martingale on M is also a ∇^N-martingale on N.*

Proof. If X is a martingale on M, then W is a local martingale. By (2.6.2) the condition $\Pi = 0$ implies that E is also a local martingale. Hence X is also a martingale on N. □

The range of the second fundamental form at $x \in M$ is

$$\operatorname{Ran}\Pi_x = \{\Pi(X,Y)(x) : X, Y \in \Gamma(TM)\}.$$

We say that Π is independent of TM if $\operatorname{Ran}\Pi_x \cap T_xM = \{0\}$ at every $x \in M$. This is the case, for example, when N is a Riemannian manifold with its Levi-Civita connection and M a submanifold of N with the induced metric and connection, because in this case Π is always perpendicular to TM; see EXAMPLE 2.6.3.

Corollary 2.6.6. *Suppose that M is a submanifold of N such that its second fundamental form $\varPi$ is independent of TM. Then a ∇^N-martingale on N which lives on M is also a ∇^M-martingale on M.*

Proof. Let $W_t = M_t + A_t$ be the decomposition of W into a local martingale and a process of locally bounded variation. It is enough to show that $A = 0$, for then W is a local martingale, which means that X is ∇^M-martingale. We have

$$E_t = \int_0^t V_s^{-1} U_s dM_s + \int_0^t V_s^{-1} \left\{ U_s dA_s + \frac{1}{2} \varPi(dX_s, dX_s) \right\}.$$

Since E is a local martingale, the last term vanishes. We have $U_s dA_s \in T_{\pi U_s} M$, $\varPi(dX_s, dX_s) \in \operatorname{Ran} \varPi_{\pi U_s}$ and $U_s dA_s + \varPi(dX_s, dX_s) = 0$. Hence

$$\varPi(dX_s, dX_s) = -U_s dA_s \in \operatorname{Ran} \varPi_{\pi U_s} \cap T_{\pi U_s} M.$$

But $\varPi$ with TM are independent; hence we must have $dA_s = 0$. $\square$

Corollary 2.6.7. *Suppose that N is a Riemannian manifold and M is a strictly convex hypersurface in N. Then every martingale on N which lives on M must be a constant.*

Proof. By EXAMPLE 2.6.3 the preceding corollary applies; hence both W and E are local martingales. From (2.6.2) We have $\varPi(dX_s, dX_s) = 0$. By the strict convexity of $\varPi$, it is an easy exercise to show from this that X is constant. $\square$

Example 2.6.8. Let M be a Riemannian manifold. A semimartingale X on M is called a Brownian motion if its anti-development with respect to the Levi-Civita connection is a euclidean Brownian motion. Suppose that M is a minimal submanifold of a euclidean space $\mathbb{R}^l$. Then a Brownian motion on M is a local martingale on $\mathbb{R}^l$. This follows easily from (2.6.4), for the differential of the last term is

$$\varPi(U_t e_i, U_t e_j)\, d\left\langle W^i, W^j \right\rangle_t = \varPi(U_t e_i, U_t e_i)\, dt = H(X_t)\, dt = 0.$$

The geometric counterpart of this result is that if M is minimal in $\mathbb{R}^l$, then the coordinate functions are harmonic on M. $\square$

Finally we study the nonconfluence property of manifold-valued martingales. Suppose that X is an $\mathbb{R}^d$-valued $\mathscr{F}_*$-martingale on $[0, T]$. Then

$$X_t = \mathbb{E}\left\{ X_T | \mathscr{F}_t \right\}, \quad 0 \le t \le T.$$

This means that a martingale is determined by its terminal value at time T and the filtration $\{\mathscr{F}_t, 0 \le t \le T\}$. We will prove a local version of this result for manifold-valued martingales. Let us start with two preliminary results, both interesting in their own right.

Proposition 2.6.9. *Suppose that M is a totally geodesic submanifold of N. Then every point of M has a neighborhood O in N with the following property: If X is a ∇^N-martingale in O on the time interval $[0,T]$ such that its terminal value X_T is in the submanifold M, then $X_t \in M$ for all $t \in [0,T]$.*

Proof. We first show that if M is totally geodeisc, then the exponential maps of M and N coincide. Let $o \in M$ and consider the exponential map $\exp : T_oN \to N$ of the manifold N. Suppose that $X \in T_oM$ and C is a ∇^M-geodesic in M from o such that $\dot{C}(0) = X$. Since M is totally geodesic, C is also a geodesic in N (see PROPOSITION 2.6.1). Hence $\exp tX = C(t)$ for $0 \le t \le 1$ if X is sufficiently small, and $\exp X = C(1) \in M$. This shows that the restriction of the exponential map $\exp : T_oN \to M$ to the subspace $T_oM \subseteq T_oN$ is just the exponential map of M.

Choose a cartesian coordinate system $\{x^i, 1 \le i \le l\}$ on T_oN such that the first d coordinates span the subspace T_oM. This coordinate system on N covers a neighborhood O having the property that

$$M \cap O = \left\{x^{d+1} = 0, \dots, x^l = 0\right\} \cap O.$$

Therefore $\tilde{x} = \left\{x^1, \dots, x^d\right\}$ is a local chart on M. Let $\hat{x} = \left\{x^{d+1}, \dots, x^l\right\}$ and consider the function We consider the function

$$f(x) = f(\tilde{x}, \hat{x}) = \left(1 + |\tilde{x}|^2\right) |\hat{x}|^2.$$

In the neighborhood O, $f(x) = 0$ if and only if $x \in M$. We show that on a possibly smaller neighborhood the symmetric part h of the Hessian $\nabla^2 f$ is positive (more precisely, nonnegative) definite (cf. the proof of THEOREM 2.5.6). Let $X_i = \partial/\partial x^i$. From

$$\nabla^2 f = \nabla^2 f(X_i, X_j)\, dx^i \otimes dx^j$$

and

$$\nabla^2 f(X_i, X_j) = f_{ij} - \Gamma_{ij}^k f_k$$

we have by a straightforward calculation

$$\begin{aligned}\nabla^2 f = {} & 2\left(1 + |\tilde{x}|^2\right) \sum_{i,j=d+1}^{l} \left(\delta_{ij} - x^k \Gamma_{ij}^k\right) dx^i \otimes dx^j \\ & + 2|\hat{x}|^2 \sum_{i,j=1}^{d} (\delta_{ij} - x^k \Gamma_{ij}^k) dx^i \otimes dx^j + 4 \sum_{i=1}^{d} \sum_{j=d+1}^{l} x^i x^j dx^i \otimes dx^j.\end{aligned}$$

Note that because the second fundamental form vanishes, those of the Christoffel symbols of N whose are from the tangent vector fields on M

coincides with the corresponding Christoffel symbols of M, so we do not need to distinguish them. The symmetric part of $\nabla^2 f$ is

$$
\begin{aligned}
h = & \left(1+|\tilde{x}|^2\right) \sum_{i,j=d+1}^{l} \left[2\delta_{ij} - x^k(\Gamma_{ij}^k + \Gamma_{ji}^k)\right] dx^i \otimes dx^j \\
& + |\hat{x}|^2 \sum_{i,j=1}^{d} \left[2\delta_{ij} - x^k(\Gamma_{ij}^k + \Gamma_{ji}^k)\right] dx^i \otimes dx^j \\
& + 2\sum_{i=1}^{d} \sum_{j=d+1}^{l} x^i x^j (dx^i \otimes dx^j + dx^j \otimes dx^i) \\
& \stackrel{\text{def}}{=} S_1 + S_2 + S_3.
\end{aligned}
$$

By LEMMA 2.5.5 the Christoffel symbols satisfy $\Gamma_{jk}^i + \Gamma_{kj}^i = 0$ at o; hence there is a constant C_1 such that $|\Gamma_{ij}^k + \Gamma_{ji}^k| \le C_1|x|$ on O. Thus as symmetric quadratic forms we have

$$
\begin{aligned}
S_1 &\ge \left(2 - C_2|x|^2\right) \sum_{j=d+1}^{l} dx^j \otimes dx^j, \\
S_2 &\ge |\hat{x}|^2 \left(2 - C_2|x|^2\right) \sum_{i=1}^{d} dx^i \otimes dx^i.
\end{aligned}
$$

Using the inequality

$$
x^i x^j (dx^i \otimes dx^j + dx^j \otimes dx^i) \le 2|x^i|^2 dx^j \otimes dx^j + \frac{1}{2}|x^j|^2 dx^i \otimes dx^i,
$$

we have

$$
S_3 \ge -4|\tilde{x}|^2 \sum_{j=d+1}^{l} dx^j \otimes dx^j - |\hat{x}|^2 \sum_{i=1}^{d} dx^i \otimes dx^i.
$$

It follows that

$$
\begin{aligned}
h \ge & \left\{2 - (C_2+4)|x|^2\right\} \sum_{j=d+1}^{l} dx^j \otimes dx^j \\
& + |\hat{x}|^2 \left(1 - C_2|x|^2\right) \sum_{i=1}^{d} dx^i \otimes dx^i.
\end{aligned}
$$

Choosing O sufficiently small so that $(C_2+4)|x|^2 \le 1/2$, we obtain $h \ge 0$ on O.

Now we prove the assertion of the proposition for the neighborhood O of N constructed above. Suppose that $\{X_t, 0 \le t \le T\}$ is a martingale in O

such that $X_T \in O \cap M$. By Itô's formula we have

$$f(X_t) = f(X_0) + \text{ local martingale} + \frac{1}{2}\int_0^t \nabla^2 f(dX_s, dX_s).$$

By PROPOSITIONS 2.4.7 and 2.4.12,

$$\int_0^t \nabla^2 f(dX_s, dX_s) = \int_0^t h(dX_s, dX_s)$$

is nondecreasing. This shows that $f(X)$ is a uniformly bounded, nonnegative submartingale with terminal value $f(X_T) = 0$. It follows that $f(X_t) = 0$ for all $t \leq T$. This shows $X_t \in M$ because f vanishes only on $O \cap M$. □

Let M be a manifold with a connection ∇^M and $N = M \times M$ the product manifold. Let π_i $(i = 1, 2)$ be the projection from N to the first and second factor respectively. We will now define the product connection $\nabla^N = \nabla^M \times \nabla^M$. For any $x = (x_1, x_2) \in N$, the map

$$X \mapsto (\pi_{1*}X, \pi_{2*}X) \stackrel{\text{def}}{=} (X_1, X_2)$$

defines an isomorphism between T_xN and $T_{x_1}M \times T_{x_2}M$ (product vector space). Suppose that $X = (X_1, X_2)$ is a vector field on N. Note that in general X_i are vector fields on M. Now let $Y = (Y_1, Y_2) \in T_xN$, and $t \mapsto x_t = (x_{1t}, x_{2t})$ be a curve on N such that $x_0 = x$ and

$$Y = \dot{x}_0 = (\dot{x}_{10}, \dot{x}_{20}) = (Y_1, Y_2).$$

Along the curve we have $X = (X_1, X_2)$, where each X_i can be regarded as a vector field along $t \mapsto x_{it}$. We define

$$\nabla^N_Y X = (\nabla^M_{Y_1} X_1, \nabla^M_{Y_2} X_2).$$

We verify that ∇^N defined above is a connection. The only defining property of a connection that is not obvious directly from the defintion is

$$\nabla^N_Y (fX) = f\nabla^N_Y X + Y(f)X$$

for any $f \in C^\infty(N)$. This can be seen as follows. Along the curve $t \mapsto x_t$, we have $fX = (f(t)X_1, f(t)X_2)$ at x_t, where $f(t) = f(x_t)$. Hence the vector field along $t \mapsto x_{it}$ is $t \mapsto f(t)X_i(x_t)$, and

$$\nabla^M_{Y_i}(fX_i) = f\nabla^M_{Y_i} X_i + \dot{f}X_i.$$

We have, of course, $\dot{f} = Y(f)$. It follows that

$$\begin{aligned}\nabla^N_Y(fX) &= \left(f\nabla^M_{Y_1}X_1 + \dot{f}X_1, f\nabla^M_{Y_2}X_2 + \dot{f}X_2\right) \\ &= f\left(\nabla^M_{Y_1}X_1, \nabla^M_{Y_2}X_2\right) + Y(f)(X_1, X_2) \\ &= f\nabla^N_Y + Y(f)X.\end{aligned}$$

This shows that ∇^N is indeed a connection on the product manifold N.

Proposition 2.6.10. *Suppose that M is a differentiable manifold equipped with a connection ∇^M. Let $N = M \times M$ be the product manifold equipped with the product connection $\nabla^N = \nabla^M \times \nabla^M$. Then the diagonal map $x \mapsto (x, x)$ embeds M as a totally geodesic submanifold of N.*

Proof. Let X, Y be vector fields on M. As vector fields on the product manifold N, they are represented by (X, X) and (Y, Y) respectively. Hence, by the definition of the product connection,

$$\nabla_X^N Y = (\nabla_X^M Y, \nabla_X^M Y).$$

Thus $\nabla_X^N Y$ is tangent to M and is equal to $\nabla_X^M Y$. This immediately implies that $\Pi(X, Y) = 0$, i.e., M is totally geodesic in N. □

The following proof of the local nonconfluence property of manifold-valued martingales is an ingenious combination of PROPOSITIONS 2.6.9 and 2.6.10.

Theorem 2.6.11. *Let M be a manifold equipped with a connection. Then every point of M has a neighborhood V with the following property: If $\{X_t, 0 \le t \le T\}$ and $\{Y_t, 0 \le t \le T\}$ are two $\mathscr{F}_*$-martingales on V defined on the same filtered probability space $(\Omega, \mathscr{F}_*, \mathbb{P})$ such that $X_T = Y_T$, then $X_t = Y_t$ for all $t \in [0, T]$.*

Proof. Let $N = M \times M$ with the product connection. By PROPOSITION 2.6.10, M is totally geodesic. Let $o \in M$ and choose a neighborhood O of (o, o) in N which has the property stated in PROPOSITION 2.6.9. Choose a neighborhood V of o in M such that $V \times V \subset O$. Let the anti-developments of X and Y be W and E respectively. By assumption they are $\mathbb{R}^d$-valued local martingales. It is easy to verify that the anti-development of $Z = (X, Y)$ with respect to the product connection ∇^N is simply (W, E), which is an $\mathbb{R}^{2d}$-valued local martingale; hence Z is a ∇^N-martingale. By assumption Z lives on O and $Z_T = (X_T, X_T) \in M$. By PROPOSITION 2.6.9 $Z_t \in M$ for $0 \le t \le T$, which is equivalent to $X_t = Y_t$ for $0 \le t \le T$. □

Chapter 3

Brownian Motion on Manifolds

With this chapter we leave the general theory and concentrate almost exclusively on Brownian motion on Riemannian manifolds. It is defined as a diffusion process generated by the Laplace-Beltrami operator $\Delta_M/2$, with the proverbial $1/2$ that has baffled many geometers. Properties of the Laplace-Beltrami operator that are important to our discussion are laid out in SECTION 3.1, especially its relation to Bochner's horizontal Laplacian. This is followed by a formal introduction of Brownian motion on a Riemannian manifold in SECTION 3.2 and a discussion of several equivalent characterizations. In view of Nash's famous embedding theorem, we assume whenever convenient that our Riemannian manifold is a submanifold of a euclidean space with the induced metric. Examples of Riemannian Brownian motion are given in SECTION 3.3 with an eye to later applications.

Fix a reference point $o \in M$ and let $r(x) = d(x, o)$, the Riemannian distance from x to o. The radial process $r(X)$ of a Brownian motion X on M will play an important role in many applications. Before Brownian motion reaches the cutlocus of its starting point, the radial process is decomposed as

$$r(X_t) = r(X_0) + \beta_t + \frac{1}{2}\int_0^t \Delta_M r(X_s)\, ds,$$

where β is a one-dimensional Brownian motion. It is clear from the above relation that the behavior of the radial process depends on the growth of $\Delta_M r$ relative to r. From differential geometry it is well known that the growth can be controlled by imposing appropriate conditions on the growth of sectional and Ricci curvatures. In SECTION 3.4 we will discuss several geometric results of this kind, including the well known Laplacian comparison theorem.

Effective use of the radial process can be made only if we can go beyond the cutlocus. In SECTION 3.5 we prove Kendall's decomposition of the radial process, namely,

$$r(X_t) = r(X_0) + \beta_t + \frac{1}{2}\int_0^t \Delta_M r(X_s)\, ds - L_t, \quad 0 \le t < e(X),$$

where L is a nondecreasing process which increases only when Brownian motion is at the cutlocus. This term can be ignored when we need an upper bound for the radial process. The analytic counterpart of this decomposition of the radial process is the well known geometric fact that the distributional Laplacian of the distance function is bounded from above by its restriction within the cutlocus. As an application of the above decomposition, we prove in the last section a useful estimate on the first exit time of Brownian motion from a geodesic ball.

3.1. Laplace-Beltrami operator

Brownian motion on a Riemannian manifold M is a diffusion process generated by $\Delta_M/2$, where Δ_M is the Laplace-Beltrami operator on M, the natural generalization of the usual Laplace operator on euclidean space. In this section, we review some relevant facts concerning this operator. For this part of Riemannian geometry the books *Riemannian Geometry* by Do Carmo [**17**] and *Riemannian Geometry and Geometric Analysis* by Jost [**50**] are highly recommended.

Let M be a differentiable manifold. A Riemannian metric $ds^2 = \langle\cdot,\cdot\rangle$ on M is a symmetric, strictly positive (0,2)-tensor on M. Equivalently, it is a smooth assignment of an inner product $ds_x^2 = \langle\cdot,\cdot\rangle_x$ for each tangent space T_xM. Let $x = \{x^i\}$ be a local chart and $X_i = \partial/\partial x^i$ the partial differentiation. Then the Riemannian metric can be written as

$$ds^2 = g_{ij}dx^i dx^j, \quad g_{ij} = \langle X_i, X_j\rangle.$$

Here $dx^i dx^j$ stands for the symmetrization of $dx^i \otimes dx^j$:

$$dx^i dx^j = \frac{dx^i \otimes dx^j + dx^j \otimes dx^i}{2}.$$

The matrix $g = \{g_{ij}\}$ is positive definite at each point. A Riemannian manifold is a differentiable manifold equipped with a Riemannian metric. We will use m to denote the Riemannian volume measure on M. In local coordinates it is given by $m(dx) = \sqrt{G(x)}dx$, where $G = \det g$ and $dx = dx^1 \cdots dx^d$ is the Lebesgue measure on $\mathbb{R}^d$. It is easy to verify that the

Riemannian volume measure m is well defined, i.e., it is independent of the choice of local coordinates. For two functions f, g on M with compact support we write

$$(f, g) = \int_M f(x)g(x)m(dx).$$

For two vector fields X, Y on M with compact support, we define

$$(X, Y) = \int_M \langle X, Y \rangle_x m(dx).$$

The inner product $\langle \cdot, \cdot \rangle_x$ on the tangent space T_xM induces an inner product on its dual T_x^*M, the cotangent space: if $\theta \in T_x^*M$, then there is a unique $X_\theta \in T_xM$ such that $\theta(Y) = \langle X_\theta, Y \rangle_x$ for all $Y \in T_xM$. For $\theta, \psi \in T_x^*M$ we define

$$\langle \theta, \psi \rangle_x = \langle X_\theta, X_\psi \rangle_x.$$

If θ, ψ are two 1-forms on M with compact support, we write

$$(\theta, \psi) = \int_M \langle \theta, \psi \rangle_x m(dx).$$

On a Riemannian manifold M, there is a unique connection ∇, the Levi-Civita connection, which is compatible with the Riemannian metric, i.e.,

$$\nabla_Z \langle X, Y \rangle = \langle \nabla_Z X, Y \rangle + \langle X, \nabla_Z Y \rangle, \qquad X, Y, Z \in \Gamma(TM); \tag{3.1.1}$$

and is torsion-free:

$$T(X, Y) \stackrel{\text{def}}{=} \nabla_X Y - \nabla_Y X - [X, Y] = 0, \qquad X, Y \in \Gamma(TM),$$

where $[X, Y]$ is the Lie bracket of the vector fields X and Y. In local coordinates, the Christoffel symbols of the Levi-Civita connection are given by

$$\Gamma^i_{jk} = \frac{1}{2} g^{im} \left(g_{jm,k} + g_{mk,j} - g_{jk,m} \right), \tag{3.1.2}$$

where $\{g^{im}\}$ is the inverse matrix of $\{g_{ij}\}$ and $g_{jk,m} = \partial g_{jk} / \partial x^m$.

From now on we assume that M is a Riemannian manifold equipped with the Levi-Civita connection. On M there is an intrinsically defined second order elliptic operator, the Laplace-Beltrami operator, which generalizes the usual Laplace operator on euclidean space. On euclidean space $\Delta f = \operatorname{div}\operatorname{grad} f$. Let us define the gradient and divergence on M. The gradient $\operatorname{grad} f$ is the dual of the differential df; thus it is the unique vector field defined by the relation

$$\langle \operatorname{grad} f, X \rangle = df(X) = Xf, \qquad \forall X \in \Gamma(TM).$$

In local coordinates, we have by an easy computation

$$\nabla f = g^{ij} \frac{\partial f}{\partial x^i} \frac{\partial}{\partial x^j}.$$

The divergence $\mathrm{div}X$ of a vector field X is defined to be the contraction of the (1,1)-tensor ∇X. If $X = a^i \dfrac{\partial}{\partial x^i}$ in local coordinates, then it is easy to verify that

$$\mathrm{div}X = \frac{1}{\sqrt{G}} \frac{\partial\,(\sqrt{G}a^i)}{\partial x^i}.$$

The Laplace-Beltrami operator is

$$\Delta_M f = \mathrm{div}\,\mathrm{grad} f.$$

Combining the local expressions the gradient and divergence, we obtain the familiar local formula for the Laplace-Beltrami operator:

$$\Delta_M f = \frac{1}{\sqrt{G}} \frac{\partial}{\partial x^i} \left(\sqrt{G} g^{ij} \frac{\partial f}{\partial x^j} \right).$$

Thus Δ_M is a nondegenerate second order elliptic operator.

Proposition 3.1.1. *For any orthonormal basis* $\{X_i\}$ *of* T_xM, *we have*

$$\Delta_M f = \mathrm{trace}\, \nabla^2 f = \sum_{i=1}^{d} \nabla^2 f(X_i, X_i).$$

Proof. From the definition of $\mathrm{div}X$ we have

$$\mathrm{div}X = \sum_{i=1}^{d} \langle \nabla_{X_i} X, X_i \rangle. \tag{3.1.3}$$

Applying this identity to $X = \mathrm{grad} f$, we obtain the proposition immediately. □

An alternative way of introducing the Laplace-Beltrami operator is as follows. Let $d : C^\infty(M) \to \Gamma(T^*M)$ be the differentiation on functions. Denote its formual adjoint with respect to the pre-Hilbert norms introduced on $C^\infty(M)$ and $\Gamma(TM)$ by $\delta : \Gamma(T^*M) \to C^\infty(M)$, i.e.,

$$\int_M f\delta\theta = \int_M \langle df, \theta \rangle.$$

Using local coordinates, we can verify that $\delta\theta = -\mathrm{div}\, X_\theta$, where $X_\theta \in \Gamma(TM)$ is the dual of the 1-form $\theta \in \Gamma(T^*M)$. Therefore,

$$\Delta_M f = -\delta(df).$$

Since $\Delta_M/2$ is a second order elliptic operator on M, general theory developed in SECTION 1.3 applies. Any M-valued diffusion process generated by $\Delta_M/2$ is called a Brownian motion on M. However, in order to take full advantage of stochastic calculus, we need to generate Brownian motion as the solution of an intrinsically defined Itô type stochastic differential equation. We have seen in CHAPTER 1 that the solution of a stochastic differential equation on M of the form

$$dX_t = V_\alpha(X_t) \circ dW_t^\alpha + V_0(X_t)\,dt,$$

where W is a euclidean Brownian motion, is an L-diffusion generated by a Hörmander type second order elliptic operator

$$L = \frac{1}{2}\sum_{i=1}^{l} V_i^2 + V_0.$$

If we can write Δ_M in this form, then Browiann motion can be generated as the solution of a stochastic differential equation on the manifold. Unfortunately there is no intrinsic way of achieving this on a general Riemannian manifold. We will see later that if M is isometrically embedded in a euclidean space, then there is a way of writing Δ_M as a sum of squares associated naturally with the embedding. In general there is a lifting of Δ_M to the orthonormal frame bundle $\mathscr{O}(M)$ which has the above form, i.e., the sum of squares of $d = \dim M$ intrinsically defined vector fields on $\mathscr{O}(M)$. This is the Eells-Elworthy-Malliavin approach to Brownian motion on manifolds.

Let $\mathscr{O}(M)$ be the orthonormal frame bundle of M and $\pi : \mathscr{O}(M) \to M$ the canonical projection. Recall that the fundamental horizontal vector fields H_i (with respect to the Levi-Civita connection) are the unique horizontal vector fields on $\mathscr{O}(M)$ such that $\pi_* H_i(u) = ue_i$, where $\{e_i\}$ is the canonical basis for $\mathbb{R}^d$. Bochner's horizontal Laplacian is the second order elliptic operator on $\mathscr{O}(M)$ defined by

$$\Delta_{\mathscr{O}(M)} = \sum_{i=1}^{d} H_i^2.$$

Its relation to the Laplace-Beltrami operator is explained in the next proposition.

Proposition 3.1.2. *Bochner's horizontal Laplacian $\Delta_{\mathscr{O}(M)}$ is the lift of the Laplace-Beltrami operator Δ_M to the orthonormal frame bundle $\mathscr{O}(M)$. More precisely, let $f \in C^\infty(M)$, and $\widetilde{f} = f \circ \pi$ its lift to $\mathscr{O}(M)$. Then for any $u \in \mathscr{O}(M)$,*

$$\Delta_M f(x) = \Delta_{\mathscr{O}(M)} \widetilde{f}(u),$$

where $x = \pi u$.

Proof. We need to find the corresponding operations for grad and div in $\mathscr{O}(M)$. Recall that the scalarization of a 1-form θ is defined as $\widetilde{\theta}(u) = \{\theta(ue_i)\}$. Thus the scalarization of df is given by

$$[\widetilde{df}]_i = df(ue_i) = (ue_i)f = H_i\widetilde{f}(u),$$

that is,

$$\widetilde{df} = \left\{ H_1\widetilde{f}, \ldots, H_d\widetilde{f} \right\}.$$

The scalarization $\widetilde{\text{grad} f}$ of the dual $\text{grad} f$ of df is given by the same vector:

$$\widetilde{\text{grad} f} = \left\{ H_1\widetilde{f}, \ldots, H_d\widetilde{f} \right\}. \tag{3.1.4}$$

On the other hand, if $u \in \mathscr{O}(M)$ and $\pi u = x$, then $\{ue_i\}$ is an orthonormal basis for $T_{\pi u}M$. By (3.1.3) and PROPOSITION 2.2.1 we have, for a vector field X,

$$\text{div} X = \langle \nabla_{ue_i} X, ue_i \rangle = \langle u^{-1} \nabla_{ue_i} X, e_i \rangle = \langle H_i \widetilde{X}(u), e_i \rangle,$$

where $\widetilde{X}$ is the scalarization of X. The above relation can be written equivalently as

$$\text{div} X(x) = \sum_{i=1}^{d} (H_i \widetilde{X})^i(u).$$

It follows that

$$\Delta_M f(\pi u) = \sum_{i=1}^{d} (H_i \widetilde{\text{grad} f})^i(u) = \sum_{i=1}^{d} H_i H_i \widetilde{f}(u) = \Delta_{\mathscr{O}(M)} \widetilde{f}(u).$$

□

We have shown that $\Delta_{\mathscr{O}(M)}$ is a sum of $d = \text{dim} M$ squares of vector fields and is a lift of the Laplace-Beltrami operator Δ_M. If we permit the number of vector fields to be larger than $\text{dim} M$, then it is possible to write Δ_M itself as a sum of squares by embedding M isometrically as a submanifold of some euclidean space. Nash's embedding theorem asserts that such an embedding always exists. Let us give a precise statement of this embedding theorem.

Let M and N be Riemannian manifolds such that $\phi : M \to N$ is an embedding. We say that ϕ is an isometric embedding if

$$\langle X, Y \rangle_x = \langle \phi_* X, \phi_* Y \rangle_{\phi(x)}, \qquad X, Y \in \Gamma(TM).$$

If we regard M as a submanifold of N, then the embedding is isometric if the inner product of X and Y with respect to the metric of M is the same as that with respect to the metric of N.

Theorem 3.1.3. (Nash's embedding theorem) *Every Riemannian manifold can be isometrically embedded in some euclidean space with the standard metric.* □

We assume for the rest of this section that M is a submanifold of $\mathbb{R}^l$ with the induced metric. Let $\{\xi_\alpha\}$ be the standard orthonormal basis on $\mathbb{R}^l$. For each $x \in M$, let $P_\alpha(x)$ be the orthogonal projection of ξ_α to $T_x M$. Thus P_α is a vector field on M.

Theorem 3.1.4. *We have* $\Delta_M = \sum_{\alpha=1}^{l} P_\alpha^2$.

Proof. In this proof we will use $\widetilde{\nabla}$ to denote the standard covariant differentiation in the ambient space $\mathbb{R}^l$. Define

$$\nabla_X Y = \text{ the projection of } \widetilde{\nabla}_X Y \text{ to } T_x M, \qquad X, Y \in \Gamma(TM).$$

Then it is easy to verify that ∇ is the Levi-Civita connection on M. Let $f \in C^\infty(M)$. Since the vector field $\mathrm{grad} f$ is tangent to M, we have

$$\mathrm{grad} f = \sum_{\alpha=1}^{l} \langle \mathrm{grad} f, \xi_\alpha \rangle \xi_\alpha = \sum_{\alpha=1}^{l} \langle \mathrm{grad} f, P_\alpha \rangle P_\alpha.$$

We now take the divergence of both sides. On the left side we have $\Delta_M f$ by definition; on the right side we use the formula

$$\mathrm{div}(hX) = Xh + h\,\mathrm{div} X$$

for $h \in C^\infty(M)$ and $X \in \Gamma(TM)$. The result is

$$\Delta_M f = \sum_{\alpha=1}^{l} P_\alpha P_\alpha f + \sum_{\alpha=1}^{l} (\mathrm{div} P_\alpha) P_\alpha f.$$

The theorem follows if we show that the last term vanishes. To see this, we first recall that from the definition of divergence,

$$\mathrm{div} P_\alpha = \sum_{i=1}^{d} \langle \nabla_{X_i} P_\alpha, X_i \rangle,$$

where $\{X_i\}$ is any orthonormal basis of $T_x M$. We take a special one to simplify the computation. Let $\{x^i\}$ be a normal coordinate system of M near x induced by the exponential map and let $X_i = \partial/\partial x^i$. Since the connection is torsion-free, we have shown in PROPOSITION 2.5.5 that the Christoffel symbols vanish at x, namely $\nabla_{X_i} X_j = 0$ at x. This means that in $\mathbb{R}^l$ the covariant differentiation $\widetilde{\nabla}_{X_i} X_j$ is perpendicular to M. Now, using the fact that ∇ is compatible with the metric, we have

$$\begin{aligned}
\langle \nabla_{X_i} P_\alpha, X_i \rangle &= X_i \langle P_\alpha, X_i \rangle, && \nabla_{X_i} X_i = 0 \\
&= X_i \langle \xi_\alpha, X_i \rangle, && X_i \text{ is tangent to } M \\
&= \langle \xi_\alpha, \widetilde{\nabla}_{X_i} X_i \rangle, && \widetilde{\nabla}_{X_i} \xi_\alpha = 0.
\end{aligned}$$

It follows that

$$\begin{aligned}\sum_{\alpha=1}^{l}(\mathrm{div}P_\alpha)P_\alpha &= \sum_{\alpha=1}^{l}\langle\xi_\alpha, \widetilde{\nabla}_{X_i}X_i\rangle P_\alpha \\ &= \text{the projection of } \sum_{\alpha=1}^{l}\langle\xi_\alpha, \widetilde{\nabla}_{X_i}X_i\rangle\xi_\alpha \\ &= \text{the projection of } \widetilde{\nabla}_{X_i}X_i \\ &= 0.\end{aligned}$$

This completes the proof. □

Finally, we record the following identity for future reference.

Corollary 3.1.5. *With the same notations as above, we have*

$$\Delta_M f = \sum_{\alpha=1}^{l}\nabla^2 f(P_\alpha, P_\alpha).$$

Proof. Let $\{X_i\}$ be an orthonormal basis for T_xM. From $X_i = \langle X_i, P_\alpha\rangle P_\alpha$ and PROPOSITION 3.1.1,

$$\Delta_M f = \sum_{i=1}^{d}\nabla^2 f(X_i, X_i) = \sum_{i=1}^{d}\sum_{\alpha,\beta=1}^{l}\nabla^2(P_\alpha, P_\beta)\langle X_i, P_\alpha\rangle\langle X_i, P_\beta\rangle.$$

We have

$$\sum_{\beta=1}^{l}\sum_{i=1}^{d}\langle X_i, P_\alpha\rangle\langle X_i, P_\beta\rangle P_\beta = \sum_{\beta=1}^{l}P_\beta\langle P_\alpha, P_\beta\rangle = P_\alpha.$$

The desired identity follows immediately. □

3.2. Brownian motion on manifolds

We assume that M is a Riemannian manifold equipped with the Levi-Civita connection ∇, and Δ_M the Laplace-Beltrami operator on M. We have shown in CHAPTER 1 that given a probability measure μ on M, there is a unique $\Delta_M/2$-diffusion measure $\mathbb{P}_\mu$ on the filtered measurable space $(W(M), \mathscr{B}_*)$ (the path space over M). Any $\Delta_M/2$-diffusion measure on $W(M)$ is called a Wiener measure on $W(M)$. In general, an M-valued stochastic process X is a measurable map (random variable) $X : \Omega \to W(M)$ defined on some measurable space $(\Omega, \mathscr{F})$. Rougly speaking, Brownian motion on M is any M-valued stochastic process X whose law is a Wiener measure on the path space $W(M)$. In the next proposition we give several equivalent definitions of Brownian motion on M.

Proposition 3.2.1. *Let $X : \Omega \to W(M)$ be a measurable map defined on a probability space $(\Omega, \mathscr{F}, \mathbb{P})$. Let $\mu = \mathbb{P} \circ X_0^{-1}$ be its initial distribution. Then the following statements are equivalent.*

(1) *X is a $\Delta_M/2$-diffusion process (a solution to the martingale problem for $\Delta_M/2$ with respect to its own filtration $\mathscr{F}_*^X$), i.e.,*

$$M^f(X)_t \stackrel{\text{def}}{=} f(X_t) - f(X_0) - \frac{1}{2}\int_0^t \Delta_M f(X_s)\, ds, \quad 0 \le t < e(X),$$

is an $\mathscr{F}_*^X$-local martingale for all $f \in C^\infty(M)$.

(2) *The law $\mathbb{P}^X = \mathbb{P} \circ X^{-1}$ is a Wiener measure on $(W(M), \mathscr{B}(W(M)))$, i.e., $\mathbb{P}^X = \mathbb{P}_\mu$.*

(3) *X is a $\mathscr{F}_*^X$-semimartingale on M whose anti-development is a standard euclidean Brownian motion.*

An M-valued process X satisfying any of the above conditions is called a (Riemannian) Brownian motion on M.

Proof. (1) $\Longleftrightarrow$ (2). This is discussed in SECTION 1.3.

(3) $\Longrightarrow$ (1). Both statements assume that X is an $\mathscr{F}_*^X$-semimartingale on M. Let U be a horizontal lift of X and W the corresponding anti-development. Then we have

$$dU_t = H_i(U_t) \circ dW_t. \tag{3.2.1}$$

Let $f \in C^\infty(M)$, and $\widetilde{f} = f \circ \pi$ its lift to $\mathscr{O}(M)$. Applying Itô's formula to $\widetilde{f}(U_t)$, we have

$$f(X_t) = f(X_0) + \int_0^t H_i \widetilde{f}(U_s)\, dW_s^i + \frac{1}{2}\int_0^t H_i H_j \widetilde{f}(U_s)\, d\langle W^i, W^j\rangle_s.$$

If W is a euclidean Brownian motion we have $\langle W^i, W^j\rangle_t = \delta^{ij}t$ and, by PROPOSITION 3.1.2,

$$\sum_{i=1}^d H_i^2 \widetilde{f}(u) = \Delta_{\mathscr{O}(M)} \widetilde{f}(u) = \Delta_M f(x), \qquad x = \pi u.$$

Hence

$$M^f(X)_t = \int_0^t H_i \widetilde{f}(U_s)\, dW_s^i$$

is a local martingale.

(1) $\Longrightarrow$ (3). We can prove this either by calculation in local coordinates or by embedding M as a submanifold of $\mathbb{R}^N$. If we ignore the technicality of passing from local to global (the process of "patching up") the first method is in fact simpler. We leave the local approach as an exercise and assume

that M is a submanifold of $\mathbb{R}^l$. Again let $f^\alpha(x) = x^\alpha, \alpha = 1, \dots, N$, be the coordinate functions and $\widetilde{f^\alpha} = f^\alpha \circ \pi$ their lifts to $\mathscr{O}(M)$. By assumption,

$$X_t^\alpha = X_0^\alpha + M_t^\alpha + \frac{1}{2}\int_0^t \Delta_M f^\alpha(X_s)\, ds, \tag{3.2.2}$$

where M^α is a local martingale. On the other hand, from (3.2.1) we have

$$X_t^\alpha = X_0^\alpha + \int_0^t H_i \widetilde{f^\alpha}(U_s)\, dW_s^i + \frac{1}{2}\int_0^t \nabla^2 f^\alpha(dX_s, dX_s). \tag{3.2.3}$$

Note that $H_i H_j \widetilde{f}(u) = \nabla^2 f(ue_i, ue_j)$. We now check that

$$\int_0^t \nabla^2 f(dX_s, dX_s) = \int_0^t \Delta_M f(X_s)\, ds. \tag{3.2.4}$$

By $dX_t = P_\alpha(X_t) \circ dX_t^\alpha$ (LEMMA 2.3.3) and PROPOSITION 2.4.8 we have

$$\int_0^t \nabla^2 f(dX_s, dX_s) = \int_0^t \nabla^2 f(P_\alpha, P_\beta)\, d\langle X^\alpha, X^\beta \rangle.$$

On the other hand, by (1.3.8) in the proof of THEOREM 1.3.6 the assumption that X is a $\Delta_M/2$-diffusion implies that

$$d\langle X^\alpha, X^\beta \rangle_t = d\langle M^\alpha, M^\beta \rangle_t = \Gamma(f^\alpha, f^\beta)\, dt. \tag{3.2.5}$$

But it is clear that

$$\Gamma(f^\alpha, f^\beta) = \langle \nabla f^\alpha, \nabla f^\beta \rangle = \langle P_\alpha, P_\beta \rangle.$$

From COROLLARY 3.1.5 we also have

$$\sum_{\alpha,\beta=1}^{l} \nabla^2 f(P_\alpha, P_\beta)\langle P_\alpha, P_\beta \rangle = \sum_{\alpha=1}^{l} \nabla^2 f(P_\alpha, P_\alpha) = \Delta_M f.$$

Hence

$$\begin{aligned}\int_0^t \nabla^2 f(dX_s, dX_s) &= \int_0^t \nabla^2 f(P_\alpha, P_\beta)\langle P_\alpha, P_\beta \rangle ds \\ &= \int_0^t \Delta_M f(X_s)\, ds.\end{aligned}$$

This proves (3.2.4)

Having identified the last terms of (3.2.2) and (3.2.3), we can also identify the terms in the middle,

$$M_t^\alpha = \int_0^t H_i f^\alpha(U_s)\, dW_s^i.$$

It remains to show that this implies that W is a Brownian motion. We want to solve W from this relation and show that it is a euclidean Brownian motion by Lévy's criterion. Since $H_i\widetilde{f^\alpha}(u) = \langle \xi_\alpha, ue_i\rangle$, where $\{\xi^\alpha\}$ is the standard basis for the ambient space $\mathbb{R}^l$, we can write the above identity as

$$dM_t^\alpha = \langle \xi_\alpha, U_t e_i\rangle \, dW_t^i.$$

Multiplying both sides by $\langle \xi_\alpha, U_t e_j\rangle$ and using the fact that

$$\sum_{\alpha=1}^l \langle \xi_\alpha, ue_i\rangle\langle \xi_\alpha, ue_j\rangle = \langle ue_i, ue_j\rangle = \delta_{ij},$$

we have

$$dW_t^j = \langle \xi_\alpha, U_t e_j\rangle \, dM_t^\alpha.$$

For the quadratic variations of M^α we have from (3.2.2) and (3.2.5),

$$d\langle M^\alpha, M^\beta\rangle_t = d\langle X^\alpha, X^\beta\rangle_t = \langle P_\alpha, P_\beta\rangle dt.$$

It is now clear that W is a local martingale whose quadratic variation is given by

$$\sum_{\alpha,\beta=1}^l \langle \xi_\alpha, U_t e_i\rangle\langle \xi_\beta, U_t e_j\rangle\langle P_\alpha, P_\beta\rangle = \langle U_t e_i, U_t e_j\rangle = \delta^{ij}.$$

We have therefore shown that the anti-development W of X is a euclidean Brownian motion. □

We make two minor extensions of the concept of Brownian motion. First, in many applications, it is useful to take a broader point of view by introducing a filtration into the definition. In most cases we can take the filtration to be $\mathscr{F}_*^X$ generated by the process itself, but there are situations where it is more convenient to consider a larger filtration with respect to which the strong Markov property still holds. We say that a $W(M)$-valued random variable defined on a filtered probability space $(\Omega, \mathscr{F}_*, \mathbb{P})$ is an $\mathscr{F}_*$-Brownian motion if it is a Brownian motion in the sense of Proposition 3.2.1 and is strong Markov with respect to the filtration $\mathscr{F}_*$. Recall that this means that for any $\mathscr{F}_*$-stopping time τ and any nonnegative $f \in \mathscr{B}(W(M))$,

$$\mathbb{E}\{f(X_{\tau+*})|\mathscr{F}_\tau\} = \mathbb{E}_{X_\tau} f \quad \text{on } \{\tau < \infty\},$$

where $X_{\tau+*} = \{X_{\tau+t}, t \geq 0\}$ is the shift process. Second, for convenience we may allow Brownian motion to be defined up to a stopping time. More precisely, on a filtered probability space $(\Omega, \mathscr{F}_*, \mathbb{P})$ an M-valued process X defined on the interval $[0, \tau)$ for an $\mathscr{F}_*$-stopping time τ is a Brownian motion up to time τ if its anti-development W is a local martingale up to τ whose quadratic variation is $\langle W, W^\dagger\rangle_t = I dt$ (Lévy's criterion). Equivalently, there are a $\mathscr{G}_*$-Brownian motion Y on a (possibly different) filtered probability space $(\Pi, \mathscr{G}_*, \mathbb{Q})$ and a $\mathscr{G}_*$-stopping time σ such that $(\tau, X_{\tau\wedge *})$ has the same distribution as $(\sigma, Y_{\sigma\wedge *})$.

If X is a Brownian motion on M, then its horizontal lift U is called a horizontal Brownian motion. Since $dU_t = H_i(U_t) \circ dW_t^i$ for a euclidean Brownian motion W, a horizontal Brownian motion is a $\Delta_{\mathscr{O}(M)}/2$-diffusion on $\mathscr{O}(M)$, where $\Delta_{\mathscr{O}(M)}$ is Bochner's horizontal Laplacian on $\mathscr{O}(M)$.

Proposition 3.2.2. *Let $U : \Omega \to W(\mathscr{O}(M))$ be a measurable map defined on a probability space $(\Omega, \mathscr{F}, \mathbb{P})$. Then the following statements are equivalent.*

(1) *U is a $\Delta_{\mathscr{O}(M)}/2$-diffusion with respect to its own filtration $\mathscr{F}_*^U$.*

(2) *U is a horizontal $\mathscr{F}_*^U$-semimartingale whose projection $X = \pi U$ is a Brownian motion on M.*

(3) *U is a horizontal $\mathscr{F}_*^U$-semimartingale on M whose anti-development is a standard euclidean Brownian motion.*

An $\mathscr{O}(M)$-valued process U satisfying any of the above conditions is called a horizontal Brownian motion on $\mathscr{O}(M)$.

Proof. Exercise. □

When M is a submanifold of $\mathbb{R}^l$, PROPOSITION 3.1.4 provides a way of constructing Brownian motion by solving a stochastic differential equation on M driven by an l-dimensional euclidean Brownian motion:

$$dX_t = P_\alpha(X_t) \circ dW^\alpha, \qquad X_0 \in M. \tag{3.2.6}$$

The solution is a diffusion generated by $\frac{1}{2}\sum_{\alpha=1}^l P_\alpha^2 = \frac{1}{2}\Delta_M$; hence it is a Brownian motion on M. This way of producing a Brownian motion on M has the advantage of being explicit and without resorting to the orthonormal frame bundle. Since the dimension of the driving Brownian motion is the dimension of the ambient space, which is usually larger than the dimension of the manifold, the driving Brownian motion W contains some extra information beyond what is provided by the Brownian motion X on the manifold. The method is extrinsic because it depends on the embedding of M into $\mathbb{R}^l$.

3.3. Examples of Brownian motion

It's time to leave general theory and see a few explicit examples of Brownian motion on manifolds. Whenever the details are not given in an example, the reader should regard it as an invitation to work them out as an exercise.

Example 3.3.1. (Brownian motion on a circle) The simplest compact manifold is the circle

$$\mathbb{S}^1 = \left\{ e^{i\theta} : 0 \le \theta < 2\pi \right\} \subseteq \mathbb{R}^2.$$

Let W be a Brownian motion on $\mathbb{R}^1$. Then Brownian motion on $\mathbb{S}^1$ is given by $X_t = e^{iW_t}$. The anti-development of X is just W.

Example 3.3.2. (Brownian motion on a sphere) Let

$$\mathbb{S}^d = \left\{ x \in \mathbb{R}^{d+1} : |x|^2 = 1 \right\}$$

be the d-sphere embedded in $\mathbb{R}^{d+1}$. The projection to the tangent sphere at x is given by

$$P(x)\,\xi = \xi - \langle \xi, x \rangle x, \qquad x \in \mathbb{S}^d, \quad \xi \in \mathbb{R}^{n+1}.$$

Hence the matirx $P(x)$ is

$$P(x)_{ij} = \delta_{ij} - x_i x_j.$$

By (3.2.6) Brownian motion on $\mathbb{S}^d$ is the solution of the equation

$$X_t^i = X_0^i + \int_0^t (\delta_{ij} - X_s^i X_s^j) \circ dW_s^j, \quad X_0 \in \mathbb{S}^d.$$

This is Stroock's representation of spherical Brownian motion.

Our next example is Brownian motion on a radially symmetric manifold. Such a manifold often serves as models for comparing Brownian motions on different manifolds. The important conclusion here is that the radial part of Brownian motion on such a manifold is a diffusion process on $\mathbb{R}_+$ generated by the radial Laplacian, whereas the angular part is a Brownian motion on the sphere $\mathbb{S}^{d-1}$ independent of the radial process, but running on a new time clock defined by the radial process. Thus probabilistic properties of such a Brownian motion are essentially determined by its radial process.

The geometric setting is as follows. In the polar coordinates $(r, \theta) \in \mathbb{R}_+ \times \mathbb{S}^{d-1}$ determined by the exponential map at its pole, the metric of a radially symmetric manifold has the special form

$$ds^2 = dr^2 + G(r)^2 d\theta^2,$$

where $d\theta^2$ denotes the standard Riemannian metric on $\mathbb{S}^{d-1}$ and G is a smooth function on an interval $[0, D)$ satisfying $G(0) = 0,\ G'(0) = 1$. Manifolds of constant curvature K are examples of such manifolds, where

$$G(r) = \begin{cases} \dfrac{\sin \sqrt{K} r}{\sqrt{K}}, & K \geq 0, \\[2ex] \dfrac{\sinh \sqrt{-K}}{\sqrt{-K}}, & K < 0. \end{cases} \tag{3.3.1}$$

The Laplace-Beltrami operator has the form

$$\Delta_M = L_r + \frac{1}{G(r)^2} \Delta_{\mathbb{S}^{d-1}}, \tag{3.3.2}$$

where L_r is the radial Laplacian

$$L_r = \left(\frac{\partial}{\partial r}\right)^2 + (d-1)\frac{G'(r)}{G(r)}\frac{\partial}{\partial r}.$$

Example 3.3.3. (Brownian motion on a radially symmetric manifold) Let $X_t = (r_t, \theta_t)$ be a Brownian motion on a radially symmetric manifold M written in polar coordinates. Applying the martingale property of the Brownian motion X to the distance function function $f(r, \theta) = r$ and using (3.3.2), we have

$$r_t = r_0 + \beta_t + \frac{d-1}{2}\int_0^t \frac{G'(r_s)}{G(r_s)} ds, \tag{3.3.3}$$

where β is a local martingale whose quadratic variation is

$$\langle \beta, \beta \rangle_t = \int_0^t \Gamma(r, r)(X_s)\, ds.$$

From

$$\Gamma(r, r) = |\nabla r|^2 = 1$$

we have $\langle \beta, \beta \rangle_t = t$; hence β is a 1-dimensional Brownian motion. It follows that the radial process is a diffusion process generated by the radial Laplacian L_r.

We now look at the angular process. Let $f \in C^\infty(\mathbb{S}^{d-1})$. Then, from (3.3.2),

$$f(\theta_t) = f(\theta_0) + M_t^f + \frac{1}{2}\int_0^t \frac{\Delta_{\mathbb{S}^{d-1}} f(\theta_s)}{G(r_s)^2} ds \tag{3.3.4}$$

for a local martingale M^f. Define a new time scale

$$l_t = \int_0^t \frac{ds}{G(r_s)^2}, \tag{3.3.5}$$

and let $\{\tau_t\}$ be the inverse of $\{l_t\}$. Let $z_t = \theta_{\tau_t}$. Then (3.3.4) becomes

$$f(z_t) = f(z_0) + M_{\tau_t}^f + \frac{1}{2}\int_0^t \Delta_{\mathbb{S}^{d-1}} f(z_s)\, ds. \tag{3.3.6}$$

Since $t \mapsto M_{\tau_t}^f$ is still a local martingale (adapted to the time-changed filtration), we see that the time-changed angular process $t \mapsto z_t = \theta_{\tau_t}$ is a Brownian motion on $\mathbb{S}^{d-1}$. We now prove the claim:

$$\{r_t\} \text{ and } \{z_t\} \text{ are independent.} \tag{3.3.7}$$

Note that these two processes are adapted to different filtrations. We compute the co-variation

$$\langle M^f, \beta \rangle_t = \int_0^t \Gamma(f, r)(X_s)\, ds = \int_0^t (\nabla f \cdot \nabla r)(X_s)\, ds.$$

Because f is a function of θ alone, from the special form of the metric we have $\nabla f \cdot \nabla r = 0$; hence $\langle M^f, \beta\rangle_t = 0$. Applying this to the coordinate functions $f^\alpha(x) = x^\alpha$ on the sphere $\mathbb{S}^{d-1} \subseteq \mathbb{R}^d$ (standard embedding), we have $\langle \theta^\alpha, \beta\rangle_t = 0$ for every $\alpha = 1, \ldots, d$. Let $\{u_t\}$ be a horizontal lift of $\{\theta_t\}$ and $\{w_t\}$ the anti-development. The latter can be expressed in terms of θ; in fact, (see PROPOSITION 2.3.8)

$$w_t = \int_0^t u_s^{-1} P_\alpha(\theta_s) \circ d\theta_s^\alpha.$$

Hence,

$$\langle w, \beta\rangle_t = \int_0^t u_s^{-1} P_\alpha(\theta_s)\, d\langle \theta^\alpha, \beta\rangle_s = 0.$$

On the other hand, since $\{\theta_{\tau_t}\}$ is a Brownian motion on the sphere, $\{w_{\tau_t}\}$ is a euclidean Brownian motion, which implies in turn that $\{w_t\}$ is a local martingale. It follows from LEMMA 3.3.4 below that $\{w_{\tau_t}\}$ is independent of $\{\beta_t\}$. Now $\{z_t\}$ is the stochastic development of $\{w_{\tau_t}\}$, and $\{r_t\}$ is the solution of a stochastic differential equation (3.3.3) driven by β; hence $\{z_t\}$ is is independent of $\{r_t\}$. This proves the claim (3.3.7)

We can now construct a Brownian motion on a radially symmetric manifold as a warped product. Let $\{r_t\}$ be a diffusion process generated by the radial Laplacian L_r and $\{z_t\}$ an independent Brownian motion on $\mathbb{S}^{d-1}$. Define l as in (3.3.5). Then $t \mapsto X_t = (r_t, z_{l_t})$ is a Brownian motion on the radially symmetric manifold. $\square$

Let us write out the lemma used the above example.

Lemma 3.3.4. *Let $M = \{M^\alpha\}$ be l real-valued local martingales mutually orthogonal in the sense that $\langle M^\alpha, M^\beta\rangle_t = 0$ for $\alpha \neq \beta$. Suppose that each $\langle M^\alpha\rangle$ is strictly increasing. Let $\{\tau_t^\alpha\}$ be its inverse and $W_t^\alpha = M^\alpha_{\tau_t^\alpha}$. Then $\{W^\alpha\}$ are l independent 1-dimensional Brownian motions.*

Proof. Let us assume $l = 2$ for simplicity. We will use the martingale representation theorem for 1-dimensional Brownian motion. This theorem can be proved from the well known the Cameron-Martin-Maruyama theorem; see SECTION 8.5.

Fix a $t > 0$ and let $X^\alpha \in \mathscr{F}_t^{W^\alpha}$, $\alpha = 1, 2$, be square integrable. We need to show that

$$\mathbb{E}(X^1 X^2) = \mathbb{E}X^1 \cdot \mathbb{E}X^2.$$

By the martingale representation theorem, there is an $\mathscr{F}_*^{W^\alpha}$-adapted process H^α such that

$$X^\alpha = \mathbb{E}X^\alpha + \int_0^t H_s^\alpha\, dW_s^\alpha.$$

From this we have

$$\mathbb{E}(X^1X^2) = \mathbb{E}X^1 \cdot \mathbb{E}X^2 + \mathbb{E}\left[\int_0^t H_s^1\, dW_s^1 \int_0^t H_s^2\, dW_s^2\right].$$

Thus it is enough to show that the last term vanishes. Let

$$N_t^\alpha = \int_0^t H^\alpha_{\langle M^\alpha\rangle_s} dM_s^\alpha.$$

Then N^1 and N^2 are martingales adapted to the same filtration, and

$$\mathbb{E}\left[\int_0^t H_s^1 dW_s^1 \int_0^t H_s^2 dW_s^2\right] = \mathbb{E}\left(N^1_{\tau_t^1} N^2_{\tau_t^2}\right).$$

Both τ_t^1 and τ_t^2 are stopping times; hence, letting $\tau = \min\left\{\tau_t^1, \tau_t^2\right\}$, we have

$$\mathbb{E}\left(N^1_{\tau_t^1} N^2_{\tau_t^2}\right) = \mathbb{E}\left(N_\tau^1 N_\tau^2\right) = \mathbb{E}\int_0^\tau H_s^1 H_s^2\, d\langle M^1, M^2\rangle_s = 0.$$

The first equality is a consequence of the following general fact: if M is an $\mathscr{F}_*$-martingale and σ, τ are two stopping times, then, assuming proper integrability conditions are satisfied, we have

$$\mathbb{E}\left[M_\sigma | \mathscr{F}_\tau\right] = M_{\min\{\sigma,\tau\}};$$

hence, if N is another martingale, then

$$\mathbb{E}\left(M_\sigma N_\tau\right) = \mathbb{E}\left(M_{\min\{\sigma,\tau\}} N_{\min\{\sigma,\tau\}}\right).$$

The proof is completed. □

It is often useful to have a stochastic differential equation of Brownian motion in local coordinates. Information about Brownian motion can often be gained by writing it in a judiously chosen coordinate system.

Example 3.3.5. (Brownian motion in local coordinates) The equation for a horizontal Brownian motion on $\mathscr{O}(M)$ is

$$dU_t = H_i(U_t) \circ dW_t^i,$$

where W is a d-dimensional euclidean Brownian motion. We have shown in PROPOSITION 2.1.3 that locally the horizontal vector fields are given by

$$H_i(u) = e_i^j X_j - e_i^j e_m^l \Gamma_{jl}^k(x) X_{km}, \tag{3.3.8}$$

where

$$X_i = \frac{\partial}{\partial x^i}, \qquad X_{km} = \frac{\partial}{\partial e_m^k}.$$

Hence the equation for $U_t = \left\{X_t^i, e_j^i(t)\right\}$ is

$$\begin{cases} dX_t^i = e_j^i(t) \circ dW_t^j, \\ de_j^i(t) = -\Gamma_{kl}^i(X_t) e_j^l(t) e_m^k(t) \circ dW_t^m. \end{cases} \tag{3.3.9}$$

We can find an equation for the Brownian motion X itself. From the first equation we have

$$dX_t^t = e_j^i(t)\, dW_t^j + \frac{1}{2} d\left\langle e_j^i, dW^j \right\rangle_t. \tag{3.3.10}$$

If we let $dM_t^i = e_j^i(t)\, dW_t^j$ be the martingale part, then

$$d\left\langle M^i, M^j \right\rangle_t = \sum_{k=1}^{d} e_k^i(t) e_k^j(t)\, dt.$$

At a frame u, by definition $ue_l = e_l^i X_i$. Hence

$$\delta_{lm} = \langle ue_l, ue_m \rangle = e_l^i g_{ij} e_m^j,$$

or $ege^\dagger = I$ in matrix notation. This shows that

$$e^\dagger e = g^{-1} \quad \text{or} \quad \sum_{k=1}^{d} e_k^i e_k^j = g^{ij}.$$

Now we have

$$d\left\langle M^i, M^j \right\rangle_t = g^{ij}(X_t)\, dt.$$

If σ is the positive definite matrix square root of g^{-1}, then

$$B_t = \int_0^t \sigma(X_s)^{-1} dM_s$$

is a euclidean Brownin motion, and we have

$$dM_t = \sigma(X_s)\, dB_s.$$

From the second equation of (3.3.9) the last term in (3.3.10) becomes

$$d\left\langle e_j^i, dW^j \right\rangle_t = -\Gamma_{kl}^i(X_t) e_m^l(t) e_m^k(t) = -g^{lk}(X_t)\Gamma_{kl}^i(X_t).$$

Therefore the equation for the Brownian motion X in local coordinates is

$$dX_t^i = \sigma_j^i(X_t)\, dB_t^j - \frac{1}{2} g^{lk}(X_t)\Gamma_{kl}^i(X_t)\, dt, \tag{3.3.11}$$

where B is a d-dimensional euclidean Brownian motion.

We can also obtain the equation for X directly from its generator in local coordinates

$$\Delta_M f = \frac{1}{\sqrt{G}} \frac{\partial}{\partial x^j}\left(\sqrt{G} g^{ij} \frac{\partial f}{\partial x^i}\right) = g^{ij} \frac{\partial}{\partial x^i} \frac{\partial f}{\partial x^j} + b^i \frac{\partial f}{\partial x^i},$$

where

$$b^i = \frac{1}{\sqrt{G}} \frac{\partial(\sqrt{G} g^{ij})}{\partial x^j}.$$

It is an easy exercise to show that $b^i = g^{jk}\Gamma_{jk}^i$. Using this, we verify that the generator of the solution of (3.3.11) is indeed $\Delta_M/2$.

3.4. Distance function

Using the exponential map based at a point o, we can introduce polar coordinates (r, θ) in a neighborhood of o. In these coordinates, Brownian motion can be decomposed into the radial process $r_t = r(X_t)$ and the angular process $\theta_t = \theta(x_t)$. In several applications of Brownian motion we will discuss in this book, the radial process plays an important role. When we try to separate the martingale part and the bounded variation part by applying Itô's formula to the distance function r, we face the problem that, although $r(x)$ is smooth if x is close to o, it is in general not so on the whole manifold. More specifically, $r(x)$ is not smooth on the cutlocus of o. Thus we have two issues to deal with. First, we want to know locally how the distance function reflects the geometry of the manifold, and hence the behavior of Brownian motion. Second, we want to know how to describe the singularity of the distance function at the cutlocus, both analytically and probabilistically.

We start our investigation of the radial process by reviewing some basic analytic properties of the distance function. For the geometric topics in this section we recommend *Comparison Theorems in Differential Geometry* by Cheeger and Ebin [**9**].

For simplicity we assume that M is geodesically complete. In this case, by the Hopf-Rinow theorem every geodesic segment can be extended in both directions indefinitely and every pair of points can be connected by a distance-minimizing geodesic. For each unit vector $e \in T_oM$, there is a unique geodesic $C_e : [0, \infty) \to M$ such that $\dot{C}_e(0) = e$. The exponential map $\exp : T_oM \to M$ is

$$\exp te = C_e(t).$$

If we identify T_oM with $\mathbb{R}^d$ by an orthonormal frame, the exponential map becomes a map from $\mathbb{R}^d$ onto M. For small t, the geodesic $C_e[0, t]$ is the unique distance-minimizing geodesic between its endpoints. Let $t(e)$ be the largest t such that the geodesic $C_e[0, t]$ is distance-minimizing from $C_e(0)$ to $C_e(t)$. Define

$$\widetilde{C}_o = \{t(e)e : e \in T_oM, \ |e| = 1\}.$$

Then the cutlocus of o is the set $C_o = \exp \widetilde{C}_o$. Sometimes we also call $\widetilde{C}_o$ the cutlocus of o. The set within cutlocus is the star-shaped domain

$$\widetilde{E}_o = \{te \in T_oM : \ e \in T_oM, \ 0 \le t < t(e), \ |e| = 1\}.$$

On M the set within cutlocus is $E_o = \exp \widetilde{E}_o$.

Theorem 3.4.1. *(i) The map* $\exp : \widetilde{E}_o \to E_o$ *is a diffeomorphism.*

(ii) The cutlocus C_o *is a closet subset of measure zero.*

(iii) If $x \in C_y$*, then* $y \in C_x$.

(iv) E_o *and* C_o *are disjoint and* $M = E_0 \cup C_o$. □

Let

$$i_o = \inf\{t(e) : e \in T_oM,\ |e| = 1\}$$

be the injectivity radius at o. Then $B(i_o)$ (centered at o) is the largest geodesic ball on which the exponential map is a diffeomorphism onto its image.

If $(r, \theta) \in \mathbb{R}_+ \times \mathbb{S}^{d-1}$ are the polar coordinates on $\mathbb{R}^d$, then through the exponential map they become the polar coordinates on M centered at o, which cover the region E_o within the cutlocus. The set they do not cover is the cutlocus C_o, a set of measure zero. The radial function $r(x) = d(x, o)$ is smooth on $M \backslash C_o$ (we discount its well behaved singularity at o) and Lipschitz on all of M. Furthermore, $|\nabla r| = 1$ everywhere on $E_o = M \backslash C_o$.

If X is a Brownian motion on M starting within E_o, then, before it hits the cutlocus C_o,

$$r(X_t) = r(X_0) + \beta_t + \frac{1}{2}\int_0^t \Delta_M r(X_s)\, ds, \quad t < T_{C_o}, \tag{3.4.1}$$

where T_{C_o} is the first hitting time X of the cutlocus C_o and β is a martingale. It is in fact a Brownian motion, because

$$\langle \beta, \beta \rangle_t = \int_0^t \Gamma(r, r)(X_s)\, ds = \int_0^t |\nabla r(X_s)|^2\, ds = t.$$

The relation (3.4.1) reveals an important principle: the behavior of the radial process is controlled by the Laplacian of the distance function $\Delta_M r$. If we can bound $\Delta_M r$ by a known function of r, then we will be able to control $r(X_t)$ by comparing it with a one-dimensional diffusion process. The Laplacian comparison theorem in Riemannian geometry does just that. Before proving this theorem, we first review several basic geometric concepts invovled in this theorem.

The Riemannian curvature tensor is defined by

$$R(X, Y)Z = \nabla_X \nabla_Y Z - \nabla_Y \nabla_X Z - \nabla_{[X,Y]} Z,$$

where X, Y, Z are vector fields on M. The sectional curvature is the quadratic form

$$K(X, Y) = \langle R(X, Y)Y, X \rangle.$$

If X and Y are orthogonal unit vectors spanning the two-dimensional plane σ in the tangent space, then $K(\sigma) \stackrel{def}{=} K(X, Y)$ is the sectional curvature of the plane σ. It is the usual Gaussian curvature of the 2-dimensional submanifold $\exp \sigma$ of M at o. The Ricci curvature tensor is obtained from the curvature tensor by contraction:

$$\mathrm{Ric}(X, Y) = \sum_{i=1}^d \langle R(X, X_i)X_i, Y \rangle,$$

where $\{X_i\}$ is an orthonormal basis. Note that

$$\mathrm{Ric}(X,X) = \sum_{i=1}^{d} K(X, X_i).$$

The fully contracted curvature tensor is the scalar curvature

$$S = \sum_{i=1}^{d} K(X_i, X_i).$$

The behavior of $\Delta_M r$ will be described in terms of the growth rate of the sectional and Ricci curvatures.

We will use $K_M(x)$ to denote the set of sectional curvatures at x:

$$K_M(x) = \{K(\sigma) : 2\text{-dimensional plane } \sigma \subseteq T_x M\}.$$

We say that the sectional curvature at x is bounded from above by K^2 and write $K_M(x) \le K^2$ if $K(\sigma) \le K^2$ for every 2-dimensional plane σ in $T_x M$. A similar remark applies to the set of Ricci curvatures

$$\mathrm{Ric}_M(x) = \{\mathrm{Ric}(X,X) : X \in T_x M, |X| = 1\}.$$

Introduce the following two functions:

$$\begin{cases} \kappa_1(r) \ge \sup\{K_M(x) : r(x) = r\}, \\ \kappa_2(r) \le \inf\{\mathrm{Ric}_M(x) : r(x) = r\}\,(d-1)^{-1}. \end{cases} \tag{3.4.2}$$

Thus $\kappa_1(r)$ is an upper bound of all sectional curvatures on $\partial B(r)$, and $\kappa_2(r)$ is a lower bound of all Ricci curvatures on $\partial B(r)$. Note that when M has constant sectional curvature K, its Ricci curvature is $(d-1)K$, hence the factor $(d-1)^{-1}$.

For a given function κ on an interval $[0, D)$, the Jacobi equation of κ is

$$G''(r) + \kappa(r)G(r) = 0, \qquad G(0) = 0, \quad G'(0) = 1.$$

Let G_i be the solution of the Jacobi equation for κ_i. In the constant curvature case, it is given in (3.3.1).

Theorem 3.4.2. (Laplacian comparison theorem) *With the notations introduced above the following inequalities holds within the cutlocus:*

$$(d-1)\frac{G_1'(r)}{G_1(r)} \le \Delta_M r \le (d-1)\frac{G_2'(r)}{G_2(r)}.$$

Proof. We sketch the proof. Fix $x \in M$ within the cutlocus of o and let γ be the unique geodesic from o to x. Let $\{X_i\}$ be an orthonormal frame at x such that $X_1 = X_r$. The frame is defined on the geodesic γ by parallel translation. Thus X_1 is the tangent field of the geodesic, which will also be denoted by T. We have

$$\Delta_M r = \sum_{i=1}^{d} \nabla^2 r(X_i, X_i).$$

From the definition of the Hessian of a function, it is an easy exercise to show that $\nabla^2 f(X, X)$ is equal to the second derivative of f along the geodesic with the initial direction X; hence each $\nabla^2 r(X_i, X_i)$ is the second variation of the distance function along the Jacobi field J_i along the geodesic γ with the boundary values $J_i(0) = 0$ and $J_i(r) = X_i$. By the second variation formula we have

$$\nabla^2(X_i, X_i) = I(J_i, J_i),$$

where $I(J, J)$ is the index form defined by

$$I(J, J) = \int_\gamma |\nabla_T J|^2 - \langle R(J, T)T, J\rangle.$$

For the upper bound of $\Delta_M r$, we define a new vector field along the geodesic γ by

$$K_i(s) = \frac{G_2(s)}{G_2(r)} X_i.$$

It has the same boundary values as J_i. Hence by the index lemma (stated in LEMMA 6.7.1) $I(J_i, J_i) \le I(K_i, K_i)$. Using this inequality, we can write

$$\begin{aligned}
\Delta_M r &\le \sum_{i=2}^{d} \int_\gamma |\nabla_T K_i|^2 - \langle R(K_i, T)T, K_i\rangle \\
&= \frac{1}{G_2(r)^2} \int_0^r \left\{(d-1)G_2'(s)^2 - G_2(s)^2 \operatorname{Ric}(T, T)\right\} ds \\
&\le \frac{d-1}{G_2(r)^2} \int_0^r \left\{G_2'(s)^2 - G_2(s)^2 \kappa_2(s)\right\} ds \\
&= (d-1)\,\frac{G_2'(r)}{G_2(r)}.
\end{aligned}$$

The proof of the upper bound is similar; it is in fact easier because J_i is orthogonal to T along γ and we can directly apply the upper bound of the sectional curvature on the index forms. □

Remark 3.4.3. Let $X_r = \partial/\partial r$ be the radial vector field. From the proof of the above theorem, it is clear that we only need

$$\begin{cases}
\kappa_1(r) \ge \sup\left\{K(X, X_r) : X, X_r \in T_x M,\ X \perp X_r,\ |X| = 1,\ r(x) = r\right\}, \\
\kappa_2(r) \le \inf\left\{\operatorname{Ric}(X_r, X_r) : X_r \in T_x M,\ r(x) = r\right\} (d-1)^{-1}.
\end{cases}$$

Thus $\kappa_1(r)$ is an upper bound of all radial sectional curvatures on $\partial B(r)$, and $\kappa_2(r)$ is a lower bound of the radial Ricci curvature on $\partial B(r)$. This remark should be kept in mind when we consider radially symmetric manifolds.

It follows that an upper bound on the sectional curvature yields a lower bound on $\Delta_M r$, whereas a lower bound on the Ricci curvature yields an upper bound on $\Delta_M r$. In view of (3.4.1) the probabilistic implication of these geometric facts is that negative curvature helps to push Brownian motion away from its starting point.

The following corollaries are immediate.

Corollary 3.4.4. *Suppose that on a geodesic ball $B(R)$ within the cutlocus the sectional curvature is bounded from above by K_1^2 and the Ricci curvature is bounded from below by $-(d-1)K_2^2$. Then*

$$(d-1)K_1 \cot K_1 r \le \Delta_M r \le (d-1)K_2 \coth K_2 r.$$

□

Corollary 3.4.5. *We have*

$$\Delta_M r = \frac{d-1}{r} + O(r).$$

In particular, $r\Delta_M r$ is uniformly bounded on any compact subset of M within the cutlocus. □

Remark 3.4.6. From COROLLARY 3.4.5 the dominating term of the drift of the radial process is $(d-1)/2r$. This implies, as in the euclidean case, that, for dimensions 2 and higher, Brownian motion never returns to its starting point; see the proof of PROPOSITION 3.5.3.

3.5. Radial process

The results in the last section give a good picture of the behavior of $\Delta_M r$ at smooth points. These estimates on $\Delta_M r$ can be used to control the radial process within the cutlocus using (3.4.1). But in many applications we need to deal with the radial process beyond the cutlocus. A close look at the simplest case $M = \mathbb{S}^1$ reveals that there should be a continuous additive functional supported on the cutlocus which helps push the radial process towards the origin:

$$r(X_t) = r(X_0) + \beta_t + \frac{1}{2}\int_0^t \Delta_M r(X_s)\, ds - L_t, \tag{3.5.1}$$

where L should be a nondecreasing process which increases only when X is on the cutlocus. Our task in this section is to prove this decomposition.

We start with an observation. Because the cutlocus C_o has Riemannian volume measure zero, the amount of time Brownian motion spends on C_o has

Lebesgue measure zero, for if $p_M(t,x,y)$ is the transition density function of the Brownian motion, then

$$\mathbb{E}_x \int_0^{e(X)} I_{C_o}(X_s)\,ds = \int_0^\infty ds \int_{C_o} p_M(t,x,y)\,dy = 0.$$

We will discuss the transition density function (the heat kernel) in the next chapter. According to COROLLARY 3.4.5 $r\Delta_M r$ is uniformly bounded on any compact subset of M. This together with the facts that X does not spend time on C_o and that X does not visit the starting point o shows that the integral on the right side of (3.5.1) is well defined up to the explosion time $e(X)$.

We now come to the main result of this section. If pressed for time and energy, you can safely skip the (not so short) proof of THEOREM 3.5.1 (but not the statement itself) without interrupting the understanding of the rest of the book. What you will miss is an opportunity to appreciate the crucial part played by the simple triangle inequality for the distance function in the decomposition of thc radial process.

Theorem 3.5.1. *Suppose that X is a Brownian motion on a Riemannian manifold M. Let $r(x) = d(x,o)$ be the distance function from a fixed point $o \in M$. Then there exist a one-dimensional euclidean Brownian motion β and a nondecreasing process L which increases only when $X_t \in C_o$ (the cutlocus of o) such that*

$$r(X_t) = \beta_t + \frac{1}{2}\int_0^t \Delta_M r(X_s)\,ds - L_t, \quad t < e(X).$$

Proof. It is enough to prove the decomposition before the first exit time of an arbitrarily large geodesic ball. Thus for simplicity we will assume the injectivity radius

$$i_M = \inf\{i_x : x \in M\}$$

is strictly positive. Suppose that $x \notin B(o; i_M/2)$ and let $\gamma : [0, r(x)] \to M$ be a distance-minimizing geodesic joining o and x. Then the geodesic $\gamma : [i_M/4, r(x)] \to M$ from $\gamma(i_M/4)$ to x can be extended beyond $\gamma(i_M/4)$ and remains distance-minimzing; hence $\gamma(i_M/4)$ does not belong to the cutlocus of x, i.e., $\gamma(i_M/4) \notin C_x$; by symmetry, $x \notin C_{\gamma(i_M/4)}$ (see THEOREM 3.4.1 (iv)). Since the cutlocus of a point is closed, there exists an ϵ_0 such that the ball $B(x;\epsilon_0)$ does not intersect $C_{\gamma(i_M/4)}$. Furthermore we can choose ϵ_0 such that it has the above property for all $x \in M\backslash B(o; i_M/2)$. We fix such an ϵ_0 for the rest of the proof.

Let $\epsilon \le \min\{\epsilon_0, i_M\}/4$, and

$$D_\epsilon = \{x \in M : d(x, C_o) < \epsilon\}$$

the ϵ-neighborhood of the cutlocus C_o. Now let $\{\sigma_n\}$ be the successive times X moves by a distance of ϵ, i.e., $\sigma_0 = 0$ and

$$\sigma_n = \inf\left\{t > \sigma_{n-1} : d\left(X_t, X_{\sigma_{n-1}}\right) = \epsilon\right\}.$$

For a fixed t, we let $t_n = t \wedge \sigma_n$ and $x_n = X_{t_n}$; then

$$r(X_t) - r(X_0) = \sum_{n\geq 1} \left\{r(x_n) - r(x_{n-1})\right\}.$$

For each term, we consider two cases depending on whether or not the starting point $x_{n-1} \in D_\epsilon$.

Case 1. $x_{n-1} \notin D_\epsilon$. In this case $X_s \notin C_o$ for $t \in [t_{n-1}, t_n]$. Since $r(x)$ is smooth away from C_o, we have

$$r(x_n) - r(x_{n-1}) = \beta_{t_n} - \beta_{t_{n-1}} + \frac{1}{2}\int_{t_{n-1}}^{t_n} \Delta_M r(X_s)\, ds. \tag{3.5.2}$$

Case 2. $x_{n-1} \in D_\epsilon$. In this case let $\gamma : [0, r(x_{n-1})] \to M$ be a distance-minimizing geodesic joining o and x_{n-1}. Let $y_{n-1} = \gamma(i_M/4)$, and let $r^*(x) = d(x, y_{n-1})$ be the distance function based at y_{n-1}. The reader is encouraged to draw a picture of the various points introduced so far in order to understand the situation. Now we have

$$\begin{aligned} r(x_n) &\leq r^*(x_n) + \frac{i_M}{4}, \\ r(x_{n-1}) &= r^*(x_{n-1}) + \frac{i_M}{4}; \end{aligned}$$

the first relation follows from the triangle inequality for the triangle $oy_{n-1}x_n$ and $r(y_{n-1}) = i_M/4$, and the second relation holds because the three points o, y_{n-1}, and x_{n-1} lie on the distance-minimizing geodesic γ. Hence

$$r(x_n) - r(x_{n-1}) \leq r^*(x_n) - r^*(x_{n-1}). \tag{3.5.3}$$

This simple-looking inequality is the key step of the proof. Since $r(x_{n-1}) \geq i_M/2$ and $\epsilon \leq \epsilon_0/4$, the ball $B(x_{n-1}; \epsilon)$ does not intersect the cutlocus of y_{n-1}; thus X_s is within the cutlocus of y_{n-1} for $t \in [t_{n-1}, t_n]$. Thus the distance function r^* (based at y_{n-1}) is smooth for this range of time, and we have

$$\begin{aligned} &r^*(x_n) - r^*(x_{n-1}) \\ &\qquad = \beta^n_{t_n - t_{n-1}} + \frac{1}{2}\int_0^{t_n - t_{n-1}} \Delta_M r^*(X_{s+t_{n-1}})\, ds \end{aligned} \tag{3.5.4}$$

for some Brownian motion β^n. In fact,

$$\beta^n_{t_n} - \beta^n_{t_{n-1}} = \int_{t_{n-1}}^{t_n} H_i \tilde{r}^*(U_s)\, dW^i_s,$$

where W is the anti-development of X.

Putting the two cases together, we have

$$r(X_t) = r(X_0) + \beta_t + \frac{1}{2}\int_0^t \Delta_M r(X_s)\,ds - L_\epsilon(t) + R_\epsilon(t), \tag{3.5.5}$$

where

$$L_\epsilon(t) = \sum_{n=1}^{\infty} \lambda_n I_{D_\epsilon}(X_{t_{n-1}}) \tag{3.5.6}$$

with (see (3.5.3))

$$\lambda_n = r^*(X_{t_n}) - r^*(X_{t_{n-1}}) - r(X_{t_n}) + r(X_{t_{n-1}}) \geq 0$$

and

$$\begin{aligned}
R_\epsilon(t) &= \sum_{n\geq 1}\left[\beta^n_{t_n - t_{n-1}} - (\beta_{t_n} - \beta_{t_{n-1}})\right] I_{D_\epsilon}(X_{t_{n-1}}) \\
&\quad + \frac{1}{2}\sum_{n\geq 1}\left[\int_{t_{n-1}}^{t_n}\{\Delta_M r^*(X_s) - \Delta_M r(X_s)\}\,ds\right] I_{D_\epsilon}(X_{t_{n-1}}) \\
&\stackrel{\text{def}}{=} m_\epsilon(t) + b_\epsilon(t).
\end{aligned}$$

We now show that $R_\epsilon(t) \to 0$ as $\epsilon \to 0$.

In the sum $m_\epsilon(t)$, the summands are orthogonal because they are stochastic integrals with respect to the Brownian motion W on disjoint time intervals. Hence we have

$$\begin{aligned}
\mathbb{E}|m_\epsilon(t)|^2 &= \sum_{n=1}^{\infty}\mathbb{E}|\beta^n_{t_n - t_{n-1}} - (\beta_{t_n} - \beta_{t_{n-1}})|^2 I_{D_\epsilon}(X_{t_{n-1}}) \\
&\leq 2\sum_{n=1}^{\infty}\mathbb{E}\left\{|\beta^n_{t_n - t_{n-1}}|^2 + |\beta_{t_n} - \beta_{t_{n-1}}|^2\right\} I_{D_\epsilon}(X_{t_{n-1}}) \\
&= 4\sum_{n=1}^{\infty}\mathbb{E}|t_n - t_{n-1}| I_{D_\epsilon}(X_{t_{n-1}}).
\end{aligned}$$

We observe that since $X_{t_{n-1}} \in D_\epsilon$, during the time interval $[t_{n-1}, t_n]$, the process X_s stays within 2ϵ from the cutlocus C_o, i.e., $X_s \in D_{2\epsilon}$. Therefore as $\epsilon \to 0$,

$$\mathbb{E}|m_\epsilon(t)|^2 \leq 4\int_0^t I_{D_{2\epsilon}}(X_s)\,ds \to 4\int_0^t I_{C_o}(X_s)\,ds = 0.$$

For the second term $b_\epsilon(t)$ of $R_\epsilon(t)$, we observe that by COROLLARY 3.4.5 $\Delta_M r$ and $\Delta_M r^*$ are uniformly bounded on $D_{2\epsilon}$ by a constant independent of ϵ. Hence as $\epsilon \downarrow 0$ we have again

$$\mathbb{E}\,|b_\epsilon(t)| \leq C\,\mathbb{E}\int_0^t I_{C_{2\epsilon}}(X_s)\,ds \to C\int_0^t I_{C_o}(X_s)\,ds = 0.$$

Now that the error term $\mathbb{E}|R_\epsilon(t)| \to 0$, we conclude from (3.5.5) that the limit $L_t = \lim_{\epsilon \to 0} L_\epsilon(t)$ exists and

$$r(X_t) = r(X_0) + \beta_t + \frac{1}{2}\int_0^t \Delta_M r(X_s)\, ds - L_t.$$

From (3.5.6), $L_\epsilon(t)$ increases by a nonnegative amount each time t crosses a t_{n-1} such that $d(X_{t_{n-1}}, C_o) \le \epsilon$. Therefore the limit L_t, which is obviously continuous, is nondecreasing and can only increase when $X_t \in C_o$. □

Remark 3.5.2. In the language of Markov processes, L is a positive continuous additive functional of the Brownian motion X supported on the cutlocus C_o. By general theory, every such functional is generated by a nonnegative measure ν on its support. It can be shown that

$$\Delta_M r = \Delta_M r\big|_{M \setminus C_o} - 2\nu,$$

where the $\Delta_M r$ on the left side should be interpreted in the distributional sense. □

The importance of THEOREM 3.5.1 lies in the fact that it allows us to use the radial process beyond the cutlocus. In most applications, what we need is an upper bound on the radial process, so we can simply throw away the term L on the cutlocus. The quantitive lower bound for the exit time of Brownian motion from a geodesic ball we will prove in the next section is a case in point. Without THEOREM 3.5.1, we would have to confine ourselves to a geodesic ball within the cutlocus.

We now prove a useful comparison theorem for the radial process.

Theorem 3.5.3. *Let M be a Riemannian manifold and $r_t = r(X_t)$ the radial process of a Brownian motion X on M:*

$$r_t = \beta_t + \frac{1}{2}\int_0^t \Delta_M r(X_s)\, ds - L_t, \quad t < e(X).$$

Let κ be a continuous function on $[0, \infty)$ and G the solution of the Jacobi equation:

$$G''(r) + \kappa(r)G(r) = 0, \quad G(0) = 0, \quad G'(0) = 1. \tag{3.5.7}$$

Let ρ be the unique nonnegative solution of the equation

$$\rho_t = r_0 + \beta_t + \frac{d-1}{2}\int_0^t \frac{G'(\rho_s)}{G(\rho_s)} ds. \tag{3.5.8}$$

(i) *Suppose that*

$$\kappa(r) \ge \max\{K_M(x) : r(x) = r\}.$$

Then $e(\rho) \ge e(X)$ and $\rho_t \le r_t$ for $t < T_{C_o}$, the first hitting time of the cutlocus C_o.

(ii) *Suppose that*

$$\kappa(r) \le \inf\left\{\mathrm{Ric}_M(x) : r(x) = r\right\}.$$

Then $e(\rho) \le e(X)$ *and* $\rho_t \ge r_t$ *for all* $t < e(\rho)$.

Proof. The two parts being similar, we prove (ii), and leave the proof of (i) to the reader. We first show that (3.5.8) has a unique nonnegative solution. Let, for $y > 0$,

$$l(y) = \frac{d-1}{2}\sqrt{y}\,\frac{G'(\sqrt{y})}{G(\sqrt{y})} + 1.$$

Then from (3.5.7) we see that $l(y) = d + O(y^2)$. Define $l(0) = 0$ and extend l to $\mathbb{R}^1$ by, say, $l(y) = l(-y)$ for $y < 0$. Then l is locally Lipschitz on $\mathbb{R}^1$. Consider the equation

$$dy_t = 2\sqrt{y_t^+}\,d\beta_t + l(y_t)dt, \quad y_0 = |r_0|^2. \tag{3.5.9}$$

This equation is obtained from (3.5.8) by setting $y_t = \rho_t^2$. By the well known Yamada's theorem (see THEOREM IV.3.2 in Ikeda and Watanabe [**48**]), this equation has a unique solution. Furthermore, since $l(0) = d \ge 2$, the solution $y_t > 0$ for all $t > 0$ (loc. cit., EXAMPLE IV.8.2). It follows that $\rho_t = \sqrt{y_t}$ is the unique nonnegative solution for (3.5.8).

We now prove that $\rho_t \ge r_t$ for $t < e(\rho) \wedge e(X)$. We have

$$\frac{G_2'(r)}{G_2(r)} = \frac{1}{r} + h(r),$$

where h is locally Lipschitz on $[0, \infty)$ and $h(0) = 0$. Since we only need to show $\rho_t \ge r_t$ for r_t and ρ_t confined to an arbitrarily fixed large interval, for the sake of convenience we may assume that h is globally Lipschitz. For any smooth function f on $\mathbb{R}^1$ we have

$$\begin{aligned}\frac{2}{d-1}\cdot d\left\{f(r_t - \rho_t)\right\} &= f'(r_t - \rho_t)\left(h(r_t) - h(\rho_t)\right)dt \\ &\quad - f'(r_t - \rho_t)\left[\left(\frac{r_t - \rho_t}{r_t\rho_t}\right)dt + dL_t\right]\end{aligned}$$

If f has the properties that $f(x) = 0$ for $x \le 0$ and $0 \le f'(x) \le 1$, then the last expression on the right side is nonnegative, and we have

$$f(r_t - \rho_t) \le C\int_0^t |r_s - \rho_s|ds$$

for some constant C. Let $\{f_n\}$ be a sequence of smooth functions with the said properties such that $f_n(x) \to x^+$ as $n \uparrow \infty$. Replacing f by f_n in the above inequality and taking the limit, we have

$$(r_t - \rho_t)^+ \le C\int_0^t (r_s - \rho_s)^+\,ds.$$

It follows that $(r_t - \rho_t)^+ = 0$, which implies immediately $e(\rho) \le e(X)$ and $\rho_t \ge r_t$ for $t < e(\rho)$. The proof is completed. □

3.6. An exit time estimate

In various applications we need an upper bound for $\mathbb{P}\{\tau_1 \le T\}$, where τ_1 is the first exit time from the geodesic ball of radius 1 centered at the starting point. We have mentioned earlier that a lower bound on the Ricci curvature implies an upper bound on the escape rate for Brownian motion. Thus we expect an upper bound of the probability $\mathbb{P}\{\tau_1 \le T\}$ in terms of the lower bound of the Ricci curvature on the geodesic ball.

Theorem 3.6.1. *Suppose that $L \ge 1$ and*

$$\mathrm{Ric}_M(z) \ge -L^2 \quad \textit{for all } z \in B(x;1).$$

Let τ_1 be the first exit time of Brownian motion from $B(x;1)$. Then there is a constant C depending only on $d = \dim M$ such that

$$\mathbb{P}_x\left[\tau_1 \le \frac{C}{L}\right] \le e^{-L/2}.$$

Proof. Let $r_t = d(X_t, x)$ be the radial process. By THEOREM 3.5.1 there is a Brownian motion such that

$$r_t = \beta_t + \frac{1}{2}\int_0^t \Delta_M r(X_s)\, ds - L_t,$$

where the nondecreasing process L increases only when X_t is on C_x, the cutlocus of x. From

$$r_t^2 = 2\int_0^t r_s dr_s + \langle r, r\rangle_t$$

and $\langle r, r\rangle_t = \langle \beta, \beta\rangle_t = t$, we have

$$r_t^2 \le 2\int_0^t r_s d\beta_s + \int_0^t r_s \Delta_M r(X_s)\, ds + t. \tag{3.6.1}$$

By the Laplacian comparison theorem (COROLLARY 3.4.4), for $r \le 1$,

$$\Delta_M r \le (d-1)L \coth Lr.$$

This together with the inequality $l \coth l \le 1 + l$ for all $l \ge 0$ gives

$$r\Delta_M r \le (d-1)Lr \coth Lr \le (d-1)(1+L).$$

We now let $t = \tau_1$ in (3.6.1), and obtain

$$1 \le 2\int_0^{\tau_1} r_s d\beta_s + (2dL)\,\tau_1.$$

If we take $C = 1/8d$, then from the above inequality the event $\{\tau_1 \le C/L\}$ implies

$$\int_0^{\tau_1} r_s d\beta_s \ge \frac{3}{8}.$$

By Lévy's criterion, there is a Brownian motion W such that

$$\int_0^{\tau_1} r_s d\beta_s = W_\eta,$$

where

$$\eta = \int_0^{\tau_1} r_s^2 ds \le \tau_1 \le \frac{C}{L}.$$

Hence $\{\tau_1 \le C/L\}$ implies

$$\max_{0 \le s \le C/L} W_s \ge W_\eta \ge \frac{3}{8}.$$

The random variable on the left side is distributed as $\sqrt{C/L}|W_1|$. It follows that

$$\mathbb{P}_x\left[\tau_1 \le \frac{C}{L}\right] \le \mathbb{P}_x\left[|W_1| \ge \sqrt{\frac{9L}{8}}\right] \le e^{-L/2}.$$

□

Chapter 4

Brownian Motion and Heat Semigroup

Heat kernels and heat semigroups are objects of intensive research in geometry. Their link to stochastic analysis is based on the fact that the minimal heat kernel $p_M(t,x,y)$ is the transition density function of Brownian motion. When a geometric problem lends itself to a probabilistic interpretation, it can be investigated by stochastic techniques. In this chapter, after reviewing basic properties of heat kernels and heat semigroups, we provide two such examples. The first one concerns the conservation of the heat kernel; namely, we want to find geometric conditions under which

$$\int_M p_M(t,x,y)\,dy = 1.$$

The probabilistic connection is provided by the identity

$$\mathbb{P}_x\{t<e\} = \int_M p_M(t,x,y)\,dt.$$

Therefore we seek geometric conditions under which Brownian motion does not explode. The second example is the so-called C_0-property, also called the Feller property, which states that the heat semigroup

$$P_t f(x) = \int_M p_M(t,x,y) f(y)\,dy$$

preserves the space of continuous functions on M vanishing at infinity. This property is shown to be equivalent to

$$\lim_{r(x)\to\infty} \mathbb{P}_x\{T_K \le t\} = 0, \quad \text{for all } t>0, \quad K \text{ compact}.$$

Probabilistically, both problems boil down to controlling the radial process of Brownian motion, a topic to which we have devoted the last three sections of CHAPTER 3. Other problems of similar flavor discussed in this chapter include recurrence and transience of Brownian motion and comparison theorems for heat kernels. A theme common to all the above applications is comparing Brownian motion on a manifold with Brownian motion on a suitably chosen model manifold, which in most cases is a radially symmetric manifold. On such a manifold, Brownian motion reduces essentially to a one-dimensional diffusion for which the problem can be exactly solved (see EXAMPLE 3.3.3).

4.1. Heat kernel as transition density function

Let M be a Riemannian manifold. As before we use $\mathbb{P}_x$ to denote the law of Brownian motion (Wiener measure) on M starting from x. It is a probability measure on the filtered path space $(W(M), \mathscr{B}(W(M))_*)$. We will denote the coordinate process on this path space by X, i.e., $X_t(\omega) = \omega_t$. The transition density function $p_M(t,x,y)$ for Brownian motion is determined by

$$\mathbb{P}_x\{X_t \in C,\, t < e\} = \int_C p_M(t,x,y)\,dy, \quad C \in \mathscr{B}(M),$$

where the integal is with respect to the Riemannian volume measure. It is well known that if $M = \mathbb{R}^d$, then

$$p_{\mathbb{R}^d}(t,x,y) = \left(\frac{1}{2\pi t}\right)^{d/2} e^{-|x-y|^2/2t}.$$

That such a function exists for a general Riemannian manifold is a basic result from differential geometry. In this section we will study basic properties of the heat kernel needed for the rest of the book. The book *Eigenvalues in Riemannian Geometry* by Chavel [**8**] contains an extensive discussion on the heat kernel on a Riemannian manifold. We also recommend the excellent exposition by Dodziuk [**19**].

In what follows we will use $\mathscr{L}_M$ (or $\mathscr{L}$ if the manifold is understood) to denote the heat operator on M:

$$\mathscr{L}_M = \frac{\partial}{\partial t} - \frac{1}{2}\Delta_M.$$

We begin with the Dirichlet heat kernel $p_D(t,x,y)$ for a smooth domain on M. Its basic properties are summarized in the following theorem.

Theorem 4.1.1. *Suppose that D is a smooth, relatively compact domain on a Riemannian manifold M. There exists a unique continuous function $p_D(t,x,y)$ on $(0,\infty)\times\overline{D}\times\overline{D}$ such that*

(1) $p_D(t,x,y)$ *is infinitely differentiable and strictly positive on* $(0,\infty)\times D\times D$;

(2) *for every fixed* $y\in D$, *it satisfies the heat equation in* (t,x):

$$\mathscr{L}_x p_D(t,x,y)=0, \quad (t,x)\in(0,\infty)\times D;$$

(3) *for every fixed* $y\in D$, *it satisfies the Dirichlet boundary condition:* $p_D(t,x,y)=0$ *for* $x\in\partial D$;

(4) *for every fixed* $y\in D$ *and every bounded continuous function* f *on* D,

$$\lim_{t\downarrow 0}\int_D p_D(t,x,y)f(x)\,dx=f(y), \quad y\in D;$$

(5) $p_D(t,x,y)=p_D(t,y,x)$;

(6) $p_D(t+s,x,y)=\displaystyle\int_D p_D(t,x,z)p_D(s,z,y)\,dz$ *(Chapman-Kolmogorov equation);*

(7) $\displaystyle\int_D p_D(t,x,y)\,dy\le 1$. *The inequality is strict if* $M\backslash\overline{D}$ *is nonempty.*

Whenever necessary, we define $p_D(t,x,y)=0$ if x or y lies outside D. We want to connect $p_D(t,x,y)$ with the law of Brownian motion $\mathbb{P}_x$. First we prove an immediate consequence of the above theorem.

Proposition 4.1.2. *Under the same condition as in the above theorem, suppose further that* f *is a bounded continuous function on* D. *Then*

$$u_f(t,x)=\int_D p_D(t,x,y)f(y)\,dy$$

is the unique solution of the initial-boundary value problem

$$\begin{cases}\mathscr{L}_M u_f(t,x)=0, & (t,x)\in(0,\infty)\times\overline{D};\\ u_f(t,x)=0, & (t,x)\in(0,\infty)\times\partial D;\\ \lim_{t\downarrow 0}u_f(t,x)=f(x), & x\in D.\end{cases} \tag{4.1.1}$$

Proof. The fact that u_f is a solution of the initial-boundary value problem follows directly from the properties of the Dirichlet heat kernel stated in THEOREM 4.1.1. To prove the uniqueness we suppose that v is another solution and let $h=u_f-v$. Then h has zero boundary value and satisfies the heat equation. Now,

$$\frac{d}{dt}\int_D h(t,x)^2dx=2\int_D h(t,x)\frac{d}{dt}h(t,x)\,dx=\int_D h(t,x)\Delta_M h(t,x)\,dx.$$

Using Green's theorem, we have

$$\frac{d}{dt}\int_D h(t,x)^2dx=-\int_D|\nabla h(t,x)|^2dx\le 0.$$

Thus $\displaystyle\int_D h(t,x)^2dx$ is nonincreasing in t. But $h(0,x)=0$; hence $h(t,x)=0$ and $u_f(t,x)=v(t,x)$. $\square$

Next, we identify the Dirichlet heat kernel $p_D(t,x,y)$ as the transition density function of Brownian motion killed at the first exit time from D:

$$\tau_D = \inf\{t: X_t \notin D\}.$$

Proposition 4.1.3. *Let $p_D(t,x,y)$ be the Dirichlet heat kernel of a relatively compact domain D on a Riemannian manifold, and τ_D the first exit time of Brownian motion on M from D. Then*

$$\mathbb{P}_x\{X_t \in C,\, t<\tau_D\} = \int_C p_D(t,x,y)\,dy.$$

In other words, the heat kernel $p_D(t,x,y)$ is the transition density function of Brownian motion on M killed at ∂D.

Proof. The coordinate process X on $(W(M),\mathscr{B}_*,\mathbb{P}_x)$ is a Brownian motion on M. By Proposition 4.1.2

$$u(t,x) = \int_D p_D(t,x,y)f(y)\,dy$$

is the solution of the initial-boundary value problem (4.1.1). Since $\mathbb{P}_x$ is an $\Delta_M/2$-diffusion measure, after applying Itô's formula to $u(t-s,X_s)$ we have, for $s<t\wedge\tau_D$,

$$u(t-s,X_s) = u(t,x) \\ +\text{martingale} - \frac{1}{2}\int_0^s \mathscr{L}_M u(t-v,X_v)\,dv.$$

The last term vanishes because u is a solution to the heat equation. Letting $s=t\wedge\tau_D$ (or more precisely, letting $s=s_n=t_n\wedge\tau_{D_n}$ and $n\uparrow\infty$, where $t_n\uparrow\uparrow t$ and $\{D_n\}$ is an exhaustion of D) and taking the expectation, we have

$$\mathbb{E}_x u\left(t-t\wedge\tau_D, X_{t\wedge\tau_D}\right) = u(t,x).$$

Using the fact that u vanishes on the boundary of D, we have

$$\mathbb{E}_x\{f(X_t), t<\tau_D\} = u(t,x) = \int_D p_D(t,x,y)f(y)\,dy,$$

which is equivalent to what we wanted to prove. $\square$

Let $\{D_n\}$ be an exhaustion of M, namely a sequence of smooth, relatively compact domains of M such that $\overline{D}_n\subseteq D_{n+1}$ and $D_n\uparrow M$. For a bounded, nonnegative smooth function f on M,

$$\mathbb{P}_x\{f(X_t), t<\tau_{D_n}\} = \int_M p_{D_n}(t,x,y)f(y)\,dy. \tag{4.1.2}$$

Hence

$$\int_M \{p_{D_{n+1}}(t,x,y) - p_{D_n}(t,x,y)\} f(y)\,dy$$
$$= \mathbb{P}_x\left\{f(X_t), \tau_{D_n} \le t < \tau_{D_{n+1}}\right\} \ge 0.$$

This shows that

$$p_{D_{n+1}}(t,x,y) \ge p_{D_n}(t,x,y).$$

The heat kernel of M is defined by

$$p_M(t,x,y) = \lim_{n\to\infty} p_{D_n}(t,x,y).$$

The limit $p_M(t,x,y)$ is independent of the choice of the exhaustion $\{D_n\}$. It is called the minimal heat kernel of M. We state the main properties of $p_M(t,x,y)$ in the next theorem. Most of them are inherited directly from those of $p_{D_n}(t,x,y)$.

Theorem 4.1.4. *Suppose that M is a Riemannian manifold. Then the minimal heat kernel $p_M(t,x,y)$ has the following properties:*

(1) *$p_M(t,x,y)$ is infinitely differentiable and strictly positive on $(0,\infty) \times M \times M$;*

(2) *for every fixed $y \in M$, it satisfies the heat equation in (t,x):*
$$\mathscr{L}_x p_M(t,x,y) = 0, \quad (t,x) \in (0,\infty) \times M;$$

(3) *for every fixed $y \in M$ and every bounded continuous function f on M,*
$$\lim_{t\downarrow 0} \int_M p_M(t,x,y) f(x)\,dx = f(y);$$

(4) *$p_M(t,x,y) = p_M(t,y,x)$;*

(5) *$p_M(t+s,x,y) = \int_M p_M(t,x,z) p_M(s,z,y)\,dz$ (Chapman-Kolmogorov equation);*

(6) *$\int_M p_M(t,x,y)\,dy \le 1$.*

A function $p_M(t,x,y)$ satisfying (1)-(3) of the above theorem is called a fundamental solution of the heat operator $\mathscr{L}_M$. The following result explains the reason for calling $p_M(t,x,y)$ the *minimal* heat kernel.

Theorem 4.1.5. *Let $p(t,x,y)$ be a fundamental solution of the heat operator $\mathscr{L}_M$. Then $p_M(t,x,y) \le p(t,x,y)$.*

Proof. Let f be a bounded, nonnegative, smooth function on M with compact support. By definition

$$u(t,x) = \int p(t,x,y) f(y)\,dy$$

is a solution of the heat equation. Its initial value is f in the weak sense; that is, for any continuous function g with compact support we have

$$\begin{aligned}\int_M g(x)u(t,x)\,dx &= \int_M f(y)\int_M g(x)p(t,x,y)\,dx \\ &\to \int_M f(y)g(y)\,dy, \qquad \text{as} \quad t \downarrow 0.\end{aligned}$$

For an $\epsilon > 0$, applying Itô's formula as in the proof of PROPOSITION 4.1.3, we see that $s \mapsto u(t+\epsilon-s, X_s)$ is a local martingale up to time $t \wedge e$; hence for any relatively compact smooth domain D of M,

$$u(t+\epsilon, x) = \mathbb{E}_x u\left(t+\epsilon - t\wedge\tau_D, X_{t\wedge\tau_D}\right).$$

Since u is nonnegative, from the above relation we have

$$u(t+\epsilon,x) \geq \mathbb{E}_x\left\{u(\epsilon, X_t), t < \tau_D\right\} = \int_M p_D(t,x,y)u(\epsilon,y)\,dy.$$

Now let $\epsilon \downarrow 0$. We have

$$u(t,x) \geq \int_M p_D(t,x,y)f(y)\,dy,$$

or equivalently, for $f \geq 0$,

$$\int_M p(t,x,y)f(y)\,dy \geq \int_M p_D(t,x,y)f(y)\,dy.$$

This shows that $p(t,x,y) \geq p_D(t,x,y)$ for any relatively compact domain D. Letting $D \uparrow M$, we obtain $p(t,x,y) \geq p_M(t,x,y)$. □

From (4.1.2) and the fact that $\tau_{D_n} \uparrow e$ we have

$$\mathbb{E}_x\left\{f(X_t),\, t < e\right\} = \int_M p_M(t,x,y)f(y)\,dy,$$

or equivalently,

$$\mathbb{P}_x\left\{X_t \in C,\, t < e\right\} = \int_C p_M(t,x,y)\,dy, \quad C \in \mathscr{B}(M). \tag{4.1.3}$$

We have proved the following result.

Proposition 4.1.6. *The minimal heat kernel $p_M(t,x,y)$ is the transition density function of Brownian motion on M.*

4.2. Stochastic completeness

In this section we study the explosion time of Brownian motion on a Riemannian manifold. Specifically, we want to find geometric conditions which guarantee that Brownian motion does not explode, and give examples of (complete) Riemannian manifolds on which Brownian motion does explode.

Letting $C = M$ in (4.1.3), we have

$$\mathbb{P}_x \{t < e\} = \int_M p_M(t, x, y)\, dy.$$

A Riemannian manifold is said to be stochastically complete if for every x with probability 1 a Brownian motion starting from x does not explode, i.e.,

$$\mathbb{P}_x \{e = \infty\} = 1, \quad \forall x \in M.$$

Equivalently, M is stochastically complete if and only if the minimal heat hernel is conservative, that is, for all $(t, x) \in (0, \infty) \times M$,

$$\int_M p_M(t, x, y)\, dy - 1. \tag{4.2.1}$$

A few remarks about the above definition are in order. If (4.2.1) holds for one $t > 0$ and one $x \in M$, then it holds for all $(t, x) \in (0, \infty) \times M$. To see this, we integrate the Chapman-Kolmogorov equation for the heat kernel

$$p_M(t, x, y) = \int_M p_M(s, x, z) p_M(t - s, z, y)\, dz, \quad 0 < s < t,$$

and obtain

$$\int_M p_M(t, x, y)\, dy \leq \int_M p_M(s, x, z)\, dz \leq 1.$$

Thus (4.2.1) implies that the equality must hold throughout, and hence

$$\int_M p_M(s, z, y)\, dy = 1, \quad (s, z) \in (0, t) \times M.$$

The equality for large times can be obtained from this by repeatedly using the Chapman-Kolmogorov equation.

A relatively compact domain D which is not the whole manifold is of course not stochastically complete:

$$\mathbb{P}_x \{t < \tau_D\} = \int_D p_D(t, x, y)\, dy < 1.$$

Of course in this case D is not geodesically complete (meaning complete as a metric space with the Riemannian distance function). The plane $M = \mathbb{R}^2$ with one point removed is an example of a stochastically complete manifold which is not geodesically complete.

One motivation for studying the problem of stochastic completeness is the following L^∞-uniqueness of the heat equation for stochastically complete manifolds.

Theorem 4.2.1. *Let M be a stochastically complete Riemannian manifold. Let f be a uniformly bounded, continuous function on M. Let $u(t,x)$ be uniformly bounded solution of the initial-boundary value problem:*

$$\begin{cases} \mathscr{L}_M u = 0, & (t,x) \in \mathbb{R}_+ \times M, \\ \lim_{t\downarrow 0} u(t,x) = f(x), & x \in M. \end{cases}$$

Then $u(t,x) = \mathbb{E}_x\{f(X_t)\}$.

Proof. The process $s \mapsto u(t-s, X_s)$ is a uniformly bounded martingale up to time $t \wedge e$; hence for any bounded domain D such that $x \in D$ we have

$$\begin{aligned} u(t,x) &= \mathbb{E}_x u\left(t - t\wedge\tau_D, X_{t\wedge\tau_D}\right) \\ &= \mathbb{E}_x\left\{f\left(X_t\right)\right), t < \tau_D\} + \mathbb{E}\left\{u\left(t-\tau_D, X_{\tau_D}\right), t \geq \tau_D\right\}. \end{aligned}$$

Now let D go through an exhaustion $\{D_n\}$ of M. From $\tau_{D_n} \uparrow \infty$ and the uniform boundedness of u the last term tends to zero as $n \uparrow \infty$; hence $u(t,x) = \mathbb{E}_x f(X_t)$. □

From now on we will assume that M is a complete Riemannian manifold. Our goal is to find sufficient geometric conditions for stochastic completeness and noncompleteness. These conditions are obtained by comparing the radial process of Brownian motion on the manifold with a one-dimensional diffusion process for which a necessary and sufficient condition for explosion is known. This theme of comparison will be played many times in this and the next few sections. The crucial comparison theorem for the radial process is provided by THEOREM 3.5.3. For one-dimensional diffusion processes, the following criteria are well known.

Proposition 4.2.2. *Let $I = (c_1, c_2)$ with $-\infty \leq c_1 < c_2 \leq \infty$. Let a and b be continuous functions on I such that $a(x) > 0$ for all $x \in I$. Define, for a fixed $c \in I$,*

$$\begin{aligned} s(x) &= \int_c^x \exp\left[-2\int_c^y \frac{b(z)}{a(z)}dz\right] dy, \\ l(x) &= \int_c^x \exp\left[-2\int_c^y \frac{b(z)}{a(z)}dz\right]\left\{\int_c^y \exp\left[2\int_c^z \frac{b(\xi)}{a(\xi)}d\xi\right]\frac{dz}{a(z)}\right\} dy. \end{aligned}$$

Let $\mathbb{P}_x$ be the diffusion measure starting from x, where

$$L = \frac{1}{2}a(x)\left(\frac{d}{dx}\right)^2 + b(x)\frac{d}{dx}.$$

Then:

(i) $\mathbb{P}_x\{e=\infty\}=1$ *if and only if* $l(c_1)=l(c_2)=\infty$;

(ii) $\mathbb{P}_x\{e<\infty\}=1$ *if and only if one of the following three conditions is satisfied:*

(1) $l(c_1)<\infty$ *and* $l(c_2)<\infty$;

(2) $l(c_1)<\infty$ *and* $s(c_2)=\infty$;

(3) $l(c_2)<\infty$ *and* $s(c_1)=-\infty$.

Proof. This is THEOREM VI.3.2 in Ikeda and Watanabe [**48**]. □

We have indicated before that a lower bound on the Ricci curvature gives an upper bound on the growth of the radial process of a Brownian motion. It is then to be expected that a growth condition on the lower bound of the Ricci curvature will provide a sufficient condition for stochastic completeness. This is what we will prove in the next theorem.

Recall that the Jacobi equation for a function κ defined on $[0,\infty)$ is

$$G''(r)+\kappa(r)G(r)=0,\quad G(0)=0,\quad G'(0)=1. \tag{4.2.2}$$

Proposition 4.2.3. *Suppose that M is a complete Riemannian manifold and $o\in M$ a fixed point on M. Let $r(x)=d(x,o)$. Suppose that $\kappa(r)$ is a negative, nonincreasing, continuous function on $[0,\infty)$ such that*

$$\kappa(r)\le(d-1)^{-1}\inf\left\{\mathrm{Ric}_M(x):r(x)=r\right\}.$$

Let G be the solution of the Jacobi equation (4.2.2) for κ, and define, for a fixed $c>0$,

$$I(G)=\int_c^\infty G(r)^{1-d}dr\int_c^r G(s)^{d-1}ds. \tag{4.2.3}$$

If $I(G)=\infty$, then M is stochastically complete.

Proof. The radial process $r(X_t)$ has the decomposition

$$r(X_t)=\beta_t+\frac{1}{2}\int_0^t\Delta_M r(X_s)\,ds-L_t.$$

By THEOREM 3.4.2 we have

$$\Delta_M r\le(d-1)\frac{G'(r)}{G(r)}. \tag{4.2.4}$$

Let ρ be the unique nonnegative solution of the equation

$$\rho_t=\beta_t+\frac{d-1}{2}\int_0^t\frac{G'(\rho_s)}{G(\rho_s)}ds. \tag{4.2.5}$$

By THEOREM 3.5.3 $e(X)\ge e(\rho)$ and $r(X_t)\le\rho_t$ for all $t<e(\rho)$. By LEMMA 4.2.2 the condition on G implies that ρ does not explode. It follows that X cannot explode under the same condition. □

By converting the integral condition (4.2.3) in the above theorem into a growth condition on the Ricci curvature, we obtain a sufficient condition for stochastic completeness.

Theorem 4.2.4. *Let M be a complete Riemannian manifold. Suppose that $\kappa(r)$ is a negative, nonincreasing, continuous function on $[0,\infty)$ such that*

$$\kappa(r) \le (d-1)^{-1} \inf\left\{\mathrm{Ric}_M(x) : r(x) = r\right\}.$$

If

$$\int_c^\infty \frac{dr}{\sqrt{-\kappa(r)}} = \infty, \tag{4.2.6}$$

then M is stochastically complete.

Proof. We need to show that (4.2.6) implies that $I(G) = \infty$. The function $-\kappa(r) = G''(r)/G(r)$ is assumed to be nondecreasing. Integrating by parts, we have

$$\int_0^r G(r)^2 d\{-\kappa(s)\} = G''(r)G(r) - G'(r)^2 + 1.$$

Hence $G''(r)G(r) - G'(r)^2 \ge -1$, or

$$\left[\frac{G'(r)}{G(r)}\right]^2 \le \frac{G''(r)}{G(r)} + \frac{1}{G'(r)^2} \le -C_1\kappa(r) \tag{4.2.7}$$

for $r \ge c$ and $C_1 = 1 - \kappa(c)^{-1}G'(c)^{-2}$. Integrating by parts again, we have

$$\begin{aligned}
\int_c^r G(s)^{d-1} ds &= \frac{1}{d}\int_c^r \frac{d\left\{G(s)^d\right\}}{G'(s)} \\
&\ge \frac{1}{d}\frac{G(r)^d}{G'(r)} - \frac{1}{d}\frac{G(c)^d}{G'(c)} - \int_c^r G(s)^d\, d\left\{\frac{1}{G'(s)}\right\} \\
&\ge \frac{1}{d}\frac{G(r)^d}{G'(r)} - \frac{1}{d}\frac{G(c)^d}{G'(c)}.
\end{aligned}$$

The last inequality holds because $G'(s)$ is nondecreasing. Multiply the above inequality by $G(r)^{1-d}$ and integrating from c to ∞, we have

$$I(G) \ge \frac{1}{d\sqrt{C_1}} \int_c^\infty \frac{dr}{\sqrt{-\kappa(r)}} - C_2,$$

where

$$C_2 = \frac{1}{d}\frac{G(c)^d}{G'(c)} \int_c^\infty G(r)^{1-d} dr.$$

The last integral is finite because G grows at least exponentially. The proof is completed. □

Remark 4.2.5. In a remarkable work Grigorian [**32**] proved a stronger result: if

$$\int_c^\infty \frac{r dr}{\ln |B(r)|} = \infty,$$

then M is stochastically complete. Here $|B(r)|$ is the volume of the geodesic ball of radius r. This shows that it is the growth of the volume that lies at the heart of the problem of stochastic completeness. The fact that the Ricci curvature condition (4.2.6) implies the volume growth condition can be proved by standard volume comparison theorems in differential geometry (cf. Bishop and Crittenden [**4**]). In the next section we will meet the Ricci curvature growth condition (4.2.6) again in a different context.

Our discussion of stochastic completeness will not be complete without demonstrating a result in the opposite direction. We will assume that M is a Cartan-Hadamard manifold so that the cutlocus is empty. See SECTION 6.1 for a discussion of such manifolds.

Proposition 4.2.6. *Suppose that M is a Cartan-Hadamard manifold, i.e., a complete, simply connected manifold with nonpositive sectional curvature. Let $o \in M$ be a fixed point, and $r(x) = d(x, o)$. Let $\kappa(r)$ be a nonpositive, nonincreasing, continuous function on $[0, \infty)$ such that*

$$\kappa(r) \geq \sup \{K_M(x) : r(x) = r\}.$$

Let G be the solution of the Jacobi equation (4.2.2) for κ and define $I(G)$ as in (4.2.3). If $I(G) < \infty$, then M is not stochastically complete.

Proof. By the Cartan-Hadamard theorem, the cutlocus is empty. Hence we have

$$r(X_t) = \beta_t + \frac{1}{2} \int_0^t \Delta_M r(X_s)\, ds.$$

By THEOREM 3.4.2 we have

$$\Delta_M r \geq (d-1) \frac{G'(r)}{G(r)}.$$

Define ρ as in (4.2.5). The condition on G implies that $\mathbb{P}_x \{e(\rho) < \infty\} = 1$ by PROPOSITION 4.2.2. On the other hand, by THEOREM 3.5.3 $r(X_t) \geq \rho_t$; hence $e(X) \leq e(\rho)$. It follows that $\mathbb{P}_x \{e(X) < \infty\} = 1$ and M is not stochastically complete. □

We end this section with an example of a complete but not stochastically complete manifold.

Example 4.2.7. Let M be a Cartan-Hadamard manifold. Suppose that there is an $\epsilon \in (0, 1]$ such that

$$K_M(x) \leq -16\, r(x)^{2+\epsilon} - 6.$$

Then M is not stochastically complete. To see this, consider the function

$$G(r) = re^{r^{2(1+\epsilon)}}.$$

Then $G(0) = 0, G'(0) = 1$, and by an easy calculation,

$$\frac{G''(r)}{G(r)} \leq 6 + 16r^{2+\epsilon}.$$

Under the assumption on the sectional curvature, we can take

$$\kappa(r) = -\frac{G''(r)}{G(r)}.$$

By the preceding proposition, M is not stochastically complete if we can prove that $I(G) < \infty$. To this end we first exchange the order of integrations in the definition of $I(G)$ to obtain

$$I(G) = \int_c^\infty G(s)^{d-1} ds \int_s^\infty G(r)^{1-d} dr.$$

For the inside integral we have

$$\begin{aligned}\int_s^\infty G(r)^{1-d} dr &= \frac{-1}{(d-1)(2+\epsilon)} \int_s^\infty r^{-(d+\epsilon)} d\left\{e^{-(d-1)r^{2+\epsilon}}\right\} \\ &\leq \frac{1}{(d-1)(2+\epsilon)} \frac{G(s)^{(1-d)}}{s^{1+\epsilon}}.\end{aligned}$$

It follows that

$$I(G) \leq \frac{1}{(d-1)(2+\epsilon)} \int_c^\infty \frac{ds}{s^{1+\epsilon}} < \infty.$$

Remark 4.2.8. The curvature condition (4.2.6) is sharp. Varopoulos [**70**] showed that under certain minor technical assumptions, the condition

$$\int_c^\infty \frac{dr}{\sqrt{-\kappa(r)}} < \infty$$

implies that M is not stochastically complete.

4.3. C_0-property of the heat semigroup

We assume that M is a (geodesically) complete, noncompact Riemannian manifold. We use $BC(M)$ to denote the space of bounded continuous functions on M. The heat semigroup $\{P_t, t \geq 0\}$ is a family of bounded operators $\mathbb{P}_t : BC(M) \to BC(M)$ defined by

$$P_t f(x) = \mathbb{P}_x \{f(X_t), t < e(X)\} = \int_M p_M(t, x, y) f(y)\, dy.$$

We have

$$P_t P_s = P_{t+s}, \quad \lim_{t \downarrow 0} \mathbb{P}_t f = f, \quad f \in BC(M).$$

In other words, $\{P_t, t \geq 0\}$ is a strongly continuous semigroup on $BC(M)$. These properties follow directly from the corresponding properties of the heat kernel.

In the one-point compactification $\widehat{M} = M \cup \{\partial_M\}$, the point at infinity ∂_M is considered to be an absorbing state for Brownian motion. Let $C_0(M)$ be the space of continuous functions on $\widehat{M}$ which vanish at infinity: $\lim_{x \to \partial_M} f(x) = 0$. The heat semigroup $\{P_t\}$ is said to have the Feller property (or C_0-property) if it preserves $C_0(M)$, namely $P_t C_0(M) \in C_0(M)$ for all $t \geq 0$. Following the line of thought we have been exploring so far, in this section we will prove a sufficient geometric condition for the C_0-property.

The first step is to find an equivalent probabilistic characterization of the C_0-property, which we will eventually convert to a problem of estimating the exit time from a geodesic ball. As before, X is the coordinate process on the filtered path space $(W(M), \mathscr{B}(W(M))_*)$ and $e = e(X)$ is the liftime. Note that for a complete Riemannian manifold, $x \to \partial_M$ if and only if $r(x) \to \infty$.

Lemma 4.3.1. *Suppose that M is a complete Riemannian manifold. The heat semigroup has the C_0-property if and only if, for every compact set K of M and every fixed $t > 0$,*

$$\lim_{r(x) \to \infty} \mathbb{P}_x \{T_K \leq t\} = 0, \tag{4.3.1}$$

where $T_K = \inf\{t : X_t \in K\}$ is the first hitting time of K.

Proof. Let $\zeta_K(x) = \mathbb{E}_x e^{-T_K}$. Suppose first that the C_0-property holds. We have, for $x \in M \backslash K$,

$$\mathbb{P}_x \{T_K \leq t\} \leq e^t \, \mathbb{E}_x e^{-T_K} = e^t \, \zeta_K(x). \tag{4.3.2}$$

Take a nonnegative $f \in C_0(M)$ which is not identically zero, and let

$$G_1 f(x) = \int_0^\infty e^{-t} P_t f(x) \, dt.$$

By the C_0-property and the dominated convergence theorem,

$$\lim_{r(x) \to \infty} G_1 f(x) = 0. \tag{4.3.3}$$

Let $m = \min\{G_1 f(x) : x \in \partial K\}$. Since $G_1 f$ is continuous and strictly positive, and K is compact, we have $m > 0$. Now,

$$\begin{aligned} G_1 f(x) &= \mathbb{E}_x \int_0^e e^{-t} f(X_t)\, dt \\ &\geq \mathbb{E}_x \int_{T_K}^e e^{-t} f(X_t)\, dt \\ &= \mathbb{E}_x \left[e^{-T_K} \int_0^{e-T_K} f(X_{T_K+t})\, dt \right] \\ &= \mathbb{E}_x \left[e^{-T_K}\, \mathbb{E}_{X_{T_K}} \left(\int_0^e f(X_t) dt \right) \right] \\ &= \mathbb{E}_x \left[e^{-T_K}\, G_1 f(X_{T_K}) \right] \\ &\geq m\, \zeta_K(x). \end{aligned}$$

Here we have used the strong Markov property, which says that, conditioned on $\mathscr{B}_{T_K}$, the shifted process $\{X_{T_K+t}, t < e - T_K\}$ is a Brownian motion from X_{T_K} with life time $e - T_K$. From this and (4.3.2) we obtain the inequality

$$\mathbb{P}_x \{T_K \leq t\} \leq \frac{e^t}{m} G_1 f(x),$$

which implies (4.3.1) by (4.3.3).

For the converse, suppose that $f \in C_0(M)$, and for any $\epsilon > 0$ choose $R > 0$ such that $|f(x)| \leq \epsilon$ for $r(x) \geq R$. Let K be the closure of the geodesic ball $B(R)$, which is compact because of the (geodesic) completeness of M. For $x \in M \backslash K$,

$$\begin{aligned} |P_t f(x)| &\leq \mathbb{E}_x \{|f(X_t)|\, ; T_K \leq t\} + \mathbb{E}_x \{|f(X_t)|\, ; T_K > t\} \\ &\leq \|f\|_\infty \mathbb{P}_x \{T_K \leq t\} + \epsilon. \end{aligned}$$

Suppose that (4.3.1) holds. Then $\limsup_{r(x)\to\infty} |P_t f(x)| \leq \epsilon$. This shows that $\mathbb{P}_t f \in C_0(M)$. □

We can paraphrase the above lemma as follows. In order that the C_0-property hold, as the starting point moves to infinity, Brownian motion should have diminishing probability of returning to a fixed compact set before a prefixed time; in other words, Brownian motion should not wander away too fast. We have seen many times before that the lower bound of the Ricci curvature controls the growth of Brownian motion. Therefore the following sufficient condition for the C_0-property should not come as a surprise. Incidentally, it is the same integral test that guarantees the nonexplosion of Brownian motion.

Theorem 4.3.2. *Suppose that M is a complete Riemannian manifold. Let $o \in M$ and $r(x) = d(x, o)$. Let $\kappa(r)$ be a negative, nonincreasing, continuous* *function on* $[0,\infty)$ *such that*

$$\kappa(r) \le (d-1)^{-1} \inf\{\mathrm{Ric}_M(x) : r(x) = r\}.$$

If

$$\int_c^\infty \frac{dr}{\sqrt{-\kappa(r)}} = \infty, \tag{4.3.4}$$

then the heat semigroup on M *has the* C_0*-property.*

Proof. Without loss of generality we assume that $\kappa(r) \uparrow \infty$ as $r \uparrow \infty$. For ease of notation, let $l(r) = \sqrt{-\kappa(r)}$. The hypothesis on the Ricci curvature becomes

$$\int_c^\infty \frac{dr}{l(r)} = \infty. \tag{4.3.5}$$

By LEMMA 4.3.1 it is enough to prove (4.3.1). We may assume that $K = B(R)$, the geodesic ball of radius R centered at o. We will consider the first htting times of the concentric spheres centered at o. Define two sequences of stopping times as follows: $\sigma_0 = 0$ and

$$\begin{aligned}\tau_n &= \inf\{t > \sigma_n : d(X_t, X_0) = 1\},\\ \sigma_n &= \inf\{t \ge \tau_{n-1} : r(X_t) = r(x) - n\}.\end{aligned}$$

Intuitively, at time σ_n, Brownian motion is at the sphere of radius $r(x) - n$ from the fixed point o. The difference $\tau_n - \sigma_n$ is the amount of time Brownian motion X takes to move a distance 1, and $\sigma_{n+1} - \tau_n$ is the waiting time for the Brownian motion to reach the next sphere of radius $r(x) - (n+1)$. We throw away these waiting times and let $\theta_n = \tau_n - \sigma_n$. Then it is clear that

$$T_K \ge \sigma_{[r(x)-R]} \ge \theta_1 + \theta_2 + \cdots + \theta_{[r(x)-R]}. \tag{4.3.6}$$

($[a]$ denotes the integral part of a.) We will use THEOREM 3.6.1 to estimate the size of θ_k. During the time interval $[\sigma_k, \sigma_k + \theta_k]$, the motion is confined to the geodesic ball $B(X_{\sigma_k}; 1)$ whose center is at a distance $r(x) - k$ from the fixed point o; hence the geodesic ball is contained in the ball $B(r(x) - k + 1)$ (centered at o). By assumption, the Ricci curvature there is bounded from below by $-(d-1)\kappa(r(x) - k + 1)$. By THEOREM 3.6.1, there are constants C_1 and C_2, depending only on the dimension of the manifold, such that

$$\mathbb{P}_x\left[\theta_k \le \frac{C_1}{l(r(x) - k + 1)}\right] \le e^{-C_2 l(r(x)-k+1)}. \tag{4.3.7}$$

Since the above probability is rather small, each θ_k cannot be too small, and then by (4.3.6) the probability of $\{T_K \le t\}$ cannot be too large. The rest of the proof is simply to quantify this intuitive argument. Choose $n(x,t)$ to be the smallest integer such that

$$\sum_{k=1}^{n(x,t)} \frac{1}{l(r(x)-k+1)} \geq \frac{t}{C_1}. \tag{4.3.8}$$

By (4.3.5) such $n(x,t)$ exists for all sufficiently large $r(x)$. By the choice of $n(x,t)$,

$$\frac{t}{C_1} \geq \sum_{k=1}^{n(x,t)-1} \frac{1}{l(r(x)-k+1)} \geq \sum_{k=[r(x)]-n(x,t)+3}^{[r(x)]+1} \frac{1}{l(j)}. \tag{4.3.9}$$

By (4.3.5) again, as $r(x) \uparrow \infty$, the lower bound $[r(x)] - n(x,t) + 3$ of the last sum must also go to infinity:

$$r(x) - n(x,t) \to \infty \quad \text{as} \quad r(x) \uparrow \infty. \tag{4.3.10}$$

This implies that $[r(x) - R] \geq n(x,t)$ for all sufficiently large $r(x)$, and the following sequence of inclusions holds:

$$\begin{aligned} \{T_K \leq t\} \subseteq \left\{ \sum_{k=1}^{[r(x)-R]} \theta_k \leq t \right\} &\subseteq \left\{ \sum_{k=1}^{n(x,t)} \theta_k \leq t \right\} \\ &\subseteq \bigcup_{k=1}^{n(x,t)} \left\{ \theta_k \leq \frac{C_1}{l(r(x)-k+1)} \right\}. \end{aligned}$$

Note that the last inclusion is a consequence of (4.3.8). Now we can estimate the probability by (4.3.7):

$$\mathbb{P}_x \{T_K \leq t\} \leq \sum_{k=1}^{n(x,t)} e^{-C_2 l(r(x)-k+1)}. \tag{4.3.11}$$

We use (4.3.9) to estimate the last sum, and obtain

$$\begin{aligned} \mathbb{P}_x \{T_K \leq t\} &\leq q(x,t) \sum_{k=1}^{n(x,t)-1} \frac{1}{l(r(x)-k+1)} + e^{-C_2 l(r(x)-n(x,t)+1)} \\ &\leq q(x,t) \cdot \frac{t}{C_1} + e^{-C_2 l(r(x)-n(x,t)+1)}, \end{aligned}$$

where

$$q(x,t) = \sup \left\{ e^{-c_2 l(r)} l(r) : r \geq r(x) - n(x,t) \right\}.$$

From (4.3.10) and the hypothesis that $l(r) \uparrow \infty$ as $r \uparrow \infty$, we have $q(x,t) \to 0$. Hence $\mathbb{P}_x \{T_K \leq t\} \to 0$ as $r(x) \to \infty$. By LEMMA 4.3.1, the C_0-property holds for the manifold. □

4.4. Recurrence and transience

It is well known that euclidean Brownian motion is recurrent in dimensions 1 and 2 and transient in dimensions 3 and higher. In this section we consider the same property for Brownian motion on a Riemannian manifold. The problem of recurrence and transience of Brownian motion is in general more difficult than both that of nonexplosion and that of C_0-property. Understandably our discussion in this section is rather incomplete. We will restrict ourselves to some general remarks and to several results whose proofs are in line with the general level of this book in terms of their complexity.

Definition 4.4.1. *Let M be a Riemannian manifold and $\mathbb{P}_x$ the law of Brownian motion starting from $x \in M$.*

(i) *A set K on M is called recurrent for a path $X \in W(M)$ if there is a sequence $t_n \uparrow\uparrow e(X)$ (strictly increasing to the explosion time) such that $X_{t_n} \in K$; otherwise K is called transient for X, or equivalently, there exists $T < e(X)$ such that $X_t \notin K$ for all $t \geq T$.*

(ii) *A set K of M is called recurrent if for all $x \in M$,*

$$\mathbb{P}_x \{X \in W(M) : K \text{ is recurrent for } X\} = 1;$$

it is called transient if for all $x \in M$,

$$\mathbb{P}_x \{X \in W(M) : K \text{ is transient for } X\} = 1.$$

(iii) *M is called parabolic (or Brownian motion is recurrent on M) if every nonempty open set of M is recurrent; it is called hyperbolic (or Brownian motion is transient on M) if every compact set of M is transient.*

Two simple observations follow immediately from the definitions. First, Brownian motion on M is transient if

$$\mathbb{P}_x \left\{ \lim_{t\uparrow\uparrow e(X)} X_t = \partial_M \right\} = 1,$$

for all $x \in M$. Here ∂_M is the point at infinity and the convergence is understood in the one-point compactification $\widehat{M} = M \cup \{\partial_M\}$. Second, if M is parabolic, then it is also stochastically complete.

Define

$$T_K = \inf\{t < e : X_t \in K\} = \text{ the first hitting time of } K,$$
$$\tau_D = \inf\{t < e : X_t \notin D\} = \text{ the first exit time from } D.$$

For testing recurrence and transience of Brownian motion, we will take

$$\mathscr{K}(M) = \text{relatively compact, connected, nonempty open sets on } M \text{ with smooth, nonempty boundary.}$$

Let $K \in \mathscr{K}(M)$ and

$$h_K(x) = \mathbb{P}_x \{T_K < e\}.$$

This hitting probability function holds the key to the problem of recurrence and transience. We start with two preliminary results.

Lemma 4.4.2. *For any relatively compact D such that $M\backslash\overline{D}$ is nonempty, we have $\sup_{x\in D}\mathbb{E}_x\tau_D < \infty$. In particular, Brownian motion always exits D in finite time.*

Proof. By enlarging D we may assume that D is connected with smooth boundary. Let $p_D(t,x,y)$ be the Dirichlet heat kernel. Then we have (see THEOREM 4.1.1)

$$\alpha \stackrel{\text{def}}{=} \sup_{x\in D} \int_D p_D(1,x,y)\,dy < 1;$$

hence for any $x \in D$,

$$\mathbb{P}_x\{\tau_D > 1\} = \int_D p_D(1,x,y)\,dy \le \alpha < 1.$$

Using the Markov property, we have

$$\mathbb{P}_x\{\tau_D > n\} = \mathbb{E}_x\left\{\mathbb{P}_{X_{n-1}}[\tau_D > 1]; \tau_D > n-1\right\} \le \alpha\mathbb{P}_x\{\tau_D < n-1\}.$$

By induction $\mathbb{P}_x\{\tau_D > n\} \le \alpha^n$, and

$$\mathbb{E}_x\tau_D \le \sum_{n=0} \mathbb{P}_x\{\tau_D > n\} \le \frac{1}{1-\alpha} < \infty.$$

□

The following lemma gives the well known probabilistic representation of the solution of a Dirichlet boundary value problem.

Lemma 4.4.3. *Let G be a relatively compact open set on M with smooth boundary. Then for any $f \in C(\partial G)$ the unique solution of the Dirichlet problem*

$$\begin{cases} \Delta_M u = 0 & \text{on } G, \\ u = f & \text{on } \partial G, \end{cases}$$

is given by

$$u(x) = \mathbb{E}_x f(X_{\tau_G}).$$

Proof. In this proof we take for granted the result from the theory of partial differential equations that the Dirichlet problem stated above is uniquely solvable in $C^2(G)\cap C(\overline{G})$. Let u be the solution and $x \in G$. It is easy to verify that $s \mapsto u(X_s)$ is a martingale up to τ_G. Hence

$$u(x) = \mathbb{E}_x u(X_{t\wedge\tau_G}).$$

Letting $t \uparrow \infty$ and using the fact that $\mathbb{P}_x \{\tau_G < \infty\} = 1$ by LEMMA 4.4.2, we have $u(x) = \mathbb{E}_x f(X_{\tau_G})$. □

Basic prperties of the hitting probability function h_K are set out in the following proposition.

Proposition 4.4.4. *The function h_K is harmonic (i.e., $\Delta_M h_K = 0$) on $M\backslash K$ and continuous on $\overline{M\backslash K}$. There are two possibilities: either $h_K(x) = 1$ for all $x \in M$, or $0 < h_K(x) < 1$ for all $x \in M\backslash K$.*

Proof. Let $D = M\backslash K$ for simplicity. Let $x \in D$, and let B be a small geodesic ball centered at x and disjoint from K. Then $\tau_B < T_K$, and by the strong Markov property at τ_B we have

$$h_K(x) = \mathbb{E}_x \mathbb{P}_{\tau_B} \{T_K < e\} = \mathbb{E}_x h_K(\tau_B).$$

By PROPOSITION 4.4.3 h_K on B is the solution of a Dirichlet problem; hence it is harmonic on B. This shows that h_K is harmonic on D.

Next, we show that $h_K \in C(\overline{D})$. Let $\{D_n\}$ be an exhaustion of M such that each D_n has a smooth boundary and contains K. Then

$$u_n(x) = \mathbb{P}_x \{T_K < \tau_{D_n}\}$$

is the solution of the Dirichlet problem on $D_n\backslash K$ with boundary value 1 on ∂K and 0 on ∂D_n. Hence u_n is continuous on the closure of $D_n\backslash K$. Since $u_n \uparrow h_K$, we see that h_K, as the limit of a sequence of increasing continuous functions, is lower semicontinuous on $D = M\backslash K$ and $h_K(x) = 1$ for $x \in \partial K$. It follows that, for any $x \in \partial K$,

$$\liminf_{y \to x} h_K(y) \geq h_K(x) = 1.$$

This together with the fact that $h_K(y) \leq 1$ implies that

$$\lim_{y \to x} h_K(y) = h_K(x) = 1,$$

which shows that $h_K \in C(\overline{D})$.

Since $0 \leq h_K \leq 1$, by the maximum principle for harmonic functions, if h_K is equal to 1 somewhere in D, then it is identically 1. Otherwise, we must have $h_K(x) < 1$ everywhere on D. If $h_K(x) = 0$ somewhere, then by the maximum principle again, it has to be identically 0, which contradicts the fact that h_K is continuous on $\overline{D}$ and takes value 1 on ∂D. □

The next proposition gives a criterion for recurrence or transience in terms of the hitting probability function h_K.

Proposition 4.4.5. *Let $K \in \mathscr{K}(M)$. If $h_K \equiv 1$, then K is recurrent. If $h_K < 1$ on $M\backslash K$, then K is transient. Thus a set $K \in \mathscr{K}(M)$ is either recurrent or transient.*

Proof. Suppose that $h_K \equiv 1$ and let

$$S_K = \{X \in W(M) : K \text{ is recurrent for } X\}. \tag{4.4.1}$$

We need to show that $\mathbb{P}_x S_K = 1$ for all $x \in M$. Let $\{D_n\}$ be an exhaustion of M and τ_n the first exit time from D_n. We may assume that $x \in D_n$ and $K \subseteq D_n$ for all n. Let

$$S_n = \{T_K \circ \theta_{\tau_n} < e\} = \{X \text{ visits } K \text{ after } \tau_n\}.$$

[θ_t is the shift operator in $W(M)$.] Then $S_n \downarrow S_K$ because τ_n strictly increases to the lifetime e. By the strong Markov property,

$$\mathbb{P}_x S_n = \mathbb{E}_x \mathbb{P}_{X_{\tau_n}} \{T_K < e\} = \mathbb{E}_x h_K(X_{\tau_n}) = 1.$$

Hence $\mathbb{P}_x S_K = 1$ and K is recurrent.

Now suppose that $h_K < 1$ on $M \backslash K$, and pick a relatively compact open set O such that $\overline{K} \subseteq O$. Then

$$\alpha \stackrel{\text{def}}{=} \inf \{h_K(z) : z \in \partial O\} < 1.$$

Define two sequences of stopping times: $\zeta_0 = 0$ and

$$\begin{aligned} \sigma_n &= \{t \geq \zeta_n : X_t \in K\} = T_K \circ \theta_{\zeta_n} + \zeta_n; \\ \zeta_n &= \{t \geq \sigma_{n-1} : X_t \notin O\} = \tau_O \circ \theta_{\sigma_{n-1}} + \sigma_{n-1}. \end{aligned}$$

Let S_K be defined in (4.4.1) as before. If X visits K infinitely often before the explosion time e, then for every n it has to visit K after ζ_n unless $\zeta_n = e$, i.e.,

$$S_K \subseteq \{\sigma_n < e\} \cup \{\zeta_n = e\}.$$

Because O is relatively compact, $\zeta_n < e$ unless $e = \infty$. Hence $\zeta_n = e$ implies that $\zeta_n = \infty$. But this cannot happen, for, according to LEMMA 4.4.2, with probability one X exits from O in finite time. Thus we have

$$\mathbb{P}_x S_K \leq \mathbb{P}_x \{\sigma_n < e\}.$$

On the other hand,

$$\begin{aligned} \mathbb{P}_x \{\sigma_n < e\} &= \mathbb{P}_x \{T_K \circ \zeta_n + \zeta_n < e\} \\ &\leq \mathbb{E}_x \left\{\mathbb{P}_{X_{\zeta_n}} [T_K < e] ; \zeta_n < e\right\} \\ &\leq \alpha \, \mathbb{P}_x \{\sigma_{n-1} < e\}. \end{aligned}$$

The last inequality holds because $X_{\zeta_n} \in \partial O$ if $\zeta_n < e$. By induction we have

$$\mathbb{P}_x S_K \leq \mathbb{P}_x \{\sigma_n < e\} \leq \alpha^{n-1} \downarrow 0 \quad \text{as } n \uparrow \infty.$$

This proves that K is transient. □

Corollary 4.4.6. *If M is compact, then Brownian motion is recurrent.*

Proof. If $h_K < 1$ on $D \stackrel{\text{def}}{=} M\backslash K$, it has to attain a mimimun somewhere in the open set D, which contradicts the fact that h_K is a harmonic function in D □

Although it does not follow immediately from the definition, recurrence or transience is independent of the choice of the testing set, as is shown in the next proposition.

Proposition 4.4.7. *If Brownian motion is recurrent for a set $K \in \mathscr{K}(M)$, then it is recurrent for every set in $\mathscr{K}(M)$.*

Proof. Let us argue in words for a change. Let K be recurrent and $F \in \mathscr{K}(M)$ another testing set. In view of COROLLARY 4.4.6 we may assume that M is noncompact and it is enough to show that F is also recurrent under the assumption that $\overline{K} \cap \overline{F}$ is empty. We have

$$\alpha \stackrel{\text{def}}{=} \inf \{h_F(x) : x \in K\} > 0.$$

If Brownian motion starts from a point $x \in K$, then it hits F at least with probability α. Because K is recurrent, for any $T > 0$ all paths come back to K sometime after T, and after that at least with probability α they will reach F. Hence

$$\mathbb{P}_x \{X \text{ visits } F \text{ after } T\} \geq \alpha.$$

If F were transient, this probability would have gone to zero when $T \uparrow \infty$. Therefore F must be recurrent. □

We can also characterize recurrence and transience in terms of the Green function

$$G_M(x, y) = \int_0^\infty p_M(t, x, y)\, dt.$$

It is well defined for all $(x, y) \in M \times M$, but may be infinite.

Proposition 4.4.8. *If M is parabolic, then $G_M(x, y) = \infty$ for all $(x, y) \in M \times M$. If M is hyperbolic, then $G_M(x, \cdot)$ is locally integrable for all $x \in M$.*

Proof. Suppose that M is parabolic, i.e., Brownian motion on M is recurrent. Let $K \in \mathscr{K}(M)$, and let O be a relatively compact open set such that $\overline{K} \subseteq O$. We have

$$G_M(x, y) \geq \int_1^\infty p_M(t, x, y)\, dt = \int_0^\infty dt \int_M p_M(t, x, z) p_M(1, z, y)\, dz.$$

Let

$$m = \inf \{p_M(1, z, y) : z \in K\}.$$

Then $m > 0$ and

$$G_M(x, y) \geq m \int_0^\infty dt \int_K p_M(t, x, y)\, dy = m\, \mathbb{E}_x \int_0^\infty I_K(X_t)\, dt.$$

Define the stopping times τ_n, σ_n as in the proof of PROPOSITION 4.4.5. By LEMMA 4.4.2, $\sigma_{n-1} < e$ implies $\zeta_n < \infty$ (see the secon half of the proof of PROPOSITION 4.4.5), and since K is recurrent, $\zeta_n < \infty$ implies $\sigma_n < e$. It follows from $\sigma_0 = T_K < \infty$ by induction that all σ_n and ζ_n are finite, and we have

$$\begin{aligned} G_M(x,y) &\geq m \sum_{n=1}^{\infty} \mathbb{E}_x \int_{\sigma_n}^{\zeta_{n+1}} I_K(X_t)dt \\ &= m \sum_{n=1}^{\infty} \mathbb{E}_x \mathbb{E}_{X_{\sigma_n}} \int_0^{\tau_O} I_K(X_t)dt. \end{aligned} \tag{4.4.2}$$

Now $X_{\sigma_n} \in K$ and the function

$$\mathbb{E}_y \int_0^{\tau_O} I_K(X_t)\, dt = \int_0^{\infty} dt \int_K p_O(t,y,z)\, dz, \quad y \in K,$$

has a positive lower bound on K because the Dirichlet heat kernel $p_O(t,y,z)$ is continuous and strictly positive on K. It follows that $G_M(x,y) = \infty$.

Now suppose that M is hyperbolic, i.e., Brownian motion is transient. For $K \in \mathscr{K}(M)$, we have

$$\int_K G_M(x,y)\, dy = \mathbb{E}_x \int_0^{e} I_K(X_t)\, dt.$$

Take the set O and the stopping times σ_n, ζ_n as before, and write

$$\int_K G_M(x,y)\, dy = \sum_{n=0}^{\infty} \mathbb{E}_x \int_{\zeta_n}^{\zeta_{n+1}} I_K(X_t)\, dt. \tag{4.4.3}$$

Using the Markov property at σ_n, we see that the terms to be summed

$$\begin{aligned} \mathbb{E}_x \int_{\zeta_n}^{\zeta_{n+1}} I_K(X_t)\, dt &= \mathbb{E}\left[\mathbb{E}_{X_{\sigma_n}} \int_0^{\tau_O} I_K(X_t)\, dt; \sigma_n < e\right] \\ &\leq \mathbb{E}_x \left\{\mathbb{E}_{X_{\sigma_n}} \tau_O; \sigma_n < e\right\} \\ &\leq a\, \mathbb{P}_x \left\{\sigma_n < e\right\}, \end{aligned} \tag{4.4.4}$$

where

$$a \stackrel{\text{def}}{=} \max\left\{\mathbb{E}_z \tau_O : z \in O\right\} < \infty$$

by LEMMA 4.4.2. Now

$$\mathbb{P}_x \left\{\sigma_n < e\right\} = \mathbb{E}_x \left\{\mathbb{P}_{\zeta_n}\left[T_K < e\right]; \zeta_n < e\right\} \leq b\, \mathbb{P}_x \left\{\zeta_n < e\right\},$$

where

$$b \stackrel{\text{def}}{=} \max\left\{h_K(z) : z \in \partial O\right\} < 1.$$

The last inequality holds because K is transient (see PROPOSITION 4.4.5). Also, by LEMMA 4.4.2,

$$\mathbb{P}_x \left\{\zeta_n < e\right\} = \mathbb{P}_x \left\{\sigma_{n-1} < e\right\}.$$

Hence, by induction,

$$\mathbb{P}_x\{\sigma_n < e\} \le b\,\mathbb{P}_x\{\sigma_{n-1} < e\} \le b^n. \tag{4.4.5}$$

It follows from (4.4.3)–(4.4.5) that

$$\int_K G_M(x,y)\,dy \le a\sum_{n=0}^{\infty} b^n = \frac{a}{1-b} < \infty.$$

This shows that $G_M(x,y)$ is locally integrable. □

We now turn to geometric conditions for recurrence and transience. The approach here is the same as in the study of stochastic completeness in SECTION 4.2, namely, we compare the radial process with a one-dimensional diffusion and then use the known necessary and sufficient condition for recurrence and transience for one-dimensional diffusions. Let us first state these one dimensional results.

Proposition 4.4.9. *Let $I = (c_1, c_2)$ with $-\infty \le c_1 < c_2 \le \infty$. Suppose that a and b are continuous functions on I such that $a(x) > 0$ for all $x \in I$. Let*

$$L = \frac{1}{2}a(x)\left(\frac{d}{dx}\right)^2 + b(x)\frac{d}{dx}$$

and X an L-diffusion. For a fixed $c \in (c_1, c_2)$ define

$$s(x) = \int_c^x \exp\left[-2\int_c^y \frac{b(z)}{a(z)}dz\right] dy.$$

(i) *If $s(c_1) = -\infty$ and $s(c_2) = \infty$, then X is recurrent;*

(ii) *If $s(c_1) = -\infty$ and $s(c_2) < \infty$, then X is transient and*

$$\mathbb{P}\left\{\lim_{t\uparrow e} X_t = c_2\right\} = 1.$$

Proof. This can be proved using the criteria in PROPOSITION 4.4.5, which in fact holds for any diffusion process generated by a strictly elliptic operator on a manifold; see THEOREM VI.3.1 in Ikeda and Watanabe [**48**] for details. □

Proposition 4.4.10. *Suppose that M is a complete Riemannian manifold without cutlocus. Let $o \in M$ be a fixed point, and $r(x) = d(x, o)$. Suppose that $\kappa(r)$ is a continuous function on $[0, \infty)$ such that*

$$\kappa(r) \ge \sup\{K_M(x) : r(x) = r\}.$$

Let G be the solution of the Jacobi equation (4.2.2) for κ. If

$$\int_c^\infty G(r)^{1-d} dr < \infty,$$

then Brownian motion on M is transient.

Proof. The proof is similar to that of THEOREM 4.2.6 for stochastic completeness. Retaining the notations there, we have $r(X_t) \geq \rho_t$. By the hypothesis of the theorem and PROPOSITION 4.4.9, $\rho_t \to \infty$ as $t \uparrow e(\rho)$. Hence we also have $r(X_t) \to \infty$ as $t \uparrow e(X)$, which implies that X must also be transient. □

Proposition 4.4.11. *Suppose that M is a complete Riemannian manifold and $o \in M$ a fixed point on M. Let $r(x) = d(x, o)$. Suppose that $\kappa(r)$ is a continuous function on $[0, \infty)$ such that*

$$\kappa(r) \leq (d-1)^{-1} \inf \left\{ \mathrm{Ric}_M(x) : r(x) = r \right\}.$$

Let G be the solution of the Jacobi equation (4.2.2) for κ. If

$$\int_c^\infty G(r)^{1-d} dr = \infty,$$

then M is recurrent.

Proof. The proof is similar to that of THEOREM 4.2.4 for stochastic noncompleteness. Retaining the notations there, we have $r(X_t) \leq \rho_t$. By PROPOSITION 4.4.9 ρ is recurrent. Hence X is also recurrent. □

For surfaces ($d = 2$) we can convert the integral test in the above propositions into curvature conditions.

Theorem 4.4.12. (i) *Let M be a complete Riemannian surface. Suppose that there exists an $r_0 > 0$ such that the Gaussian curvature*

$$K_M(x) \geq -\frac{1}{r(x)^2 \ln r(x)}, \qquad r(x) \geq r_0.$$

Then M is recurrent.

(ii) *Let M be a complete Riemannian surface without cutlocus. Suppose that there exist positive ϵ and r_0 such that the Gaussian curvature*

$$K_M(x) \leq -\frac{1+\epsilon}{r(x)^2 \ln r(x)}, \qquad r(x) \geq r_0.$$

Then M is transient.

Proof. (i) Fix an $r_1 > \min\{r_0, 2\}$ and choose a smooth function $\kappa(r)$ such that

$$K_M(x) \geq \kappa(r(x)) \quad \text{for all } x \in M,$$

and

$$\kappa(r) = -\frac{1}{r^2 \ln r} \quad \text{for all } r > r_1.$$

Such a function clearly exists by the hypothesis on the curvature $K_M(x)$. By PROPOSITION 4.4.10 it is enough to show that

$$\int_{r_1}^{\infty} \frac{dr}{G(r)} = \infty.$$

For this purpose we compare G with the function $G_1(r) = Cr \ln r$. Choose C large enough so that

$$G_1(r_1) > G(r_1), \quad G_1'(r_1) > G(r_1).$$

With this choice of C we claim that

$$G_1(r) \geq G(r) \quad \text{for all } r \geq r_1.$$

This will complete the proof, because then

$$\int_{r_1}^{\infty} \frac{dr}{G(r)} \geq \int_{r_1}^{\infty} \frac{dr}{G_1(r)} = \infty.$$

To prove the claim we let $h(r) = G_1(r) - G(r)$. Suppose that r_2 is the first zero of h, if it exists. From

$$h''(r) = -\kappa(r)h(r) \geq 0$$

we have $h''(r) > 0$ on $[r_1, r_2)$. Now h' is increasing and $h'(r_1) > 0$; hence $h'(r) > 0$ and h is also increasing on $[r_1, r_2]$. But $h(r_1) > 0$, and this means that r_2 cannot be a zero of $h(r)$. This contradiction shows that $h(r) > 0$ for all $r \geq r_1$.

(ii) The proof of this part is similar, but with a different comparison function. Choose a smooth, nonpositive function $\kappa(r)$ such that

$$\kappa(r) \begin{cases} = 0, & r \leq r_0; \\ \geq -(1+\epsilon)/r^2 \ln r, & r_0 < r < r_1; \\ = -(1+\epsilon)/r^2 \ln r, & r \geq r_1. \end{cases}$$

By the hypothesis on the Gaussian curvature, we have $K_M(x) \leq \kappa(r(x))$ for all $x \in M$. By PROPOSITION 4.4.11 it is enough to show that

$$\int_{r_2}^{\infty} \frac{dr}{G(r)} < \infty.$$

We compare G with $G_1(r) = Cr(\ln r)^{(1+\epsilon/2)}$ with a positive C, and see that it is enough to show that $G_1(r) \leq G(r)$ for sufficiently large r. An explicit computation shows that

$$\frac{G_1''(r)}{G_1(r)} \leq \frac{G''(r)}{G(r)}$$

for sufficiently large r, say $r \geq r_2$. The choice of r_2 is independent of C. We can fix a small C such that

$$G_1(r_2) \leq G(r_2), \quad G_1'(r_2) \leq G'(r_2).$$

Now the same argument as in Part (i) shows that $G_1(r) \le G(r)$ for all $r > r_2$. □

For $d \ge 3$, the situation is more complicated because the curvature function $\kappa(r)$ does not always determine whether Brownian motion is recurrent or transient. For example, in dimension $d = 3$, Brownian motion is recurrent on a flat cylinder, but transient on $\mathbb{R}^3$, but both manifolds have zero curvature. We can, however, still make some positive statements. Consider a typical curvature function

$$\kappa(r) = \frac{c}{r^2}.$$

If $c > 1/4$, then the general solution of the Jacobi equation is

$$G(r) = C_1\sqrt{r}\sin\left(\sqrt{4c-1}\ln r + C_2\right).$$

Thus G cannot be strictly positive for all r. This means that if there is a positive ϵ such that

$$\mathrm{Ric}_M(x) \ge \frac{1+\epsilon}{4}\cdot\frac{d-1}{r(x)^2}, \qquad r(x) \ge r_0,$$

then M is compact, so Brownian motion is recurrent.

For $c = 1/4$, we have

$$G(r) = C_1\sqrt{r} + C_2\sqrt{r}\ln r.$$

Thus if M is noncompact without cutlocus, has the dimension $d \ge 4$, and

$$K_M(x) \le \frac{1}{4}\cdot\frac{1}{r(x)^2}, \qquad r(x) \ge r_0,$$

then Brownian motion is transient on M.

For $0 < c < 1/4$, we write $c = \alpha(1-\alpha)$ for $0 < \alpha < 1/2$. In this case

$$G(r) = C_1 r^\alpha + C_2 r^{1-\alpha}.$$

Hence, if M is noncompact without cutlocus, has the dimension $d > 1 + 1/\alpha$, and

$$K_M(x) \le \frac{\alpha(1-\alpha)}{r(x)^2}, \qquad r(x) \ge r_0,$$

then Brownian motion is transient on M.

In the range $c \le 0$, a solution of the Jacobi equation may be bounded. In order to eliminate this case, we assume that M is a Cartan-Hadamar manifold, i.e, a complete, simply connected manifold of nonpositive curvature. Under this assumption, the solution of the Jacobi equation grows at least as fast as r. This implies that Brownian motion is transient on all Cartan-Hadamard manifolds of dimension $d \ge 3$.

4.5. Comparison of heat kernels

Let M be a complete Riemannian manifold. In this section we compare the heat kernel $p_M(t,x,y)$ with the heat kernel on a space of constant curvature. Let K be a constant and $\mathbb{M}_K$ a simply connected, d-dimensional Riemannian manifold of constant curvature K (space form). We denote the heat kernel on $\mathbb{M}_K$ by $p^K(t,a,b)$. By symmetry, $p^K(t,a,b)$ is a function of t and the distance between a and b; hence there is a function $p^K(t,r)$ such that

$$p^K(t,a,b) = p^K\left(t, d_{\mathbb{M}_K}(a,b)\right).$$

When $K>0$, $\mathbb{M}_K$ is a euclidean sphere of radius $1/\sqrt{K}$, and we adopt the convention that $p^K(t,r) = p^K(t,\pi/\sqrt{K})$ if $r \geq \pi/\sqrt{K}$.

Theorem 4.5.1. *Suppose that M is a complete Riemannian manifold and $x \in M$. Let K be a positive number such that*

$$K_M(z) \leq K \quad \textit{for all } z \in B(x;\pi/\sqrt{K}).$$

Then for all $(t,y) \in (0,\infty)\times M$,

$$p_M(t,x,y) \leq p^K\left(t, d_M(x,y)\right).$$

Proof. Let $r(z) = d_M(x,z)$. We have, under the probability $\mathbb{P}_y$,

$$r(X_t) = d(x,y) + \beta_t + \frac{1}{2}\int_0^t \Delta_M r(X_s)\,ds - L_t, \quad t<e.$$

On $\mathbb{M}_K$, the solution for the Jacobi equation with constant curvature K is $G(\rho) = \sin\sqrt{K}\rho/\sqrt{K}$. Hence if $\rho(b) = d_{\mathbb{M}_K}(o,b)$ is the distance function on $\mathbb{M}_K$, then

$$\theta_K(\rho) \stackrel{\text{def}}{=} \Delta_{\mathbb{M}_K}\rho = (d-1)\frac{G'(\rho)}{G(\rho)} = (d-1)\sqrt{K}\cot\sqrt{K}\rho.$$

The radial process of a Brownian motion on $\mathbb{M}_K$ starting from a point b with distance

$$d_{\mathbb{M}_K}(o,b) = \min\left\{d_M(x,y), \pi/\sqrt{K}\right\}$$

is the solution of the equation

$$\rho_t = d_{\mathbb{M}_K}(o,b) + \beta_t + \frac{1}{2}\int_0^t \theta_K(\rho_s)\,ds - A_t,$$

where A increases only when $\rho_t = \pi/\sqrt{K}$. Of course, if the dimension is 2 and higher, we do not need to worry about the cutlocus on $\mathbb{M}_K$, because it degenerates to a point, which Brownian motion never hits. By the Laplacian comparison theorem we have $\Delta_M r \geq \theta_K(r)$. On the other hand, by the choice of K, the geodesic ball $B(x;\pi/\sqrt{K})$ is within the cutlocus of x. Thus if $r(X_t) < \rho_t \leq \pi/\sqrt{K}$, then $r(X_t) \in M\backslash C_x$ and L does not increase. This observation together with the radial process comparison THEOREM 3.5.3 implies that $r(X_t) \ge \rho_t$ for all $t < e = e(X)$. Now, for any $\epsilon > 0$,

$$\mathbb{P}_y \{r(X_t) \le \epsilon, \, t < e\} \le \mathbb{P}_b \{\rho_t \le \epsilon\}.$$

Writing this inequality in terms of the heat kernels we have

$$\int_{B(x;\epsilon)} p_M(t,z,y)\, dz \le \int_{B(o;\epsilon)} p^K(t,a,b)\, da.$$

On both M and $\mathbb{M}_K$, as $\epsilon \downarrow 0$, the volume of the geodesic ball of radius ϵ is asymptotically equivalent to $\epsilon^d |\mathbb{S}^d|/d$, the volume of a euclidean ball of radius ϵ. Dividing the above inequality by this volume and letting $\epsilon \downarrow 0$, we obtain

$$p_M(t,x,y) \le p^K(t,o,b) = p^K\left(t, d_{\mathbb{M}_K}(o,b)\right).$$

This completes the proof. □

For any fixed $x \in M$, the condition of the theorem is always satisfied if K is sufficiently large, and the theorem applies. However, if the sectional curvature of M is bounded from above by a nonpositive constant K, in order to compare $p_M(t,x,y)$ directly with $p^K(t,a,b)$, we need further assumptions on M to remove the cutlocus. This is the case if M is a Cartan-Hadamard manifold, i.e., a complete, simply connected Riemannian manifold with non-positive sectional curvature.

Theorem 4.5.2. *Suppose that M is a complete Riemannian manifold.*

(i) If M is a Cartan-Hadamard manifold and $K_M(z) \le K$ for all $z \in M$, then for all $(t,x,y) \in (0,\infty) \times M \times M$,

$$p_M(t,x,y) \le p^K(t, d_M(x,y)).$$

(ii) If $\mathrm{Ric}_M(z) \ge (d-1)K$ for all $z \in M$, then for all $(t,x,y) \in (0,\infty) \times M \times M$,

$$p_M(t,x,y) \ge p^K(t, d_M(x,y)).$$

Proof. Exercise. □

Chapter 5

Short-time Asymptotics

In this chapter we study short-time behaviors of both the heat kernel and Brownian motion. We will show by several typical results how the knowledge about the heat kernel can be used to study the behavior of Brownian motion and vice versa. Our starting point is the short-time asymptotics of the heat kernel $p_M((t, x, y)$ for near points, more precisely when x and y are within each other's cutlocus. The parametrix method for constructing the heat kernel in differential geometry yields the asymptotic expansion for the heat kernel as $t \downarrow 0$ stated in SECTION 5.1. This asymptotic expansion is the starting point of our discussion. One can obtain this expansion by probabilistic considerations (at least for the leading term, see Molchanov [**59**] and Elworthy and Truman [**22**]), but we will not discuss them here. Once we have this local expansion, probabilistic ideas can be used to study global short-time behavior of the heat kernel. In SECTION 5.2, we prove Varadhan's well known relation for the logarithmic heat kernel on a complete Riemannian manifold. In SECTION 5.3 we will give a general method for computing the leading term of the short-time asymptotics of $p(t, x, y)$ when x, y are not close to each other. This method is illustrated for the case the heat kernel on a sphere at antipodal points. The last two sections are devoted to several topics related to Brownian bridge and its relation to the heat kernel on a compact Riemannian manifold. Basic properties of Brownian bridge are discussed in SECTION 5.4. They are used in SECTION 5.5 to prove global esimtates on the first and second derivatives of the logarithmic heat kernel on a compact manifold for small times.

5.1. Short-time asymptotics: near points

Let M be a Riemannian manifold and C_x the cutlocus of $x \in M$. It is a basic fact in differential geometry that $y \in C_x$ is equivalent to $x \in C_y$ (THEOREM 3.4.1, see also Cheeger and Ebin [**9**]). The asymptotic behavior of the heat kernel $p_M(t, x, y)$ when x and y are within each other's cutlocus is given in the following theorem.

Theorem 5.1.1. *Let M be a complete Riemannian manifold and*

$$C_M = \{(x, y) \in M \times M : x \in C_y\}.$$

There are smooth functions $H_n(x, y)$ defined on $(M \times M)\backslash C_M$ with the properties

$$H_0(x, y) > 0 \quad \text{and} \quad H_0(x, x) = 1$$

the following asymptotic expansion

$$p_M(t, x, y) \sim \left(\frac{1}{2\pi t}\right)^{d/2} e^{-d(x,y)^2/2t} \sum_{n=0}^{\infty} H_n(x, y)\, t^n \tag{5.1.1}$$

holds uniformly as $t \downarrow 0$ for any compact subset of $(M \times M)\backslash C_M$. In particular, if x is within the cutlocus of y, then

$$\lim_{t \downarrow 0} t \ln p_M(t, x, y) = -\frac{d(x, y)^2}{2}.$$

Proof. We sketch the proof. If M is compact, the asymptotic expansion can be obtained by the method of parametrix (see Berger, Gauduchon, and Mazet [**3**] or Chavel [**8**]). If M is not compact, the proof is more difficult. The first thing we need to do is to localize the heat kernel. Let K be a compact subset of M and D a large, smooth, relatively compact open set containing K. Suppose that x and y are in K. By considering two possibilities $\{t < \tau_D\}$ and $\{t \geq \tau_D\}$ and using the Markov property at τ_D in the second possibility, we have

$$p_M(t, x, y) = p_D(t, x, y) + \mathbb{E}_x \{p_M(t - \tau_D, X_{\tau_D}, y);\, t \geq \tau_D\}.$$

Hence, for $t < t_0$,

$$0 \leq p_M(t, x, y) - p_D(t, x, y) \leq C(t_0, K, D)\, \mathbb{P}_x \{t \geq \tau_D\},$$

where

$$C(t_0, K, D) = \sup \{p_M(s, z_1, z_2) : s \leq t_0,\ z_1 \in K,\ z_2 \in \partial D\}.$$

By comparing with the heat kernel on a sphere with very small radius (see THEOREM 4.5.1), we can show that $C(t_0, K, D)$ is finite. For the probability $\mathbb{P}_x \{t \geq \tau_D\}$, a proof similar to the one for THEOREM 3.6.1 shows that for any μ, there are a sufficiently large D and a positive t_0 such that

$$\mathbb{P}_x \{t \geq \tau_D\} \leq e^{-\mu/t} \quad \text{for all } (t, x) \in (0, t_0) \times K.$$

It follows that

$$0 \le p_M(t,x,y) - p_D(t,x,y) \le Ce^{-\mu/t}.$$

Since D is compact and smooth, we can embed D isometrically into a compact manifold M_1, and the above argument applied to M_1 shows that

$$0 \le p_{M_1}(t,x,y) - p_D(t,x,y) \le Ce^{-\mu/t};$$

hence

$$|p_M(t,x,y) - p_{M_1}(t,x,y)| \le 2Ce^{-\mu/t}.$$

Since M_1 is compact, the method of parametrix applies and we obtain the asymptotic expansion stated in the theorem for $p_{M_1}(t,x,y)$. Now if we choose D large enough so that $\mu > 1 + d(K)^2/2$, where $d(K)$ is the diamter of K, then $p_M(t,x,y)$ and $p_{M_1}(t,x,y)$ are asymptotically equivalent, and the same expansion also holds for $p_M(t,x,y)$. □

The leading coefficient $H_0(x,y)$ has a simple geometric interpretation. Let $\exp_x : T_xM \to M$ be the exponential map based at x. Let $J(\exp_x)(Y)$ be the Jacobian of $\exp_x$ at $Y \in T_xM$ (the ratio of the volume element at $\exp_x Y$ over the euclidean volume element at Y). Then

$$H_0(x,y) = [J(\exp_x)(Y)]^{-1/2}, \quad Y = \exp_x^{-1} y.$$

Example 5.1.2. Consider the d-dimensional sphere $\mathbb{S}^d$ (the space form of constant curvature 1). The Riemannian metric in the polar coordinates on T_xM is

$$ds^2_{\mathbb{S}^d} = dr^2 + (\sin r)^2 d\theta^2,$$

where $d\theta^2$ is the standard metric on $\mathbb{S}^{d-1}$. Let $e_1, \ldots, e_{d-1}$ be an orthonormal basis for $\mathbb{S}^{d-1}$ at θ, and e_r the unit radial vector on T_xM. Then $e_r, e_1/r, \ldots, e_{d-1}/r$ form an orthonormal basis of T_xM (with the euclidean metric) at y. As vectors on M, they remain orthogonal but with lengths $1, \sin r, \ldots, \sin r$, respectively. Hence the Jacobian of the exponential map $\exp_x$ at y is $(\sin r/r)^{d-1}$, where $r = d(x,y)$, and

$$H_0(x,y) = \left[\frac{r}{\sin r}\right]^{(d-1)/2}.$$

By THEOREM 5.1.1 we have

$$p_{\mathbb{S}^d}(t,x,y) \sim \frac{e^{-d(x,y)^2/2t}}{(2\pi t)^{d/2}} \left[\frac{d(x,y)}{\sin d(x,y)}\right]^{(d-1)/2}$$

for $d(x,y) < \pi$. □

Example 5.1.3. For the d-dimensional hyperbolic space $\mathbb{H}^d$ (the space form of constant curvature -1), the metric is

$$ds^2_{\mathbb{H}^d} = dr^2 + (\sinh r)^2 d\theta^2,$$

and by an argument similar to that in EXAMPLE 5.1.2 we have

$$H_0(x,y) = \left[\frac{r}{\sinh r}\right]^{(d-1)/2}.$$

Hence

$$p_{\mathbb{H}^d}(t,x,y) \sim \frac{e^{-d(x,y)^2/2t}}{(2\pi t)^{d/2}}\left[\frac{d(x,y)}{\sinh d(x,y)}\right]^{(d-1)/2}.$$

For $\mathbb{H}^2$ and $\mathbb{H}^3$, smple explicit formulas are known:

$$p_{\mathbb{H}^2}(t,x,y) = \frac{\sqrt{2}e^{-t/8}}{(2\pi t)^{3/2}}\int_{d(x,y)}^{\infty}\frac{re^{-r^2/2t}dr}{\sqrt{\cosh r - \cosh d(x,y)}},$$

$$p_{\mathbb{H}^3}(t,x,y) = \frac{e^{-d(x,y)^2/2t}}{(2\pi t)^{3/2}}\frac{d(x,y)}{\sinh d(x,y)}e^{-t}.$$

For a more detailed discussion on these formulas, see Chavel [**8**].

We can give a precise description of the short-time behavior of the first exit time of a geodesic ball. Again let i_K be the minimum of the injectivity radii i_x on K.

Proposition 5.1.4. *Let K be a compact subset of M and $r < i_K$, the injectivity radius of K. Then there is a strictly positive smooth function $c_r(x)$ on K such that as $t \downarrow 0$,*

$$\mathbb{P}_x\{\tau_r \le t\} \sim \frac{c_r(x)}{t^{(d-2)/2}}e^{-r^2/2t},$$

uniformly on K.

Proof. Choose $r_1 \in (r, i_K)$ and let $A = \{y \in M : r \le d(y,x) \le r_1\}$. By the Markov property at τ_r we have

$$\mathbb{P}_x\{X_t \in A\} = \mathbb{E}_x\{S(X_\tau, t-\tau_r); \tau_r \le t\},$$

where

$$S(z,s) = \mathbb{P}_z\{X_s \in A\} = \int_A p_M(s,z,y)\,dy.$$

From THEOREM 5.1.1 we have for $z \in \partial A$, as $s \downarrow 0$,

$$S(z,s) = \frac{1+O(s)}{(2\pi s)^{d/2}}\int_A e^{-d(z,y)^2/2s}dy \to \frac{1}{2}.$$

The last limit can be proved by a local computation. Hence

$$\mathbb{P}_x\{\tau_r < t\} \sim 2\mathbb{P}_x\{X_t \in A\} = 2\int_A p_M(t,x,y)\,dy,$$

and, by THEOREM 5.1.1 again,

$$\mathbb{P}_x\{\tau_r < t\} = \frac{2+O(t)}{(2\pi t)^{d/2}}\int_A e^{-d(x,y)^2/2t}dy.$$

We use polar coordinates $y = (u, \theta)$ centered at x to calculate the last integral. The volume elements have the form

$$dy = f(u, \theta; x)\, du d\theta,$$

where $d\theta$ is the volume element on $\mathbb{S}^{d-1}$ and the coefficient $f(u, \theta; x)$ depends smoothly on x. We have

$$\begin{aligned}\int_A e^{-d(x,y)^2/2t} dy &= \int_r^{r_1} e^{-u^2/2t} \left(\int_{\mathbb{S}^{d-1}} f(u, \theta; x) d\theta \right) du \\ &\sim \left(\int_{\mathbb{S}^{d-1}} f(r, \theta; x)\, d\theta \right) \int_r^{\infty} e^{-u^2/2t} du \\ &\sim \left(\int_{\mathbb{S}^{d-1}} f(r, \theta; x)\, d\theta \right) \frac{t}{r}\, e^{-r^2/2t}.\end{aligned}$$

The desired result follows immediately. □

In the case $M = \mathbb{R}^d$, the heat kernel is

$$p_{\mathbb{R}^d}(t, x, y) = \left(\frac{1}{2\pi t} \right)^{d/2} e^{-|x-y|^2/2t},$$

and the volume element is $dy = r^{d-1} dr d\theta$. Hence we have explicitly

$$\mathbb{P}_x \{\tau_r \le t\} \sim \frac{2|\mathbb{S}^{d-1}|}{\pi} \left(\frac{r^2}{2\pi t} \right)^{(d-2)/2} e^{-r^2/2t}.$$

5.2. Varadhan's asymptotic relation

The local asymptotic relation (5.1.1) implies that if x and y are not on each other's cutlocus, then

$$\lim_{t \downarrow 0} t \ln p_M(t, x, y) = -\frac{1}{2}\, d(x, y)^2.$$

We will show in this section that this asymptotic relation holds for any pair of points on a complete Riemannian manifold.

Theorem 5.2.1. *Let M be a complete Riemannian manifold and $p_M(t, x, y)$ the minimal heat kernel on M. Then, uniformly on every compact subset of $M \times M$, we have*

$$\lim_{t \downarrow 0} t \ln p_M(t, x, y) = -\frac{1}{2}\, d(x, y)^2.$$

We will devote the rest of this section to the proof of this result. It will be divided into PROPOSITION 5.2.4 for the upper bound and PROPOSITION 5.2.5 for the lower bound.

Let K be a compact subset of M and i_K the injectivity radius of K. The proof of the upper bound is based on the following fact, which follows directly from PROPOSITION 5.1.4: Uniformly for $x \in K$ and $r \le i_K/2$,

$$\lim_{t \to 0} t \ln \mathbb{P}_x \{\tau_r < t\} = -\frac{1}{2} r^2. \tag{5.2.1}$$

Lemma 5.2.2. *Le τ be a nonnegative random variable such that*

$$\mathbb{P}\{\tau < t\} \le e^{-a^2/2t}$$

for some positive constant a and all $t \le t_0$. Let $b > 0$. Then for any $\eta \in (0,1)$, there exists a positive t_1, depending on t_0, b, and η, such that for all $t \le t_1$,

$$\mathbb{E}\left\{e^{-b^2/2(t-\tau)}; \tau < t\right\} \le e^{-(1-\eta)(a+b)^2/2t}.$$

Proof. Integrating by parts, we have

$$\begin{aligned}
&\mathbb{E}\left\{e^{-b^2/2(t-\tau)}; \tau < t\right\} \\
&\quad = \int_0^t e^{-b^2/2(t-s)} d\,\mathbb{P}\{\tau < s\} \\
&\quad = \frac{b^2}{2} \int_0^t \frac{e^{-b^2/2(t-s)}}{(t-s)^2} \mathbb{P}\{\tau < s\}\, ds \\
&\quad \le \frac{C(\eta)}{b^2} \int_0^t \exp\left\{-\frac{1-\eta}{2}\left(\frac{a^2}{s} + \frac{b^2}{t-s}\right)\right\} ds.
\end{aligned}$$

Here $C(\eta)$ is a constant depending only on η. Since

$$\frac{a^2}{s} + \frac{b^2}{t-s} \ge \frac{(a+b)^2}{t},$$

the integrand in the last integral is bounded by $te^{-(1-\eta)(a+b)^2/2t}$, hence the lemma. □

Let $r > 0$, and let σ_n be the succesive times when Brownian motion moves by a distance r:

$$\begin{aligned}
\sigma_0 &= 0, \\
\sigma_1 &= \tau_r = \inf\{t > 0 : d(X_0, X_t) = r\}, \\
\sigma_n &= \inf\left\{t > \tau_{n-1} : d(X_{\tau_{n-1}}, X_t) = r\right\}.
\end{aligned}$$

We consider the probability

$$I_n(t, x) = \mathbb{P}_x\{\sigma_n < t\}.$$

Lemma 5.2.3. *For any fixed* $\eta \in (0,1)$, $R > 0$, *and a compact subset* K *of* M, *there exist positive* $t_0 = t_0(\eta, R, K)$ *and* r_0 *such that for all* $x \in K$, $r \le r_0$, $n \le R/r$, *and* $t \le t_0$,

$$\mathbb{P}_x \{\sigma_n < t\} \le e^{-(1-\eta)(nr)^2/2t}.$$

Proof. We argue by induction. Since M is complete and K is relatively compact, the set

$$K_R = \{z \in M : d(z, K) \le R\}$$

is also relatively compact. Let $r_0 = i_{K_R}/2$. The case $n = 1$ is just (5.2.1). For $n > 1$, by the Markov property,

$$I_n(t,x) = \mathbb{E}_x \{I_{n-1}(t - \tau_r, X_{\tau_r}); \tau_r < t\}.$$

The induction step follows by using (5.2.1) and LEMMA 5.2.2. □

We can now prove the upper bound in THEOREM 5.2.1.

Proposition 5.2.4. *Let* $K \subseteq M$ *be compact. We have, uniformly for* $(x,y) \in K \times K$,

$$\limsup_{t \to \infty} t \ln p_M(t,x,y) \le -\frac{1}{2} d(x,y)^2.$$

Proof. Take r_0 as in LEMMA 5.2.3 and $r < r_0$, and let $n = [d(x,y)/r]$. Let σ_n be defined as before. Since it takes at least n steps of distance r to go from x to y, we have

$$p_M(t,x,y) = \mathbb{E}_x \left\{p_M(t - \sigma_{n-1}, X_{\sigma_{n-1}}, y); \sigma_{n-1} < t\right\}.$$

Because $d(X_{\sigma_{n-1}}, y) \ge r$, we have $p_M(t - \sigma_{n-1}, X_{\sigma_{n-1}}, y) \le C$ by THEOREM 4.5.1, with a constant C depending on K and r. Hence

$$p_M(t,x,y) \le C\, \mathbb{P}_x \{\sigma_{n-1} < t\}.$$

Using LEMMA 5.2.3, we have

$$\limsup_{t \downarrow 0} t \ln p_M(t,x,y) \le -\frac{1}{2} (n-1)^2 r^2 \overset{r \downarrow 0}{\longrightarrow} -\frac{1}{2} d(x,y)^2.$$

This completes the proof. □

We now prove the lower bound in THEOREM 5.2.1.

Proposition 5.2.5. *Let* $K \subseteq M$ *be compact. We have, uniformly on* $(x,y) \in K \times K$,

$$\liminf_{t \to 0} t \ln p_M(t,x,y) \ge -\frac{1}{2} d(x,y)^2$$

Proof. Let $r > 0$, and let $C : [0, l] \to M$ be a smooth curve of unit speed joining x and y with length $l \le d(x, y) + r$. Let D be a relatively compact open set containing the curve C. For a positive integer n, define n geodesic balls of radius r along the curve C as follows:

$$B_i = B(z_i; r), \quad z_i = C(il/n).$$

We choose r small enough so that all geodesic balls are contained in D. Using the Chapman-Kolmogorov equation for the heat kernel, we have

$$\begin{aligned}(5.2.2) \qquad & p_M(t, x, y) \\ & = \int_{M \times \cdots \times M} p_M\left(\frac{t}{n}, x, x_1\right) \cdots p_M\left(\frac{t}{n}, x_{n-1}, y\right) dx_1 \cdots dx_{n-1} \\ & \ge \int_{B_1 \times \cdots \times B_{n-1}} p_M\left(\frac{t}{n}, x, x_1\right) \cdots p_M\left(\frac{t}{n}, x_{n-1}, y\right) dx_1 \cdots dx_{n-1}.\end{aligned}$$

If $x_{i-1} \in B_{i-1}$ and $x_i \in B_i$, then

$$d(x_{i-1}, x_i) \le 3r + \frac{d(x, y)}{n}.$$

Hence by fixing a sufficiently large n, for all sufficiently small r we can apply the short-time asymptotics for the heat kernel (Theorem 5.1.1) to obtain the lower bounds

$$\begin{aligned} p_M\left(\frac{t}{n}, x_{i-1}, x_i\right) & \ge C\left(\frac{n}{t}\right)^{d/2} e^{-nd(x_{i-1}, x_i)^2/2t} \\ & \ge C\left(\frac{n}{t}\right)^{d/2} e^{-(d(x,y)+3rn)^2/2nt},\end{aligned}$$

and

$$p_M\left(\frac{t}{n}, x, x_1\right) \cdots p_M\left(\frac{t}{n}, x_{n-1}, y\right) \ge C^n \left(\frac{t}{n}\right)^{nd/2} e^{-(d(x,y)+3rn)^2/2t}.$$

Using this in (5.2.2), we have

$$\liminf_{t \downarrow 0} t \ln p_M(t, x, y) \ge -\frac{1}{2}\left(d(x, y) + 3rn\right)^2 \xrightarrow{r \downarrow 0} -\frac{1}{2} d(x, y)^2.$$

□

With Propositions 5.2.4 and 5.2.5 we have completed the proof of Theorem 5.2.1.

As an application of Varadhan's asymptotic relation for logarithmic heat kernel, we prove a similar asymptotic result for the exit time from a smooth open set.

Theorem 5.2.6. *Let D be a connected open D with smooth boundary Then for any $x \in D$*

$$\lim_{t \downarrow 0} t \ln \mathbb{P}_x \{\tau_D < t\} = -\frac{1}{2} d(x, \partial D)^2.$$

Proof. Let us prove the lower bound first. Let $y \in \partial D$ be such that $d(x, y) = d(x, \partial D)$, and $B = B(y; r)$ a small geodesic ball centered at y. Take a point $z \in B \cap (M \backslash D)$ but not on the boundary $B \cap \partial D$. Then Brownian motion which lands at z at time t must first pass through ∂D; hence by the Markov property at τ_D,

$$p_M(t, x, z) = \mathbb{E}_x \{p_M(t - \tau_D, X_{\tau_D}, z); \tau_D < t\}.$$

Since $d(z, \partial D) > 0$, by THEOREM 4.5.1 $p_M(t - \tau_D, X_{\tau_D}, z) \le C$. Hence by THEOREM 5.2.1 we have

$$\liminf_{t \downarrow 0} t \ln \mathbb{P}_x \{\tau_D \le t\} \ge \liminf_{t \downarrow 0} t \ln p_M(t, x, z) \ge -\frac{1}{2} d(x, z)^2.$$

By the choice of the point z, we have $d(x, z) \le d(x, \partial D) + r$. Letting $r \downarrow 0$, we obtain

$$\liminf_{t \downarrow 0} t \ln \mathbb{P}_x \{\tau_D < t\} \ge -\frac{1}{2} d(x, \partial D)^2.$$

For the upper bound, we fix a small $r > 0$ and let $n = [d(x, \partial D)/r]$. The ball $K = B(x; (n-1)r) \subseteq D$ is relatively compact and

$$\mathbb{P}_x \{\tau_D < t\} \le \mathbb{P}_x \{\sigma_n < t\}.$$

By LEMMA 5.2.3 we have

$$\limsup_{t \downarrow 0} t \ln \mathbb{P}_x \{\tau_D < t\} \le -\frac{1}{2} (nr)^2 \xrightarrow{r \downarrow 0} -\frac{1}{2} d(x, \partial D)^2.$$

□

5.3. Short-time asymptotics: distant points

THEOREM 5.1.1 describes precisely the short-time behavior of $p_M(t, x, y)$ when x and y are not on each other's cutlocus. However, if $x \in C_y$, the problem becomes rather complicated and the asymptotic behavior of the heat kernel depends very much on the geometry of the space

$$\Gamma_{x,y} = \text{distance-minimizing geodesics joining } x \text{ and } y.$$

In this section we describe a general method for studying the asymptotics of $p_M(t, x, y)$ for distant points, and illustrate this method by computing the leading term of the heat kernel of a sphere at antipodal points.

For the sake of simplicity, we assume that M is compact. In this case $\Gamma_{x,y}$ is a compact subset in the space of paths from x to y. We are interested in its middle section

$$\Gamma^{1/2}_{x,y} = \left\{ C\left(\frac{d(x,y)}{2}\right) : C \in \Gamma_{x,y} \right\}.$$

It is a closed set on M, and there is an obvious one-to-one correspondence between $\Gamma^{1/2}_{x,y}$ and $\Gamma_{x,y}$ itself. For example, if x and y are antipodal points on the d-sphere $\mathbb{S}^d$, then $\Gamma^{1/2}_{x,y}$ is the great sphere $\mathbb{S}^{d-1}$ perpendicular to the minimal geodesics joining x and y. This is a rather ideal situation because $\Gamma^{1/2}_{x,y}$ has the structure of a smooth Riemannian manifold, namely that of $\mathbb{S}^{d-1}$.

It is clear that $\Gamma^{1/2}_{x,y}$ is disjoint from the cutloci C_x and C_y of x and y, hence if ϵ is sufficiently small we have $d(O_\epsilon, C_x \cup C_y) > 0$, where

$$O_\epsilon(x,y) = \left\{ z \in M : d\left(z, \Gamma^{1/2}_{x,y}\right) < \epsilon \right\}$$

is the ϵ-neighborhood of $\Gamma^{1/2}_{x,y}$.

Lemma 5.3.1. *Suppose that M is a compact Riemannian manifold. Let $O_\epsilon(x,y)$ be the ϵ-neighborhood of $\Gamma^{1/2}_{x,y}$ and choose ϵ sufficiently small so that*

$$\inf_{x,y\in M} d(O_\epsilon(x,y), C_x \cup C_y) > 0.$$

Then there exist positive λ and C such that for all $(t,x,y) \in (0,1) \times M \times M$,

$$p_M(t,x,y) = \{1 + e(t,x,y)\} \int_{O_\epsilon} p_M\left(\frac{t}{2}, x, z\right) p_M\left(\frac{t}{2} z, y\right) dz, \tag{5.3.1}$$

with $|e(t,x,y)| \le Ce^{-\lambda/t}$.

Proof. By the Chapman-Kolmogorov equation,

$$p_M(t,x,y) = \left\{ \int_{O_\epsilon} + \int_{M\backslash O_\epsilon} \right\} p_M\left(\frac{t}{2}, x, z\right) p_M\left(\frac{t}{2}, z, y\right) dz.$$

By THEOREM 5.2.1 we have, uniformly in $z \in M\backslash O_\epsilon$,

$$\lim_{t\downarrow 0} t \ln \left[p_M\left(\frac{t}{2}, x, z\right) p_M\left(\frac{t}{2}, z, y\right) \right] = -\left[d(x,z)^2 + d(z,y)^2 \right].$$

Now the function

$$z \mapsto d(x,z) + d(z,y) - d(x,y)$$

has a strictly positive minimum on $M\backslash O_\epsilon$. Hence there exists an $\epsilon_1 > 0$ such that

$$d(x,z)^2 + d(z,y)^2 \ge \frac{1}{2}\left[d(x,z) + d(z,y)\right]^2 \ge \frac{1}{2}\left[d(x,y) + \epsilon_1\right]^2.$$

Decreasing ϵ_1 if necessary, we see that there is a $\lambda > 0$ such that for all sufficiently small t,

$$\int_{M\setminus O_\epsilon} p_M\left(\frac{t}{2}, x, z\right) p_M\left(\frac{t}{2}, z, y\right) dz \\ \leq e^{-[d(x,y)+\epsilon]^2/2t} \leq p_M(t, x, y)\, e^{-\lambda/t}.$$

Here in the second step we have used

$$\lim_{t\to 0} t \ln p_M(t, x, y) = -\frac{1}{2}\, d(x, y)^2.$$

The result follows immediately. □

Because $d(O_\epsilon, C_x \cup C_y) > 0$, we can use the local expansion in THEOREM 5.1.1 for the $p_M(t/2, x, z)$ and $p_M(t/2, y, z)$ in (5.3.1), which yields the following formula for computing the leading term of $p_M(t, x, y)$ as $t \downarrow 0$.

Theorem 5.3.2. *Suppose that M is a compact Riemannian manifold. Then there exists a constant C such that for all $(x, y, t) \in M \times M \times (0, 1)$,*

$$p_M(t, x, y) = \{1 + f(t, x, y)\} \times \\ \frac{e^{-d(x,y)^2/2t}}{(\pi t)^d} \int_{O_\epsilon} H_0(x, z) H_0(z, y) e^{-E_{x,y}(z)/2t} dz, \tag{5.3.2}$$

where $|f(t, x, y)| \leq Ct$ and

$$E_{x,y}(z) = 2d(x, z)^2 + 2d(z, y)^2 - d(x, y)^2.$$

Proof. This follows immediately from LEMMA 5.3.1 and THEOREM 5.1.1. □

It is clear that the above theorem is effective only when we can calculate explicitly the integral in (5.3.2). Let us illustrate by a typical example.

Example 5.3.3. We return to the case of the d-sphere (EXAMPLE 5.1.2) and compute $p_{\mathbb{S}^d}(t, N, S)$ when N and S are a pair of antipodal points. We take N as the origin and $r = d(N, z)$. The middle section $\Gamma_{N,S}^{1/2}$ is the great sphere $\mathbb{S}^{d-1}$ determined by the equation $r = \pi/2$. We have $d(z, S) = \pi - r$ and

$$E_{N,S}(z) = 2r^2 + 2(\pi - r)^2 - \pi^2 = 4\left(r - \frac{\pi}{2}\right)^2. \tag{5.3.3}$$

We have shown that in this case,

$$H_0(x, y) = \left[\frac{d(x, y)}{\sin d(x, y)}\right]^{(d-1)/2}, \quad d(x, y) < \pi.$$

The volume element on the sphere is $\sin^{d-1} r dr d\theta$. By THEOREM 5.3.2 we have

$$p_{\mathbb{S}^d}(t, N, S) \sim \frac{e^{-\pi^2/2t}}{(\pi t)^d} \int_{\mathbb{S}^{d-1}} d\theta \int_{|r-\pi/2|<\epsilon} e^{-2(r-\pi/2)^2/t} h(r)\, dr,$$

where

$$h(r) = \left[\frac{r}{\sin r} \frac{\pi - r}{\sin(\pi - r)} \right]^{(d-1)/2} \sin^{d-1} r.$$

By the Laplace approximation (see Copson [**12**]),

$$\int_{|r-\pi/2|<\epsilon} e^{-2(r-\pi/2)^2/t} h(r)\, dr \sim \sqrt{\frac{\pi t}{2}}\, h\left(\frac{\pi}{2}\right).$$

It follows that

$$p_{\mathbb{S}^{d-1}}(t, N, S) \sim \frac{|\mathbb{S}^{d-1}|}{\sqrt{2\pi}\, 2^{d-1}} \frac{e^{-\pi^2/2t}}{t^{(2d-1)/2}}. \qquad \square$$

From THEOREM 5.3.2 and the above example we see that the aysmptotic behavior of the heat kernel depends very much on the function $E_{x,y}(z)$ near the set $\Gamma^{1/2}_{x,y}$. Even without knowing precise geometric structure of this set, we can still draw some general conclusions from the asymptotic formula in THEOREM 5.3.2. We observe that, because $\Gamma^{1/2}_{x,y}$ does not intersect $C_x \cup C_y$, the function $E_{x,y}(z)$ is smooth in a neighborhood of $\Gamma^{1/2}_{x,y}$. It is everywhere nonnegative:

$$E_{x,y}(z) \geq [d(x,z) + d(z,y)]^2 - d(x,y)^2 \geq 0,$$

and vanishes on $\Gamma^{1/2}_{x,y}$. Let us now investigate $E_{x,y}(z)$ near a point $z_0 \in \Gamma^{1/2}_{x,y}$. Since $E_{x,y}(z_0) = 0$, it attains a minimum at z_0. Hence in any local chart $z = \left\{ z^i \right\}$ centered at z_0, there is a constant C such that

$$E_{x,y}(z) \leq C \sum_{i=1}^{d} |z^i|^2$$

in a neighborhood of z_0.

We now find a lower bound for $E_{x,y}(z)$. For a point z sufficiently close to z_0, we take the first coordinate to be

$$z^1 = \frac{d(x,y)}{2} - d(x,z).$$

Applying the triangle inequality to the geodesic triangle xzy, we have

$$d(z,y) \geq d(x,y) - d(x,z) = \frac{d(x,y)}{2} + z^1.$$

It follows that

$$E_{x,y}(z) \geq 2\left|\frac{d(x,y)}{2} - z^1\right|^2 + 2\left|\frac{d(x,y)}{2} + z^1\right|^2 - d(x,y)^2 = 4|z^1|^2.$$

This lower bound is attained when x and y are a pair of antipodal points on a sphere (see (5.3.3)).

To summarize, we have shown that for every $z_0 \in \Gamma_{x,y}^{1/2}$ there is a local chart $z = \{z^i\}$ centered at z_0 and a constant C such that

$$4\left|z^1\right| \leq E_{x,y}(z) \leq C\sum_{i=1}^{d}\left|z^i\right|^2 \tag{5.3.4}$$

in a neighborhood of z_0. Now we are in position to prove the following general bounds for the heat kernel on a compact Riemannian manifold.

Theorem 5.3.4. *Let M be a compact Riemannian manifold. There exist positive constants C_1 and C_2 such that for all $(t, x, y) \in (0, 1) \times M \times M$,*

$$\frac{C_1}{t^{d/2}}\, e^{-d(x,y)^2/2t} \leq p_M(t, x, y) \leq \frac{C_2}{t^{(2d-1)/2}}\, e^{-d(x,y)^2/2t}.$$

Proof. Pick a point $z_0 \in \Gamma_{x,y}^{1/2}$ and a local chart on a small ball $B = B(z_0; \epsilon)$ centered at z_0 such that the upper bound in (5.3.4) holds. Both $H_0(x, z)$ and $H_0(z, y)$ are continuous and strictly positive on B. Hence by THEOREM 5.3.2 and the upper bound in (5.3.4) there is a constant C_3 such that

$$p_M(t, x, y) \geq \frac{C_3}{t^d} e^{-d(x,y)^2/2t} \int_B e^{-C|z|^2/2t} dz.$$

Using local coordinates, it is easy to show that

$$\int_B e^{-C|z|^2/2t} dz \sim \text{const.} \times t^{d/2}.$$

It follows that

$$p_M(t, x, y) \geq \frac{C_1}{t^{d/2}} e^{-d(x,y)^2/2t}.$$

The last inequality follows from analyzing the integral in the middle in local coordinates.

We use the lower bound in (5.3.4) to prove the upper bound. Since $\Gamma_{x,y}^{1/2}$ is compact, there are a finite number of neighborhoods $\{O_l, 1 \leq l \leq N\}$ covering $\Gamma_{x,y}^{1/2}$ such that each one of them has a local chart on which the lower bound in (5.3.4) holds. By THEOREM 5.3.2 there is a constant C_4 such that

$$p_M(t, x, y) \leq \frac{C_4}{t^d}\, e^{-d(x,y)^2/2t} \sum_{l=1}^{N} \int_{O_i} e^{-2|z^1|^2/t}\, dz.$$

Again using local coordinates we can prove that

$$\int_{O_i} e^{-2|z^1|^2/t}\, dz \sim \text{const.} \times t^{1/2}.$$

Hence,

$$p_M(t, x, y) \le \frac{C_2}{t^{(2d-1)/2}}\, e^{-d(x,y)^2/2t}.$$

The proof is completed. □

Corollary 5.3.5. *Suppose that M is a compact Riemannian manifold. Then there exist positive constants C_1 and C_2 such that for all $(t, x, y) \in (0,1) \times M \times M$,*

$$\frac{C_1}{t^{d/2}} e^{-d(x,y)^2/2t} \le p_M(t, x, y) \le \frac{C_2}{t^{d/2}}. \tag{5.3.5}$$

Proof. The lower bound is the same as in the theorem. If $d(x, y) \le i_M/2$ (the injectivity radius), the upper bound follows from the local asymptotic expansion of the heat kernel in in THEOREM 5.1.1; if $d(x,y) \ge i_M/2$, it follows from the upper bound in the above theorem and the fact that $\sup_{t \ge 0} t^{-(d-1)/2} e^{-i_M^2/8t} < \infty$. □

5.4. Brownian bridge

In this section we assume that M is a compact Riemannian manifold. A Brownian bridge from x to y in time T is obtained from Brownian motion starting from x by conditioning on those paths that are at y at time T. As the transition density function of Brownian motion motion, the heat kernel measures the probability that Brownian motion starting from x will be at y at time t; therefore it is not surprising that Brownian bridge should play a role in the study of the heat kernel.

Consider the bridge space

$$L_{x,y;T}(M) = \{\omega \in W(M) : \omega_0 = x,\ \omega_T = y\}.$$

The special case $L_{x;T}(M) = L_{x,x;T}(M)$ is the loop space based at x. The law of the Brownian bridge from x to y in time T is a probability measure $\mathbb{P}_{x,y;T}$ on $L_{x,y;T}(M)$ defined roughly by

$$\mathbb{P}_{x,y;T}(C) = \mathbb{P}_x\{C | X_T = y\}, \qquad C \in \mathscr{B}(W(M)).$$

We will call $\mathbb{P}_{x,y;T}$ the Wiener measure on $L_{x,y;T}(M)$. It is sometimes helpful to regard $\mathbb{P}_{x,y;T}$ as a measure on the path space $W(M)$ concentrated on the subset $L_{x,y;T}(M)$.

We first address the question of the existence of $\mathbb{P}_{x,y;T}$. Let us first proceed intuitively and calculate the finite-dimensional marginal distributions of $\mathbb{P}_{x,y;T}$. Suppose that $s < T$ and let F be a nonnegative function on $W(M)$

measurable with respect to $\mathscr{B}_s$, and f a nonnegative measurable function on M. Then, by the definition of conditional probabilities,

$$\mathbb{E}_x\{f(X_T)F(X)\} = \mathbb{E}_x\{f(X_T)\mathbb{E}_{x,X_T;T}F(X)\}.$$

For the left side, using the Markov property at time s and the fact that $p_M(t,x,y)$ is the transition density for X, we have

$$\begin{aligned}\mathbb{E}_x\{f(X_T)F(X)\}] &= \mathbb{E}_x\{F(X)\mathbb{E}_{X_s}f(X_{T-s})\}\\ &= \mathbb{E}_x\left[F(X)\int_M p_M(T-s,X_s,y)f(y)dy\right].\end{aligned}$$

Hence

$$\begin{aligned}&\int_M \mathbb{E}_x\{F(X)p_M(T-s,X_s,y)\}f(y)dy\\ &\qquad = \int_M \mathbb{E}_{x,y;T}\{F(X)\}p_M(T,x,y)f(y)dy.\end{aligned}$$

This being true for all nonnegative measurable f, we have, for all $F \in \mathscr{B}_s$,

$$\mathbb{E}_{x,y;T}\{F(X)\} = \frac{\mathbb{E}_x\{F(X)p_M(T-s,X_s,y)\}}{p_M(T,x,y)}, \quad 0 \le s < T. \tag{5.4.1}$$

The above formula shows that $\mathbb{P}_{x,y;T}$, as a measure on $W(M)$, is absolutely continuous with respect to $\mathbb{P}_x$ on $\mathscr{B}_s$ for any $s < T$, and the Radon-Nikodym derivative is given by

$$\left.\frac{d\,\mathbb{P}_{x,y;T}}{d\,\mathbb{P}_x}\right|_{\mathscr{B}_s} = \frac{p_M(T-s,X_s,y)}{p_M(T,x,y)} \stackrel{\text{def}}{=} e_s. \tag{5.4.2}$$

It follows that $\{e_s, 0 \le s < T\}$ is a positive local martingale under the probability $\mathbb{P}_x$.

If the probability measure $\mathbb{P}_{x,y;T}$ exists, then from (5.4.2) it is easy to see that under $\mathbb{P}_{x,y;T}$, the joint density function of $X_{s_1},\ldots,X_{s_l}$, $0 = s_0 < s_1 < \cdots < s_l < s_{l+1} = T$, is

$$p_M(T,x,y)^{-1}\prod_{i=0}^{l} p_M(s_{i+1}-s_i,x_i,x_{i+1}). \tag{5.4.3}$$

$[x_0 = x,\ x_{l+1} = y.]$

Theorem 5.4.1. *Let M be a compact Riemannian manifold. Then there is a probability measure $\mathbb{P}_{x,y;T}$ on $L_{x,y;T}$ whose finite-dimensional marginal distributions are given by (5.4.3).*

Proof. This can be proved in the usual manner using Kolmogorov's extension theorem and Kolmogorov's criterion for sample path continuity (see

Chung [**11**]). In order to apply Kolmogorov's criterion, it is enough to show that there are positive α and β such that

$$\mathbb{E}_{x,y;T}\, d(X_s, X_t)^\alpha \le C(t-s)^{1+\beta}, \qquad 0 \le s \le t \le T.$$

By (5.4.3) we have

$$\mathbb{E}_{x,y;T} d(X_s, X_t)^\alpha = \frac{1}{p(T,x,y)} \int_M p_M(s,x,x_1)\, dz_1$$
$$\times \int_M d(x_1,x_2)^\alpha p_M(t-s,x_1,x_2) p_M(T-t,x_2,y)\, dx_2.$$

Without loss of generality we may assume that $0 \le s \le t \le 2T/3$. Then the last factor $p_M(T-t,x_2,y)$ is bounded by a constant. Using the upper bound for the heat kernel (5.3.4), we have

$$\int_M d(x_1,x_2)^\alpha p_M(t-s,x_1,x_2)\, dx_2$$
$$\le \frac{C_1}{(t-s)^{(2d-1)/2}} \int_M d(x_1,x_2)^\alpha e^{-d(x_1,x_2)^2/2t} dx_2.$$

Using polar coordinates it is easy to show that

$$\int_M d(x_1,x_2)^\alpha e^{-d(x_1,x_2)^2/2t} dx_2 \le C_2(t-s)^{(\alpha+d)/2}.$$

It follows that there is a constant C such that

$$\mathbb{E}_{x,y;T}\, d(X_s,X_t)^\alpha \le C(t-s)^{(\alpha-d+1)/2}.$$

Therefore it is enough to take $\alpha > d+1$ and $\beta = (\alpha-d-1)/2$. □

We will need the following estimate.

Lemma 5.4.2. *Fix $(T,x,y) \in (0,\infty)\times M \times M$ and a positive N. There is a constant $C = C(x,y,T,N)$ such that*

$$\mathbb{E}_{x,y;T}\, d(X_t,x)^{2N} \le Ct^N$$

for all $t \in [0,T]$.

Proof. It is enough to show the inequality for $t \le T/2$. We have

$$\mathbb{E}_{x,y;T}\, d(X_t,x)^{2N} =$$
$$\frac{1}{p_M(T,x,y)} \int_M d(z,x)^{2N} p_M(t,x,z) p_M(T-t,z,y)\, dz.$$

The factor $p_M(T-t,z,y)$ is uniformly bounded because $t \le T/2$. Fix a positive r and consider the integral on $B(x;r)$ and $M\backslash B(x;r)$ separately. If r is sufficiently small, by THEOREM 5.1.1 we have

$$p_M(t,x,z) \le \frac{C_1}{t^{d/2}} e^{-d(x,z)^2/2t}.$$

On $M\backslash B(x;r)$, letting T_r be the first hitting time of $B(x;r)$ we have

$$\begin{aligned} p_M(t,x,z) &= \mathbb{E}_z \left\{ p_M\left(t - T_r, x, X_{T_r}\right); T_r < t \right\} \\ &\le \sup\left\{ p_M(s,x,z) : 0 \le s \le t,\ d(x,z) = r \right\}. \end{aligned}$$

Hence, by THEOREM 5.1.1 again we have

$$p_M(t,x,z) \le \frac{C_2}{t^{d/2}} e^{-r^2/2t}.$$

It follows that there is a constant C_3 such that

$$\mathbb{E}_{x,y;T}\, d(X_t,x)^{2N} \le \frac{C_3}{t^{d/2}} \int_{B(x;r)} d(z,x)^{2N} e^{-d(z,x)^2/2t} dz + \frac{C_3}{t^{d/2}} e^{-r^2/2t}.$$

The integral can be estimated easily in polar coordinates, and we have

$$\int_{B(x;r)} d(z,x)^{2N} e^{-d(z,x)^2/2t} dz \le C_4\, t^{N+d/2}.$$

This completes the proof. □

Brownian bridge has the following symmetry property under time reversal.

Proposition 5.4.3. *Let $\mathbb{P}_{x,y;T}$ be the law of the Brownian bridge from x to y in time T. Then under $\mathbb{P}_{x,y;T}$ the process $\{X_{T-s},\ 0 \le s \le T\}$ has the law $\mathbb{P}_{y,x;T}$, i.e., it is a Brownian bridge from y to x in time T.*

Proof. This follows from the finite-dimensional marginal distributions given above and the symmetry of the heat kernel. □

Next we derive a stochastic differential equation for a Brownian bridge, or more precisely, for the horizontal lift of a Brownian bridge. In the case $M = \mathbb{R}^d$, the equation is well known: under the probability measure $\mathbb{P}_{x,y;T}$, there is a d-dimensional Brownian motion b such that

$$X_t = x + b_t - \int_0^t \frac{X_s}{T-s} ds. \tag{5.4.4}$$

We first write the positive local martingale $\{e_s,\ 0 \le s < T\}$ (see (5.4.2)) in the form of an exponential martingale. Let us compute the stochastic differential of $\ln e_s$. Using the fact that $p_M(t,x,y)$ satisfies the heat equation in (t,x) for fixed y, we have, after an easy computation, an equation for $\ln p_M$:

$$\frac{\partial}{\partial t} \ln p_M = \frac{1}{2} \Delta \ln p_M + \frac{1}{2} |\nabla \ln p_M|^2.$$

This equation can be lifted to the orthonormal frame bundle $\mathscr{O}(M)$. Let $\widetilde{p}_M(s,u,y) = p_M(s,\pi u,y)$ be the lift of $p_M(s,x,y)$ to the orthonormal frame bundle $\mathscr{O}(M)$. The above equation for $\ln p_M(t,x,y)$ becomes

$$\frac{\partial}{\partial t}\ln\widetilde{p}_M = \frac{1}{2}\Delta_{\mathscr{O}(M)}\ln\widetilde{p}_M + \frac{1}{2}|\nabla^H \ln\widetilde{p}_M|^2, \tag{5.4.5}$$

where

$$\nabla^H \ln\widetilde{p}_M = \{H_1\ln\widetilde{p}_M, \ldots, H_d\ln\widetilde{p}_M\}$$

is the horizontal gradient of $\ln\widetilde{p}$ and $\Delta_{\mathscr{O}(M)}$ the horizontal Laplacian on $\mathscr{O}(M)$ (see SECTION 3.1). Let U be a horizontal lift of the coordinate process X on $W(M)$. We know that U is a horizontal Brownian motion under the probability $\mathbb{P}_x$; hence its anti-devlopment W is a d-dimensional euclidean Brownian motion and

$$dU_s = \sum_{i=1}^n H_i(U_s)\circ dW_s^i.$$

We now apply Itô's formula to

$$\ln e_s = \ln\widetilde{p}_M(T-s,U_s,y) - \ln p_M(T,x,y).$$

Using (5.4.5), it is easy to verify that

$$d\ln e_s = \langle V_s, dW_s\rangle - \frac{1}{2}|V_s|^2\,ds,$$

where

$$V_s \stackrel{\text{def}}{=} \nabla^H\ln\widetilde{p}_M(T-s,U_s,y).$$

Note that the gradient is always taken with respect to the first space variable.

Putting things together, we have obtained the following formula for the Radon-Nikodym derivative of Brownian bridge with respect to Brownian motion:

$$\left.\frac{d\,\mathbb{P}_{x,y;T}}{d\,\mathbb{P}_x}\right|_{\mathscr{B}_s} = \exp\left[\int_0^s \langle V_u, dW_u\rangle - \frac{1}{2}\int_0^s |V_u|^2\,du\right].$$

Now, by Girsanov's theorem (see THEOREM 8.1.2), under the probability $\mathbb{P}_{x,y;T}$, the process

$$b_s = W_s - \int_0^s V_\tau d\tau, \quad 0\le s<T,$$

is a Brownian motion. We now summarize.

Theorem 5.4.4. *Let $\mathbb{P}_{x,y;T}$ be the law of the Brownian bridge from x to y of time length T. Then there is a Brownian motion $\{b_s, 0\le s<T\}$ such that the horizontal lift of the Brownian bridge X satisfies the stochastic differential equation*

$$dU_s = H_i(U_s)\circ\left\{db_s^i + H_i\ln\widetilde{p}_M(T-s,U_s,y)\,ds\right\}. \tag{5.4.6}$$

In other words, the anti-development of the Brownian bridge X is

$$W_s = b_s + \int_0^s U_\tau^{-1} \nabla \ln p_M(T-\tau, X_\tau, y)\, d\tau. \tag{5.4.7}$$

Remark 5.4.5. We have shown (basically by Girsanov's theorem) that under $\mathbb{P}_{x,y;T}$, the horizontal lift U of the coordinate pricess X is a semimartingale on $[0,T)$ and satisfies the equation (5.4.6). To show that U is a semimartingale on the closed time interval $[0,T]$ we need to show that

$$\int_0^T |\nabla \ln p_M(T-s, X_s, y)|\, ds < \infty$$

with $\mathbb{P}_{x,y;T}$-probability 1. This requires us to deal with the singularity of the logarithmic heat kernel $\nabla \ln p_M(t,x,y)$ near $t=0$. We will address this question in PROPOSITION 5.5.6 below. □

Using the explict formula for the heat kernel on $\mathbb{R}^d$, we can verify easily that (5.4.6) reduces to (5.4.4) in the euclidean case.

5.5. Derivatives of the logarithmic heat kernel

In order to deal with the singularity of the drift in (5.4.6), we need to estimate $\nabla \ln p(T,x,y)$ for T near 0. In this section we will prove universal bounds for the gradient $|\nabla \ln p_M(T,x,y)|$ and the Hessian $|\nabla^2 \ln p_M(T,x,y)|$ in the range $(0,1)\times M \times M$. A bound of this kind for the heat kernel $p_M(T,x,y)$ itself has already been proved in COROLLARY 5.3.5. The key idea in our proof is to apply Itô's formula to the processes $\nabla \ln p_M(T-t, X_t, y)$ and $\nabla^2 \ln p_M(T-t, X_t, y)$, where X is a Brownian bridge.

We introduce a few notational simplifications. For the rest of this section, we will work exclusively with the measure $\mathbb{P}_{x,y;T}$, so we simply write it as $\mathbb{P}$. The terminal point y is fixed throughout the discussion, so we usually drop it from our notations and work with the function

$$J(t,u) \stackrel{\text{def}}{=} \ln p_M(T-t, \pi u, y)$$

defined on the orthonormal frame bundle $\mathscr{O}(M)$. Note the reversal of time on the right side. Recall that the horizontal gradient is

$$\nabla^H J = \{H_1 J, \ldots, H_d J\}.$$

For a multi-index $I = i_1 \cdots i_l$ of length l, we introduce the notation

$$H_I J = H_{i_1} \cdots H_{i_l} J.$$

Let us first work with a general index I. Using the equation for the horizontal Brownian bridge U in THEOREM 5.4.4, we have

$$dH_I J(t, U_t) = \left\langle \nabla^H H_I J(t, U_t), db_t \right\rangle + \{\cdots\}\, dt, \tag{5.5.1}$$

where

$$\{\cdots\} = \frac{\partial}{\partial t} H_I J + \frac{1}{2}\Delta_{\mathscr{O}(M)} H_I J + \langle \nabla^H H_I J, \nabla^H J\rangle,$$

evaluated at (t, U_t). In terms of J, the heat equation (5.4.5) for $\ln p_M(s, z, y)$ becomes

$$\frac{\partial J}{\partial t} + \frac{1}{2}\Delta_{\mathscr{O}(M)} J = -\frac{1}{2}|\nabla^H J|^2.$$

Applying H_I to both sides of the equation and using the result in the last term of (5.5.1), we obtain

$$d\left\{H_I J(t, U_t)\right\} = \left\langle \nabla^H H_I J(t, U_t), db_t \right\rangle + E_I(t, U_t)\, ds, \tag{5.5.2}$$

where

$$E_I = \frac{1}{2}\left[\Delta_{\mathscr{O}(M)}, H_I\right] J + \langle \nabla^H H_I J, \nabla^H J\rangle - \frac{1}{2} H_I \langle \nabla^H J, \nabla^H J\rangle. \tag{5.5.3}$$

In order to avoid the singularity at $t = T$, we only use (5.5.2) up to time $t = T/2$. Integrating from 0 to $T/2$ and taking the expected value, we have

$$\mathbb{E}\, H_I J(T/2, U_{T/2}) - H_I J(0, u_0) = \mathbb{E}\int_0^{T/2} E_I(t, U_t)\, dt \tag{5.5.4}$$

Note that the second term on the right is nothing but the covariant derivative of $\ln p_M(T, x, y)$ wirtten in the orthonormal frame bundle $\mathscr{O}(M)$. The above relation is the starting point for proving the estimates for the derivatives of the logarithmic heat kernel. The necessary estimates for E_I with indices of lengths 0, 1, and 2 are set out in LEMMA 5.5.2 below. The case $E_\emptyset$ follows directly from the relevant definitions. For other cases we hope that the commutator in the first term will produce some lower order terms and that some cancellations will happen between the last two terms. Clearly we have to turn to differential geometry for computing commutators between horizontal vector fields. A reader who does not want to be interrupted by these differential geometric calculations should read the statement of LEMMA 5.5.1 and proceed directly to THEOREM 5.5.3.

On the orthonormal frame bundle, besides the d fundamental horizontal vector fields H_i, there are also $d(d-1)/2$ fundamental vertical vector fields defined as follows. Let $\mathfrak{o}(d)$ be the Lie algebra of the orthogonal group $O(d)$, i.e., the space of $(d \times d)$ antisymmetric matrices. Each element $A \in \mathfrak{o}(d)$ defines a vertical vector field V_A on $\mathscr{O}(M)$ by

$$V_A F(u) = dF(ue^{tA})/dt|_{t=0}.$$

The map $A \mapsto V_A(u)$ is an isomorphism from $\mathfrak{o}(d)$ to the vertical subspace $V_u\mathscr{O}(M)$. Let $A_{ij} \in \mathfrak{o}(d)$ be the matrix whose (i,j)th entry is $1/2$, (j,i)th entry is $-1/2$, and all other entries are zero. Then $\{A_{ij}, 1 \le i < j \le d\}$ is a basis for $\mathfrak{o}(d)$. Let $V_{ij} = V_{A_{ij}}$, the vertical vector field corresponding to the antisymmetric matrix A_{ij}. Then

$$V_A = A^{ab} V_{ab}, \quad A = \left\{A^{ab}\right\} \in \mathfrak{o}(d).$$

The $d(d-1)/2$ vertical vector fields $\{V_{ij},\ 1 \le i < j \le d\}$ are called the fundamental vertical vector fields. It is clear that $\{H_i, V_{ij}\}$ is a basis for $T_u\mathscr{O}(M)$ everywhere on $\mathscr{O}(M)$. We need to calculate commutators of these vector fields.

The solder form θ on $\mathscr{O}(M)$ is defined by

$$\theta(X)(u) = u^{-1}\pi_* X, \quad X \in T_u\mathscr{O}(M).$$

The connection form ω is an $\mathfrak{o}(d)$-valued vertical 1-form on $\mathscr{O}(M)$ defined by the following properties:

(i) $\omega(X) = 0$ if X is horizontal;

(ii) $V_{\omega(X)} = X$ if X is vertical.

From these definitions, we have, for any $X \in T\mathscr{O}(M)$,

$$X = H_{\theta(X)} + V_{\omega(X)}. \tag{5.5.5}$$

Recall the usual definition of the Riemannian curvature tensor:

$$R(X,Y)Z = \nabla_X\nabla_Y Z - \nabla_Y\nabla_X Z - \nabla_{[X,Y]}Z. \tag{5.5.6}$$

For $X, Y \in T_xM$, the linear map $R(X,Y) : T_xM \to T_xM$ is antisymmetric because the Levi-Civita connection is compatible with the metric. The curvature form Ω is an $\mathfrak{o}(d)$-valued horizontal 2-form on $\mathscr{O}(M)$ defined by

$$\Omega(X,Y)(u) = u^{-1}R(\pi_*X, \pi_*Y)\,u, \quad X, Y \in T_u\mathscr{O}(M).$$

In terms of these differential forms on $\mathscr{O}(M)$, the definitions of the torsion tensor (which is assumed to be zero here) and the curvature tensor become the first and the second structure equations:

$$\begin{aligned} d\theta &= -\,\theta \wedge \omega, \\ d\omega &= -\,\omega \wedge \omega + \Omega. \end{aligned}$$

Lemma 5.5.1. *The following commutation relations hold:*

$$\begin{cases} [H_i, H_j] = -V_{\Omega(H_i,H_j)} = -\Omega_{ij}^{ab} V_{ab}, \\ [H_i, V_{jk}] = -A_{jk}^{il} H_l. \end{cases}$$

Proof. Recall that the exterior differential of a 1-form η is defined by

$$d\eta(X,Y) = X\eta(Y) - Y\eta(X) - \eta([X,Y]),$$

where $[X,Y]$ is the bracket of the two vector fields. We rewrite the above identity as

$$\eta([X,Y]) = X\eta(Y) - Y\eta(X) - d\eta(X,Y).$$

Let $\eta = \theta$, $X = H_i$, $Y = H_j$, and use the first structure equation on the right side. We find that $\theta([H_i, H_j]) = 0$, i.e., the commutator is a vertical vector field. Using the second structure equation, we can compute $\omega([H_i, H_j])$ in the same way and obtain

$$\omega([H_i, H_j]) = -\Omega(H_i, H_j).$$

It follows from (5.5.5) that $[H_i, H_j] = -\Omega_{ij}^{ab} V_{ab}$. The second commutation relation can be proved similarly. □

We now come to the estimates of E_I.

Lemma 5.5.2. *Let E_I be defined in (5.5.3). There is a constant C such that*

$$\begin{cases} |E_\emptyset| = \dfrac{1}{2}|\nabla^H J|^2, \\ |E_i| \le C|\nabla^H J|, \\ |E_{ij}| \le C|\nabla^H J| + C|\nabla^H \nabla^H J| + C|\nabla^H \nabla^H J|^2. \end{cases} \tag{5.5.7}$$

Proof. The case $E_\emptyset$ follows directly from the definition.

For the case E_i, we first note that $V_{ab} J = 0$ because V_{ab} is a vertical and J is the lift of a function on M. Hence from the first relation in LEMMA 5.5.1 we have

$$[H_i, H_j]\, J = 0.$$

This implies that the last two terms in E_i cancel. For the first term of E_i, a simple calculation using LEMMA 5.5.1 shows that

$$[H_j^2, H_i] J = -\Omega_{ij}^{ab} V_{ab} H_i J.$$

From the definitions of V_{ab} and $H_i J$ it is easy to verify that

$$V_{ab} H_i J = A_{ab}^{ik} H_k J.$$

It follows that there is a constant such that $|E_i J| \le C|\nabla^H J|$.

The proof of the inequality for E_{ij} needs the second commutation relation in LEMMA (5.5.1), and can be carried out along the same line. This inequality is used only in estimating the second derivatives of $\ln p_M(T, x, y)$ in THEOREM 5.5.7, which is not needed for the rest of the book. For this reason we omit its proof here and leave it as an exercise to the reader. □

We are ready for a global estimate of $\nabla \ln p_M(T, x, y)$.

Theorem 5.5.3. *Let M be a compact Riemannian manifold. Then there is a constant C such that, for all $(T, x, y) \in (0,1) \times M \times M$,*

$$|\nabla \ln p_M(T, x, y)| \le C\left[\frac{d(x,y)}{T} + \frac{1}{\sqrt{T}}\right].$$

Proof. From (5.5.4) with $I = \emptyset$ and the first line in (5.5.7) we have

$$\mathbb{E}\int_0^{T/2} |\nabla^H J(t, U_t)|^2 dt = 2\mathbb{E}J\left(T/2, U_{T/2}\right) - 2J(0, u_0). \tag{5.5.8}$$

From (5.3.5) there is a constant C_1 such that the right side of (5.5.8) satisfies

$$2\mathbb{E}\left[\ln \frac{p\left(T/2, X_{T/2}, y\right)}{p_M(T, x, y)}\right] \leq C_1\left[\frac{d(x,y)^2}{T} + 1\right];$$

hence

$$\mathbb{E}\int_0^{T/2} |\nabla^H J(s, U_s)|^2 ds \leq C_1\left[\frac{d(x,y)^2}{T} + 1\right]. \tag{5.5.9}$$

Now taking $I = i$ in (5.5.4) and integarting from 0 to $T/2$, we have

$$\begin{aligned} T\nabla^H J(0, u_o) = {} & 2\mathbb{E}\int_0^{T/2} \nabla^H J(t, U_t)\, dt \\ & - 2\mathbb{E}\int_0^{T/2} \left(\frac{T}{2} - t\right) E(t, U_t)\, dt, \end{aligned}$$

where $E = (E_1, \ldots, E_d)$. From (5.5.7), there is a constant C_2 such that

$$|E(t, U_t)| \leq C_2 |\nabla^H J(t, U_t)|.$$

Hence,

$$T|\nabla^H J(0, u_o)| \leq C_3 \mathbb{E}\int_0^{T/2} |\nabla^H J(t, U_t)| ds.$$

Using the Cauchy-Schwarz inequality and (5.5.9), we obtain

$$T|\nabla^H J(0, u_o)| \leq C_4\sqrt{T}\left[\frac{d(x,y)^2}{T} + 1\right]^{1/2}.$$

Dividing by T, we obtain

$$|\nabla \ln p_M(T, x, y)| = |\nabla^H J(T, u_o)| \leq C\left[\frac{d(x,y)}{T} + \frac{1}{\sqrt{T}}\right].$$

□

Let us use our inequality for $\nabla \ln p_M(T, x, y)$ to prove a few useful facts.

Proposition 5.5.4. *There is a constant C such that for all $(T, x, y) \in (0,1) \times M \times M$,*

$$\mathbb{E}_{x,y;T}\, d(X_t, y)^2 \leq C d(x,y)^2 + C \min\{t, T - t\}, \qquad 0 \leq t \leq T.$$

Proof. Suppose first that $t \le T/2$. Let O be a neighborhood of y covered by a normal coordinate system $z = \{z^i\}$. Let f be a smooth function on M such that $f(z) = |z|^2$ for $z \in O$ and f is strictly positive outside O. It is clear that that there is a constant C_1 such that for all $z \in M$,

$$C_1^{-1} d(z,y)^2 \le f(z) \le C_1 d(z,y)^2, \quad |\nabla f(z)| \le C_1 \sqrt{f(z)}.$$

From THEOREM 5.5.3 we have

$$|\nabla \ln p_M(T-t,z,y)| \le C_2 \left[\frac{\sqrt{f(z)}}{T-t} + \frac{1}{\sqrt{T-t}} \right].$$

Hence for $t \le T/2$, there is a constant C_3 such that

$$|\langle \nabla f(z), \nabla \ln p_M(T-t,z,y)\rangle| \le C_3 \left[\frac{f(z)}{T} + 1 \right]. \tag{5.5.10}$$

From the stochastic differential equation for the Brownian bridge in THEOREM 5.4.4 we have

$$\begin{aligned} f(X_t) =& f(x) + \text{martingale} + \frac{1}{2}\int_0^t \Delta f(X_s)\, ds \\ &+ \int_0^t \langle \nabla f(X_s), \nabla \ln p_M(T-s, X_s, y)\rangle ds. \end{aligned}$$

The last term can be bounded by (5.5.10), and we obtain, for $0 \le t \le T/2$,

$$\mathbb{E} f(X_t) \le f(x) + C_4 t + \frac{C_4}{T}\int_0^t \mathbb{E} f(X_s)\, ds.$$

This implies by Gronwall's lemma that

$$\mathbb{E} f(X_t) \le C_5 \left\{ f(x) + t \right\}.$$

Note that C_5 is independent of T. The above inequality is equivalent to

$$\mathbb{E}\, d(X_t, y)^2 \le C_6 \left\{ d(x,y)^2 + t \right\}, \qquad t \le T/2.$$

This shows the assertion for $t \le T/2$. If $t \ge T/2$, using

$$d(X_t, y) \le d(X_t, x) + d(x, y)$$

and the fact that $\{X_{T-t}\, 0 \le t \le T\}$ under the probability $\mathbb{P}_{y,x;T}$ is the same Brownian bridge as $\{X_t,\, 0 \le t \le T\}$ under $\mathbb{P}_{x,y;T}$ we have

$$\begin{aligned} \mathbb{E}_{x,y;T}\, d(X_t, y)^2 &\le 2\, \mathbb{E}_{x,y;T} d(X_t, x)^2 + 2d(x,y)^2 \\ &= 2\, \mathbb{E}_{y,x;T} d(X_{T-t}, x)^2 + 2d(x,y)^2 \\ &\le C_7 \left\{ d(x,y)^2 + (T-t) \right\}. \end{aligned}$$

This completes the proof. □

Remark 5.5.5. The same line of thought can be used to show in general that

$$\mathbb{E}_{x,y;T}\, d(X_t,y)^{2N} \le C_N d(x,y)^2 + C_N \min\{t, T-t\}^N, \qquad 0 \le t \le T. \quad \square$$

We now address the question whether the horizontal lift of a Brownian bridge is a semimartingale.

Proposition 5.5.6. *Let X be the coordinate process on the path space $W(M)$ and U its horizontal lift. Then, nder the probability $\mathbb{P}_{x,y;T}$, both $\{U_s,\, 0 \le s \le T\}$ and $\{X_s,\, 0 \le s \le T\}$ are semimartingales.*

Proof. Clearly it is enough to show that U is a semimartingale. We have shown that U is a semimartingale on $[0,T)$ and satisfies the equation (5.4.6). In order to verify that U is a semimartingale on $[0,T]$, it is enough to show that

$$\mathbb{E}_{x,y;T} \int_0^T |\nabla \ln p_M(T-s, X_s, y)|\, ds < \infty. \tag{5.5.11}$$

Using our estimate for the gradient of the logarithmic heat kernel (THEOREM 5.5.3) we have

$$|\nabla \ln p_M(T-s, X_s, y)| \le C_1 \left[\frac{d(X_s,y)}{T-s} + \frac{1}{\sqrt{T-s}}\right].$$

From LEMMA 5.4.2 we have

$$\mathbb{E}_{x,y;T}\, d(X_s,y) \le C_2\sqrt{T-s}.$$

It follows that there is a constant C_3 such that

$$\mathbb{E}_{x,y;T}\, |\nabla \ln p_M(T-s, X_s, y)| \le \frac{C_3}{\sqrt{T-s}}.$$

Now it is clear that (5.5.11) holds, and the proof is completed. $\square$

We now proceed to the second derivatives of $\ln p_M(T,x,y)$. Although this estimate is not needed later, it has been proven useful in several applications not covered in this book.

Theorem 5.5.7. *Let M be a compact Riemannian manifold. There is a constant C such that, for all $(T,x,y) \in (0,1) \times M \times M$,*

$$\left|\nabla^2 \ln p\,(T,x,y)\right| \le C\left[\frac{d(x,y)}{T} + \frac{1}{\sqrt{T}}\right]^2. \tag{5.5.12}$$

Proof. The proof is similar to that of THEOREM 5.5.3. We first show that

$$\int_0^{T/2} |\nabla^H \nabla^H J(t,U_t)|^2 dt \le C_1 \left[\frac{d(x,y)}{T} + \frac{1}{\sqrt{T}}\right]^2. \tag{5.5.13}$$

Let $I = i$ and $t = T/2$ in (5.5.2). We have

$$\int_0^{T/2} \langle \nabla^H \nabla^H J(t, U_t), db_t \rangle$$
$$= \nabla^H J(T/2, U_{T/2}) - \nabla^H J(0, u_0) - \int_0^t E(s, U_s)\, ds.$$

Squaring it and taking the expectation, we have

$$\mathbb{E} \int_0^{T/2} |\nabla^H \nabla^H J(s, U_s)|^2 ds \le C_2 \{S_1 + S_2 + S_3\}, \tag{5.5.14}$$

where

$$S_1 = \mathbb{E}|\nabla^H J(T/2, U_{T/2})|^2,$$
$$S_2 = |\nabla^H J(0, u_o)|^2,$$
$$S_3 = \mathbb{E} \int_0^{T/2} |E(t, U_t)|^2 dt.$$

By THEOREM 5.5.3 (the estimate for the first derivatives of the logarithmic heat kernel) and LEMMA 5.5.4,

$$\begin{aligned} S_1 &\le C_3\, \mathbb{E} \left[\frac{d(X_{T/2}, y)}{T} + \frac{1}{\sqrt{T}} \right]^2 \\ &\le C_4 \left[\frac{\mathbb{E}\, d(X_{T/2}, y)^2}{T^2} + \frac{1}{T} \right] \\ &\le C_5 \left[\frac{d(x, y)}{T} + \frac{1}{\sqrt{T}} \right]^2. \end{aligned} \tag{5.5.15}$$

Also, by THEOREM 5.5.3,

$$S_2 \le C_6 \left[\frac{d(x, y)}{T} + \frac{1}{\sqrt{T}} \right]^2. \tag{5.5.16}$$

For S_3 we use the inequality $|E(t, U_t)| \le C_7 |\nabla^H J(t, U_t)|$ in (5.5.7) and (5.5.9) to obtain

$$S_3 \le C_8 \left[\frac{d(x, y)^2}{T} + 1 \right] \le C_9 \left[\frac{d(x, y)}{T} + \frac{1}{\sqrt{T}} \right]^2. \tag{5.5.17}$$

From (5.5.14) — (5.5.17) we obtain (5.5.13)

Now integrating (5.5.4) with $I = ij$ from 0 to $T/2$, we have

$$T \nabla^H \nabla^H J(0, u_o) = 2\mathbb{E} \int_0^{T/2} \nabla^H \nabla^H J(t, U_t)\, dt$$
$$+ 2\mathbb{E} \int_0^{T/2} \left(\frac{T}{2} - t \right) F(t, U_t)\, dt,$$

where $F = \{E_{ij}\}$. From the last line of (5.5.7) we have a bound for $|F(t, U_t)|$, and so

$$T|\nabla^H \nabla^H J(0, u_o)| \le C_3 \{K_1 + K_2 + TK_3\}, \tag{5.5.18}$$

where

$$K_1 = \mathbb{E} \int_0^{T/2} |\nabla^H J(t, U_t)| dt,$$
$$K_2 = \mathbb{E} \int_0^{T/2} |\nabla^H \nabla^H J(t, U_t)| dt,$$
$$K_3 = \mathbb{E} \int_0^{T/2} |\nabla^H \nabla^H J(t, U_t)|^2 dt.$$

Using (5.5.9), we have

$$\begin{aligned} K_1 &\le \sqrt{T} \left[\mathbb{E} \int_0^{T/2} |\overline{\nabla} J(t, U_t)|^2 dt \right]^{1/2} \\ &\le C_4 \sqrt{T} \left[\frac{d(x,y)^2}{T} + 1 \right]^{1/2} \\ &\le C_5 T \left[\frac{d(x,y)}{T} + \frac{1}{\sqrt{T}} \right]^2. \end{aligned}$$

Using (5.5.13), we have

$$\begin{aligned} K_2 &\le \sqrt{T} \left[\mathbb{E} \int_0^{T/2} |\nabla^H \nabla^H J(t, U_t)|^2 dt \right]^{1/2} \\ &\le C_7 \sqrt{T} \left[\frac{d(x,y)}{T} + \frac{1}{\sqrt{T}} \right] \\ &\le C_8 T \left[\frac{d(x,y)}{T^{3/2}} + \frac{1}{T} \right] \\ &\le C_9 T \left[\frac{d(x,y)}{T} + \frac{1}{\sqrt{T}} \right]^2. \end{aligned}$$

Here in the third step we have used the inequality

$$\frac{d(x,y)}{T^{3/2}} \le \frac{d(x,y)^2}{T^2} + \frac{1}{T}.$$

Also using (5.5.13), we have

$$K_3 \le C_{10} \left[\frac{d(x,y)}{T} + \frac{1}{\sqrt{T}} \right]^2.$$

Putting the upper bounds for K_1, K_2, and K_3 into (5.5.18), we obtain

$$|\nabla^2 \ln p_M(T,x,y)| = |\nabla^H \nabla^H J(T,u_o)| \leq C_{11} \left[\frac{d(x,y)}{T} + \frac{1}{\sqrt{T}} \right]^2 .$$

□

Remark 5.5.8. In general we can show that, for $(t,x,y) \in (0,1) \times M \times M$,

$$|\nabla^N \ln p_M(T,x,y)| \leq C_N \left[\frac{d(x,y)}{T} + \frac{1}{\sqrt{T}} \right]^N .$$

Chapter 6

Further Applications

In this chapter we study two geometric problems by probabilistic methods. The first one is the Dirichlet problem at infinity for a Cartan-Hadamard manifold M. By the Cartan-Hadamard theorem, it can be compactified by its sphere at infinity $\mathbb{S}_\infty(M)$. We reduce the solvability of the Dirichlet problem on $\overline{M} = M \cup \mathbb{S}_\infty(M)$ to the problem of angular convergence of Brownian motion on M. After explaining a general scheme for proving angular convergence in SECTION 6.1, we prove, in SECTIONS 6.2 and 6.3 respectively, the solvability of the Dirichlet problem at infinity under two typical upper bounds on the curvature:

(i) $K_M(x) \le -a^2$ for a constant $a > 0$;

(ii) $K_M(x) \le -\dfrac{C}{r(x)^2}$ for a constant $C > 2$.

In both cases we need to assume an appropriate lower bound on the growth of the Ricci curvature. If the manifold is radially symmetric, then the solvability of the Dirichlet problem at infinity can be reduced to a one-dimensional problem. This case is discussed in SECTION 6.4.

The second problem is the eigenvalue estimates for a compact Riemannian manifold with nonegative Ricci curvature by the method of coupling. It is somewhat surprising that this probabilistic method yields the sharp lower bounds for the first nonzero eigenvalue; namely, if $\mathrm{Ric}_M(x) \ge (d-1)K$ for some nonnegative constant K, then

$$\lambda_1(M) \ge \begin{cases} dK, & K > 0, \\ \dfrac{\pi^2}{d(M)^2}, & K = 0. \end{cases}$$

Here $d(M)$ is the diamter of M. Unlike geometric approaches to these estimates, the probabilistic proofs of these two cases are almost identical. The key idea, due to Chen and Wang [**10**], is to calculate the decay rate of the expected value of a suitable chosen function of the distance between the coupled Brownian motions. In SECTIONS 6.5 and 6.6 we will describe the Kendall-Cranston coupling and its relation with index forms. The two eigenvalue estimates are proved in SECTION 6.7.

6.1. Dirichlet problem at infinity

If D is a smooth domain in a Riemannian manifold and $f \in C(\partial D)$, then in LEMMA 4.4.3 we have shown that the solution to the Dirichlet boundary valued problem

$$\begin{cases} \Delta u = 0, & \text{on } D, \\ u = f, & \text{on } \partial D, \end{cases}$$

is given by $u(x) = \mathbb{E}_x f(X_{\tau_D})$. Here $\mathbb{E}_x$ is the expectation with respect to the Wiener measure (the law of Brownian motion) on $W(M)$ starting from x. The Dirichlet problem at infinity for a Cartan-Hadamard manifold has a similar representation formula. We will start with some basic facts about Cartan-Hadamard manifolds; see Jost [**50**] and Schoen and Yau [**64**] for more detailed discussion.

A Cartan-Hadamard manifold M is a complete, simply connected Riemannian manifold of nonpositive (sectional) curvature. For any fixed point $o \in M$, the exponential map $\exp : T_oM \to M$ is a diffeomorphism between T_oM and M, and we can introduce global polar coordinates on M. Two geodesic rays γ_1 and γ_2 on M are said to be equivalent if there is a constant C such that $d(\gamma_1(t), \gamma_2(t)) \le C$ for all $t \ge 0$. It can be shown that this is an equivalence relation on the set of geodesic rays. The set of equivalence classes is called the sphere at infinity $\mathbb{S}_\infty(M)$. If we fix a reference point $o \in M$, then $\mathbb{S}_\infty(M)$ can be identified with the unit sphere on the tangent space T_oM.

We now describe a topological structure on the union $\overline{M} = M \cup \mathbb{S}_\infty(M)$. Let $\theta_0 \in \mathbb{S}_\infty(M)$ be a point on the boundary. For any $o \in M$, θ_0 can be identified with a unit vector on the tangent space T_oM. Take a positive λ. We define a truncated cone

$$T(o, \theta_0, \lambda, R) = \{(r, \theta) \in M : \angle(\theta_0, \theta) < \lambda, \, r > R\}.$$

If we take $\{T(o, \theta_0, \lambda, R) : \lambda > 0, R > 0\}$ as a topological basis of $\overline{M}$ at $\theta_0 \in \mathbb{S}_\infty(M)$, then it can be shown that the resulting topology (called the cone topology) on $\overline{M}$ is independent of the reference point o, and with this topology $\overline{M}$ is a compactification of M, namely, $\overline{M}$ is compact and the natural inclusion $i : M \to \overline{M}$ maps M homeomorphically onto a dense open subset of $\overline{M}$. If (r, θ) are the polar coordinates based at o, then a squence of points $z_n = (r_n, \theta_n) \in M$ converges to a boundary point $\theta_0 \in \mathbb{S}_\infty(M)$ if and only if $r_n \to \infty$ and $\theta(z_n) \to \theta_0$ in $\mathbb{S}^{d-1}$. Thus topologically $\overline{M}$ is $\mathbb{R}^d$ compactified by its own boundary at infinity.

Given a continuous function f on $\mathbb{S}_\infty(M)$, the Dirichlet problem at infinity is to find a function $u \in C^\infty(M) \cap C(\overline{M})$ such that

$$\begin{cases} \Delta u = 0, & \text{on } M, \\ u = f, & \text{on } \mathbb{S}_\infty(M). \end{cases}$$

We say that the Dirichlet problem at infinity is solvable for M if for every $f \in C(\mathbb{S}_\infty(M))$ there is a unique solution u_f. This property of a Cartan-Hadamard manifold can be obtained under certain conditions on the curvature of M, and has been studied extensively by both analytic and probabilistic methods. By analogy with the case of a bounded domain, we will solve this problem probabilistically by showing that under suitable geometric conditions Brownian motion X on M converges to a random variable X_e on the boundary $\mathbb{S}_\infty(M)$ as $t \uparrow e(X)$. The function $u_f(x) = \mathbb{E}_x f(X_e)$ will be harmonic on M. If we can further show that the law of X_e satisfies

$$\lim_{x \to \theta_0} \mathbb{P}_x \circ X_e^{-1} = \text{point mass at } \theta_0,$$

then u_f is continuous on $\overline{M}$ and its value on the boundary is f. In particular, there are nonconstant bounded harmonic functions on M.

Proposition 6.1.1. *Let M be a Cartan-Hadamard manifold. Let $\mathbb{P}_x$ be the law of Brownian motion on M starting from x. Suppose that for any $x \in M$ we have*

$$\mathbb{P}_x \left\{ X_e = \lim_{t \uparrow e} X_t \text{ exists} \right\} = 1$$

(in the topology of $\overline{M}$) and for any $\theta_0 \in \mathbb{S}_\infty(M)$ and any neighborhood O of θ_0 in $\mathbb{S}_\infty(M)$

$$\lim_{x \to \theta_0} \mathbb{P}_x \{X_e \in O\} = 1.$$

Then the Dirichlet problem at infinity for M is uniquely solvable. For any $f \in C(\mathbb{S}_\infty(M))$, the function $u_f(x) = \mathbb{E}_x f(X_e)$ is the unique solution of the Dirichlet problem with boundary function f.

Proof. Since $u_f(x) = \mathbb{E}_x u_f(\tau_D)$ for any relatively compact open set D containing x, the function u_f is harmonic on M. We show that it is continuous on $\overline{M}$ and is equal to f on the boundary. For any $\theta_0 \in \mathbb{S}_\infty(M)$ and $\epsilon > 0$, choose a neighborhood O of θ_0 in $\mathbb{S}_\infty(M)$ such that

$$|f(\theta) - f(\theta_0)| \le \epsilon, \quad \forall \theta \in O.$$

Then

$$\begin{aligned}|u_f(x) - f(\theta_0)| &\le \mathbb{E}_x|f(X_e) - f(\theta_0)| \\ &\le \epsilon\,\mathbb{P}_x\{X_e \in O\} + 2\|f\|_\infty \mathbb{P}_x\{X_e \notin O\}.\end{aligned}$$

Letting $x \to \theta_0$, we have $\limsup_{x\to\theta_0} |u_f(x) - f(\theta_0)| \le \epsilon$. This shows that $\lim_{x\to\theta_0} u_f(x) = f(\theta_0)$. Thus u_f is a solution of the Dirichlet problem.

To prove the uniqueness, let $\{D_n\}$ be an exhaustion of M and u a solution of the Dirichlet problem at infinity with boundary function f. Then $\{u(X_{t\wedge\tau_{D_n}}),\, t \ge 0\}$ is a uniformly bounded martingale under $\mathbb{P}_x$; hence $u(x) = \mathbb{E}_x u(X_{t\wedge\tau_{D_n}})$. Letting $t \uparrow \infty$ and then $n \uparrow \infty$, and noting that $\tau_{D_n} \uparrow e$, we have

$$u(x) = \mathbb{E}_x u(X_e) = \mathbb{E}_x f(X_e) = u_f(x).$$

Thus every solution coincides with u_f. This proves the uniqueness. □

We now describe a general scheme for proving the convergence of Brownian motion on a Cartan-Hadamard manifold. We have mentioned before that the statement

$$X_t = (r_t, \theta_t) \to \theta_0 \quad \text{as } t \uparrow e$$

is equivalent to

$$r_t \to \infty \quad \text{and} \quad \theta_t \to \theta_0 \quad \text{as } t \uparrow e.$$

In the latter statement, the first part is guaranteed if Brownian motion is transient on M. The second part is the angular convergence of Brownian motion. To prove the angular convergence, we look at succesive times Brownian motion moves a distance 1:

$$\begin{aligned}\tau_1 &= \inf\{t \ge 0 : d(X_t, o) = 1\}, \\ \tau_n &= \inf\{t \ge \tau_{n-1} : d(X_t, X_{\tau_{n-1}}) = 1\}.\end{aligned}$$

$\tau_n - \tau_{n-1}$ is the amount of time spent on the nth step. The angular oscillation during the time interval $[\tau_{n-1}, \tau_n]$ is

$$\Delta\theta_n = \max_{\tau_{n-1}\le t\le\tau_n} d_{\mathbb{S}^{d-1}}(\theta(X_{\tau_{n-1}}), \theta(X_t)).$$

Proposition 6.1.2. *Let M be a Cartan-Hadamard manifold. Suppose that Brownian motion is transient:*

$$\mathbb{P}_x\{r_t \to \infty \text{ as } t \uparrow e\} = 1.$$

Then the Dirichlet problem at infinity is solvable if for any positive δ,

$$\lim_{r(x)\to\infty} \mathbb{P}_x\left\{\sum_{n=1}^{\infty} \Delta\theta_n \le \delta\right\} = 1. \tag{6.1.1}$$

Proof. First we note that that $\sum_{n=1}^{\infty} \Delta\theta_n < \infty$ implies that $\lim_{t\uparrow e} X_t = X_e$ exists. By the assumption, for any fixed positive ϵ and δ, there is an R such that

$$\mathbb{P}_z \left\{ \sum_{n=1}^{\infty} \Delta\theta_n \leq \delta \right\} \geq 1 - \epsilon, \qquad r(z) \geq R.$$

Suppose that $\delta < 1$ and let $\tau_R = \inf\{t : r_t = R\}$. Then by the Markov property at τ_R, we have

$$\begin{aligned}
&\mathbb{P}_x \left\{ X_e = \lim_{t\uparrow e} X_t \text{ exists} \right\} \\
&\geq \mathbb{P}_x \left\{ \sum_{n=1}^{\infty} \Delta\theta_n \leq 1 \right\} \\
&= \mathbb{E}_x \mathbb{P}_{X_{\tau_R}} \left\{ \sum_{n=1}^{\infty} \Delta\theta_n \leq 1 \right\} \\
&\geq 1 - \epsilon.
\end{aligned}$$

Since ϵ is arbitrary, we have

$$\mathbb{P}_x \left\{ X_e = \lim_{t\uparrow e} X_t \text{ exists} \right\} = 1, \qquad \forall x \in M.$$

Let $\theta_0 \in \mathbb{S}_\infty(M)$, and O a neighborhood of θ_0 in $\mathbb{S}_\infty(M)$ containing θ_0. Then there is a $\delta > 0$ such that

$$\{\theta \in \mathbb{S}_\infty(M) : d_{\mathbb{S}^{d-1}}(\theta, \theta_0) \leq 2\delta\} \subseteq O.$$

Note that $X_0 = x$, from which we have $\theta(X_0) = \theta(x)$. Hence,

$$d_{\mathbb{S}^{d-1}}(\theta_0, \theta_e) \leq d_{\mathbb{S}^{d-1}}(\theta(x), \theta_0) + \sum_{n=0}^{\infty} \Delta\theta_n.$$

For any $\epsilon > 0$ and the above δ, fix R as at the beginning of the proof. If x is sufficiently close to θ_0, we have $d_{\mathbb{S}^{d-1}}(\theta(x), \theta_0) \leq \delta$, and $r(x) \geq R$. Therefore,

$$\left\{ \sum_{n=0}^{\infty} \Delta\theta_n \leq \delta \right\} \subseteq \{\theta_e \in O\},$$

and

$$\mathbb{P}_x \{\theta_e \in O\} \geq \mathbb{P}_x \left\{ \sum_{n=0}^{\infty} \Delta\theta_n \leq \delta \right\} \geq 1 - \epsilon.$$

This shows that

$$\lim_{x \to \theta_0} \mathbb{P}_x \{X_e \in O\} = 1.$$

By PROPOSITION 6.1.1, the Dirichlet problem at infinity is uniquely solvable. □

The above proposition reduces the solvability of the Dirichlet problem at infinity to estimating the total oscillation $\sum_{n=1}^{\infty}\Delta\theta_n$. This can be done as follows. Let

$$J_k = \#\left\{n : r_{\tau_n} \le k\right\}$$

be the total number of steps in the geodsic ball $B(k)$. We have

$$J_k - J_{k-1} = \#\left\{n : k-1 < r_{\tau_n} \le k\right\}.$$

For each n such that $r_{\tau_n} \in B(k)\backslash B(k-1)$,

$$r_{\tau_n} \ge k-1, \qquad d(X_t, X_{\tau_{n-1}}) \le 1, \quad t \in [\tau_n, \tau_{n+1}].$$

We show below that this is the situation where the angular oscillation during the time interval $[\tau_n, \tau_{n+1}]$ can be estimated by geometric considerations (see LEMMAS 6.2.1 and 6.3.1). Let us assume that

$$\Delta\theta_n \le h_k, \qquad \text{for } k-1 \le r_{\tau_n} \ge k-1, \tag{6.1.2}$$

where h_k is decreasing in k. They will be calculated once we impose explicit upper bound on the sectional curvature of the manifold. From the inequality

$$\sum_{n=1}^{\infty}\Delta\theta_n \le \sum_{k=1}^{\infty}(J_k - J_{k-1})h_k$$

we have

$$\sum_{n=1}^{\infty}\Delta\theta_n \le \sum_{k=2}^{\infty} J_{k-1}(h_{k-1} - h_k) + \liminf_{k\to\infty} J_N h_N. \tag{6.1.3}$$

Now the problem reduces to obtaining good bounds for J_k, the number of steps in $B(k)$.

An upper bound on J_k is possible if Brownian motion goes to infinity at a fast rate. Suppose that f is a continuous, strictly increasing function, $f(0) = 0$, such that for any R

$$\lim_{r(x)\to\infty} \mathbb{P}_x\left\{r_t \ge R + f(t), \quad \forall t \ge 0\right\} = 1. \tag{6.1.4}$$

The function f will be given explicitly later. Let

$$A = \left\{r_t \ge R + f(t), \quad \forall t \ge 0\right\}.$$

On A, Brownian motion always lies outside $B(R)$; hence the sum in (6.1.3) can be limited to the range $k \ge R$. Also on A we have $r_t \ge f(t)$. This implies that

$$\text{mes}\left\{t : r_t \le k\right\} \le f^{-1}(k), \tag{6.1.5}$$

where f^{-1} is the inverse function of f and mes$\{\cdots\}$ denotes the Lebesgue measure on $[0,\infty)$. On the other hand, the total amount of time Brownian spends in $B(k)$ satisfies

$$\text{(6.1.6)} \qquad \operatorname{mes}\{t : r_t \le k\} \ge \sum_{r_{\tau_n} \le k-1} \{\tau_{n+1} - \tau_n\}.$$

If $r_{\tau_n} \le k-1$, the time difference $\tau_{n+1} - \tau_n$ can be estimated by THEOREM 3.6.1 as follows:

$$\text{(6.1.7)} \qquad \mathbb{P}_x\left[\tau_{n+1} - \tau_n \le \frac{C_1}{L_k}\middle| r_{\tau_n} \le k-1\right] \le e^{-C_2 L_k},$$

where

$$L_k \stackrel{\text{def}}{=} \sqrt{-\inf\{\operatorname{Ric}_M(x) : r(x) \le k\}}$$

is the lower bound of the Ricci curvature on $B(k)$. The explict form of L_k depends on the lower bound on the Ricci curvature we will impose on the manifold. In general, L_k goes to infinity fairly fast, so that the probability in (6.1.7) is fairly small. Now it is clearly intuitively why we can bound J_{k-1}, the number of steps in the geodesic ball $B(k-1)$. If we ignore the small probability in (6.1.7) and assume that

$$\tau_{n+1} - \tau_n > \frac{C_1}{L_k} \qquad \text{for } \tau_{\tau_n} \le k-1,$$

Then from (6.1.5) and (6.1.6) we have

$$J_{k-1} \le \frac{f^{-1}(k)L_k}{C_1}.$$

If we do not ignore the probability in (6.1.7), then the above event should occur with high probability. To be precise, let n_l be the lth step such that $r_{\tau_n} \le k-1$ and

$$A_l = \left\{\tau_{n_l+1} - \tau_{n_l} > \frac{C_1}{L_k}\right\}.$$

If the event $A \cap A_1 \cap \cdots \cap A_N \cap [J_{k-1} \ge N]$ occurs, then by (6.1.5) and (6.1.6),

$$f^{-1}(k) \ge \operatorname{mes}\{t : r_t \le l\} > \frac{C_1 N}{L_k}.$$

This cannot happen if

$$\text{(6.1.8)} \qquad N \ge \frac{f^{-1}(k)L_k}{C_1}.$$

Therefore for such N, we have

$$A \cap A_1 \cap \cdots \cap A_N \cap [J_{k-1} \ge N] = \emptyset,$$

or equivalently,

$$A \cap [J_{k-1} \ge N] \subseteq A_1^c \cup \cdots \cup A_N^c.$$

By (6.1.7) each A_i^c has probability not greater than $e^{-C_2 L_k}$. This implies

$$\mathbb{P}_x\{A \cap [J_{k-1} \ge N]\} \le N e^{-C_2 L_k}.$$

This being true for all N satisfies (6.1.8), we may replace N by $C_3 f^{-1}(k) L_k$ for a large C_3 and obtain

$$\mathbb{P}_x \left\{ A \cap \left[J_{k-1} \geq C_3 f^{-1}(k) L_k \right] \right\} \leq C_3 f^{-1}(k) L_k e^{-C_2 L_k}.$$

Summing over $k \geq R$ and taking the complement, we have

$$\begin{aligned} \mathbb{P}_x & \left\{ A \cap \left[J_{k-1} \leq C_3 f^{-1}(k) L_k, \ \forall k \geq R \right] \right\} \\ & \geq \mathbb{P}_x A - C_3 \sum_{k \geq R} f^{-1}(k) L_k e^{-C_2 L_k}. \end{aligned}$$

If the event on the left side occurs, then $J_k = 0$ for $k < R$ and, provided that $J_N h_N \to 0$ as $N \to \infty$, we have, by (6.1.3),

$$\sum_{n=1}^{\infty} \Delta\theta_n \leq C_3 \sum_{k \geq R} f^{-1}(k) L_k (h_{k-1} - h_k).$$

To summarize this part of the argument, we have

$$\mathbb{P}_x \left\{ \sum_{n=1}^{\infty} \Delta\theta_n \leq \sum_{k \geq R} \epsilon_k \right\} \geq \mathbb{P}_x \left\{ r_t \geq R + f(t), \ \forall t \geq 0 \right\} - \sum_{k \geq R} \eta_k,$$

where

$$\begin{aligned} \epsilon_k &= C_3 f^{-1}(k) L_k (h_{k-1} - h_k), \\ \eta_k &= C_3 f^{-1}(k) L_k e^{-C_2 L_{k+1}}, \\ L_k &= \sqrt{- \inf \left\{ \mathrm{Ric}_M(x) : r(x) \leq k \right\}}. \end{aligned}$$

By PROPOSITION 6.1.2 the Dirichlet problem at infinity is uniquely solvable if the following conditions are satisfied:

$$\text{(6.1.9)} \qquad \begin{cases} \lim_{r(x) \to \infty} \mathbb{P}_x \left\{ r_t \geq R + f(t), \ t \geq 0 \right\} = 1, \quad \forall R \geq 0, \\ \sum_{k=1}^{\infty} \eta_k < \infty, \\ \sum_{k=1} \epsilon_k < \infty, \\ \lim_{N \to \infty} J_N h_N = 0. \end{cases}$$

In the next two sections we will discuss two typical cases where the above conditions are satisfied.

We end this general discussion of the Dirichlet problem at infinity with a geometric comparison theorem that will be needed to estimate $\Delta\theta_n$, i.e., to calculate h_k; see (6.1.2). Again let

$$\kappa(r) \geq \sup \left\{ K_M(x) : r(x) = r \right\}.$$

Let G be the solution of the Jacobi equation

$$G''(r) + \kappa(r) G(r) = 0, \qquad G(0) = 0, \quad G'(0) = 1.$$

Theorem 6.1.3. *Let M be a Cartan-Hadamard manifold and $o \in M$. Define G as above and let N be the radially symmetric Cartan-Hadamard manifold with metric $ds^2 = dr^2 + G(r)^2 d\theta^2$ ($d\theta^2$ is the standard Riemannian metric on $\mathbb{S}^{d-1}$). Let xoy be a geodesic triangle on M with a vertex at o and $x'o'y'$ a geodesic triangle on N such that*

$$d_N(o', x') = r(x), \quad d_N(o', y') = r(y),$$

and

$$d_{\mathbb{S}^{d-1}}(\theta(x'), \theta(y')) = d_{\mathbb{S}^{d-1}}(\theta(x), \theta(y)).$$

Then we have

$$d_M(x, y) \geq d_N(x', y').$$

Proof. On M let γ be the minimal geodesic joining x and y. On the geodesic triangle xoy suppose that γ is given by $\{r(\theta), 0 \leq \theta \leq \theta_0\}$. On N, consider the curve γ' given by the same equation. Let V be the Jacobi field on M along a geodesic ray such that $V(0) = 0$ and $V(r(\theta)) = \dot{\gamma}(r(\theta))$, let V' be the corresponding Jacobi field on N. The radial sectional curvature on N is given by

$$K_N(X', X'_r) = -\frac{G''(r)}{G(r)}, \qquad X' \perp X'_r, \quad |X'| = 1.$$

By the hypothesis, each radial sectional curvature on M satisfies

$$K_M(X, X_r) \leq -\frac{G''(r)}{G(r)}, \qquad X \perp X_r, \quad |X| = 1.$$

Hence we have $K_M(X, X_r) \leq K_N(X', X'_r)$, and by the Rauch comparison theorem (see Cheeger and Ebin[**9**], pp. 29–30), we have

$$|V(\gamma(\theta))| \geq |V'(\gamma'(\theta))|, \quad \text{i.e.,} \quad |\dot{\gamma}(\theta)| \geq |\dot{\gamma}'(\theta)|.$$

Now,

$$l(\gamma) = \int_0^{\theta_0} |\dot{\gamma}(\theta)| d\theta \geq \int_0^{\theta_0} |\dot{\gamma}'(\theta)| d\theta = l(\gamma').$$

Since $d_M(x, y) = l(\gamma)$ and $d_N(x', y') \leq l(\gamma')$, the result follows. □

The above theorem will be used to bound the angle between two geodesic rays on a Cartan-Hadmard manifold M by the corresponding angle on a radially symmetric manifold constructed from the upper bound of the sectional curvature of M; see LEMMAS 6.2.1 and 6.3.1.

6.2. Constant upper bound

In this section we show that the Dirichlet problem at infinity is solvable under the following curvature condition: there are positive constants a, C, and $\eta \in (0,1)$ such that

$$-Ce^{(2-\eta)ar(x)} \le \mathrm{Ric}_M(x), \quad K_M(x) \le -a^2.$$

From the discussion in the last section, it is enough to verify (6.1.9) under these curvature conditions.

Lemma 6.2.1. *Let M be a Cartan-Hadamard manifold. Suppose that there is a positive constant a such that $K_M(x) \le -a^2$. Let $x, y \in M$ be such that $r(x), r(y) \ge r \ge 2$ and $d(x,y) \le 1$. Then*

$$d_{\mathbb{S}^{d-1}}(\theta(x), \theta(y)) \le \frac{a}{\sinh a(r-1)} \le C(a)e^{-ar}.$$

Proof. Let $\mathbb{M}_{-a^2}$ be a complete, simply connected manifold of constant curvature $-a^2$. On $\mathbb{M}_{-a^2}$ consider a geodesic triangle $x'o'y'$ such that

$$d(o', x') = r(x), \quad d(o', y') = r(y),$$

and

$$d_{\mathbb{S}^{d-1}}(\theta(x'), \theta(y')) = d_{\mathbb{S}^{d-1}}(\theta(x), \theta(y)).$$

The Riemannian metric on $\mathbb{M}_{-a^2}$ has the form

$$ds^2_{\mathbb{M}_{-a^2}} = dr^2 + G(r)^2 d\theta^2, \quad G(r) = \frac{\sinh ar}{a};$$

hence

$$d(x', y') \ge G(r-1) d_{\mathbb{S}^{d-1}}(\theta(x'), \theta(y')) \quad = G(r-1) d_{\mathbb{S}^{d-1}}(\theta(x), \theta(y)).$$

By the comparison THEOREM 6.1.3, we have $d(x', y') \le d(x,y)$. Hence

$$1 \ge G(r-1) d_{\mathbb{S}^{d-1}}(\theta(x), \theta(y)),$$

and the result follows immediately. □

We now estimate the probability that Brownian motion returns to a fixed geodesic ball.

Lemma 6.2.2. *Suppose that $K_M(x) \le -a^2$. For any $R \ge 0$ we have, for $r = r(x) \ge R$,*

$$\mathbb{P}_x \{ r_t \le R \text{ for some } t \ge 0 \} \le \cosh^{1-d} a(r-R). \tag{6.2.1}$$

Proof. By the radial process comparison THEOREM 3.5.3 we may assume that M is radially symmetric. The radial process in this case is

$$r_t = r_0 + \beta_t + \frac{d-1}{2} \int_0^t a \coth ar_s \, ds,$$

where β is a Brownian motion. The following argument is well known. The generator for the one-dimensional diffusion process $\{r_t\}$ is

$$A = \frac{1}{2}\left(\frac{d}{dr}\right)^2 + \frac{d-1}{2}\, a\coth ar\, \frac{d}{dr}.$$

Let

$$s(r) = \int_r^\infty (\sinh au)^{1-d} du.$$

Then $A\,s = 0$, i.e., s is A-harmonic. Let $\sigma_R = \inf\{t : r_t = R\}$. If $r(x) \geq R$, then $\{s(r_{t\wedge\sigma_R})\}$ is a uniformly bounded martingale. Letting $t \uparrow \infty$, we have

$$s(r) = \mathbb{E}_x s\left(r_{t\wedge\sigma_R}\right) = s(R)\mathbb{P}_x\left\{\sigma_R < \infty\right\}.$$

Hence

$$\mathbb{P}_x\left\{r_t \leq R \text{ for some } t \geq 0\right\} = \mathbb{P}_x\left\{\sigma_R < \infty\right\} = \frac{s(r)}{s(R)}.$$

On the other hand,

$$\begin{aligned}
\frac{s(r)}{s(R)} &= \frac{\int_r^\infty (\sinh au)^{1-d}\, du}{\int_R^\infty (\sinh au)^{1-d}\, du} \\
&\leq \sup_{u\geq R}\left[\frac{\sinh a(u+r-R)}{\sinh au}\right]^{1-d} \\
&\leq \cosh^{1-d} a(r-R).
\end{aligned}$$

In the last step we have used the inequality

$$\frac{\sinh(x+y)}{\sinh x} = \frac{\sinh x\cosh y + \cosh x \sinh y}{\sinh x} \geq \cosh y.$$

The result follows. □

Next we consider the rate of escape for Brownian motion.

Lemma 6.2.3. *Suppose that $K_M(x) \leq -a^2$. For any $\lambda < (d-1)a/2$ and $R \geq 0$ we have*

$$\lim_{r(x)\to\infty} \mathbb{P}_x\left\{r_t \geq R + \lambda t \quad \forall t \geq 0\right\} = 1.$$

Proof. By the radial process comparison THEOREM 3.5.3, it is enough to show the result for the process

$$r_t = r_0 + \beta_t + \frac{d-1}{2}\int_0^t a\coth ar_s\, ds.$$

Fix $\lambda_1 \in (\lambda, (d-1)a/2)$ and $\epsilon > 0$. By the law of the iterated logarithm for Brownnian motion,

$$\liminf_{t\uparrow\infty} \frac{\beta_t}{\sqrt{2t\ln\ln t}} = -1.$$

Hence there is an R_1 independent of x such that

(6.2.2) $$\mathbb{P}_x \{\beta_t \geq -(\lambda - \lambda_1)t - R_1, \ \forall t \geq 0\} \geq 1 - \epsilon.$$

Take R_2 such that

$$\frac{d-1}{2} a \coth ar \geq \lambda_1, \quad \forall r \geq R_2.$$

By LEMMA 6.2.2 for sufficiently large $r(x)$ we have

(6.2.3) $$\mathbb{P}_x \{r_t \geq R_2, \forall t \geq 0\} \geq 1 - \epsilon.$$

If the events in (6.2.2) and (6.2.3) happen simultaneously, then for all sufficiently large $r(x)$,

$$\begin{aligned} r_t &= r(x) + \beta_t + \frac{d-1}{2} \int_0^t a \coth ar_s ds \\ &\geq r(x) - (\lambda_1 - \lambda)t - R_1 + \lambda_1 t \\ &\geq R + \lambda t. \end{aligned}$$

This implies that

$$\mathbb{P}_x \{r_t \geq R + \lambda t, \quad \forall t \geq 0\} \geq 1 - 2\epsilon.$$

□

Theorem 6.2.4. *Let M be a Cartan-Hadamard manifold such that*

$$-Ce^{(2-\eta)ar(x)} \leq \mathrm{Ric}_M(x), \quad K_M(x) \leq -a^2$$

for some positive constants a, C, and $\eta \in (0,1)$. Then the Dirichlet problem at infinity on M is uniquely solvable.

Proof. By the discussion in SECTION 6.1, it is enough to show (6.1.9). We can take

$$\begin{cases} f(t) = \lambda t, & \text{by LEMMA 6.2.3;} \\ L_k = C_1 e^{(2-\eta)ak/2}, & \text{by the lower bound of the Ricci curvature;} \\ h_k = C_2 e^{-ak}, & \text{by LEMMA 6.2.1.} \end{cases}$$

It is now straightforward to verify (6.1.9). □

6.3. Vanishing upper bound

We now consider the second typical situation where the Dirichlet problem at inifnity is uniquely solvable: the upper bound of the sectional curvature approaches zero at infinity. We assume that there are positive constants $\alpha > 2$, $\beta < \alpha - 2$, and R_0 such that for all $r(x) \geq R_0$,

$$-r(x)^{2\beta} \leq \mathrm{Ric}_M(x), \quad K_M(x) \leq -\frac{\alpha(\alpha-1)}{r(x)^2}.$$

Again we need to verify (6.1.9) under this curvature assumption.

Lemma 6.3.1. *Let M be a Cartan-Hadamard manifold. Suppose that there are positive constants $\alpha \geq 1$ and $R_0 \geq 1$ such that for all $r(x) \geq R_0$,*

$$K_M(x) \leq -\frac{\alpha(\alpha-1)}{r(x)}.$$

Let $x, y \in M$ be such that $r(x), r(y) \geq 2R_0$ and $d(x,y) \leq 1$. Then there is a constant C independent of x and y such that the angle between the geodesic rays ox and oy satisfies

$$d_{\mathbb{S}^{d-1}}(\theta(x), \theta(y)) \leq \frac{C}{r(x)^\alpha}.$$

Proof. Let

$$\kappa(r) = \min\left\{ \sup_{r(x)=r} K_M(x), -\frac{\alpha(\alpha-1)}{r^2} \right\}.$$

Let G be the unique solution of the Jacobi equation

$$G''(r) + \kappa(r)G(r) = 0, \quad G(0) = 0, \quad G'(0) = 1.$$

Since $\kappa(r) = -\alpha(\alpha-1)/r^2$ for $r \geq R_0$, there are constants C_1, C_2, C_3 and $R_1 \geq R_0$ such that

$$G(r) = C_1 r^\alpha + C_2 r^{1-\alpha} \geq C_3 r^\alpha, \quad \forall r \geq R_1. \tag{6.3.1}$$

Now the proof can be completed as in the proof of LEMMA 6.2.1. □

As in the last section, we need to find the probability of Brownian motion returning to a geodesic ball.

Lemma 6.3.2. *Let M be a Cartan-Hadamard manifold. Suppose that there are positive constants $\alpha \geq 1$ and $R_0 \geq 1$ such that for all $r(x) \geq R_0$,*

$$K_M(x) \leq -\frac{\alpha(\alpha-1)}{r(x)}.$$

There is a constant C such that, for all $R \geq 1$ and $x \in M$ with $r = r(x) \geq R$,

$$\mathbb{P}_x\{r_t \leq R \text{ for some } t \geq 0\} \leq C\left(\frac{R}{r}\right)^{(d-1)\alpha-1}.$$

Proof. Define the function G as in the proof of LEMMA 6.3.1. As before, we may assume that M is radially symmetric with the metric $ds^2 = dr^2 + G(r)^2 d\theta^2$. Let

$$s(r) = \int_r^\infty G(s)^{1-d} ds.$$

By the same argument as in LEMMA 6.2.2 we have

$$\mathbb{P}_x\{r_t \leq R \text{ for some } t \geq 0\} = \frac{s(r(x))}{s(R)}.$$

We have $G(r) \sim C_1 r^\alpha$ and $s(r) \sim C_2 r^{1-\alpha(d-1)}$ as $r \uparrow \infty$, for some positive constants C_1 and C_2. The result follows immediately. □

In order to find the rate of escape for Brownian motion, we need some facts about Bessel processes. Let $Y_{q,a}$ be the Bessel process of index $q > 1$ from $a \geq 0$:

$$Y_{q,a}(t) = a + \beta_t + \frac{q}{2}\int_0^t \frac{ds}{Y_{q,a}(s)}, \tag{6.3.2}$$

where β is a one-dimensional Brownian motion. Suppose that ψ is a positive nonincreasing function such that $\int_0^\infty \psi(t)^{q-1}dt < \infty$. Then

$$\mathbb{P}\left\{\liminf_{t\to\infty} \frac{Y_{q,a}(t)}{\sqrt{t}\psi(t)} \geq 1\right\} = 1.$$

This is the well known integral test for lower functions of Bessel processes; see Shiga and Watanabe [**67**]. In particular, for any $\lambda > 0$ we have

$$\mathbb{P}\left\{\lim_{t\uparrow\infty} \frac{Y_{q,a}(t)}{t^{1/2-\lambda}} = \infty\right\} = 1. \tag{6.3.3}$$

Lemma 6.3.3. *For any $\lambda > 0$ and $R \geq 0$ we have*

$$\lim_{r(x)\to\infty} \mathbb{P}_x\left\{r_t \geq R + t^{1/2-\lambda} \quad \forall t \geq 0\right\} = 1.$$

Proof. By the comparison THEOREM 3.5.3 it is enough to assume that M is radially symmetric, as in LEMMA 6.3.2. The radial process in this case is given by

$$r_t = r_0 + \beta_t + \frac{d-1}{2}\int_0^t \frac{G'(r_s)}{G(r_s)}\,ds.$$

Fix q and q_1 such that $1 < q_1 < q < (d-1)\alpha$. For any $\epsilon > 0$, by (6.3.3) there is an R_1, dependent on ϵ but independent of x, such that

$$\mathbb{P}_x\left\{Y_{q_1,1}(t) \geq t^{1/2-\lambda} - R_1, \forall t \geq 0\right\} \geq 1 - \epsilon. \tag{6.3.4}$$

By the inequality in (6.3.1) there is an R_2 independent of ϵ such that

$$\frac{d-1}{2}\frac{G'(r)}{G(r)} \geq \frac{q}{2r}, \quad \forall r \geq R_2.$$

By LEMMA 6.3.2 for sufficiently large $r(x)$ we have

$$\mathbb{P}_x\left\{r_t \geq \frac{q\,(R + R_1 + R_2)}{q - q_1}, \quad \forall t \geq 0\right\} \geq 1 - \epsilon. \tag{6.3.5}$$

Let $X_s = r_s - (R + R_1)$. If the event in (6.3.5) holds, then

$$r_s \geq \frac{q\,(R + R_1 + R_2)}{q - q_1} \geq R_2$$

and

$$\frac{q}{r_s} \geq \frac{q_1}{r_s - (R + R_1)} = \frac{q_1}{X_s}.$$

This implies a lower bound on the drift:

$$\frac{d-1}{2}\frac{G'(r_s)}{G(r_s)} \geq \frac{q}{r_s} \geq \frac{q_1}{X_s}.$$

Now, for the process $\{X_t\}$ we have

$$\begin{aligned} X_t &= r(x) - (R + R_1) + \beta_t + \frac{d-1}{2}\int_0^t \frac{G'(r_s)}{G(r_s)}\,ds \\ &= r(x) - (R + R_1) + \beta_t + \frac{q_1}{2}\int_0^t \frac{ds}{X_s} + l_t, \end{aligned}$$

for some nondecreasing process $\{l_t\}$. Compare this with the equation

$$Y_{q_1,1}(t) = 1 + \beta_t + \frac{q_1}{2}\int_0^t \frac{ds}{Y_{q_1,1}(s)}.$$

If $r(x) \geq R + R_1 + 1$, we have $X_t \geq Y_{q_1,1}(t)$ (see the proof of THEOREM 3.5.3). Hence, if both events in (6.3.4) and (6.3.5) hold, then

$$r_t = R + R_1 + X_t \geq R + R_1 + Y_{q_1,1}(t) \geq R + t^{1/2-\lambda}.$$

It follows that, for sufficiently large $r(x)$,

$$\mathbb{P}_x\left\{r_t \geq R + t^{1/2-\lambda},\ \forall t \geq 0\right\} \geq 1 - 2\epsilon.$$

This completes the proof. □

Theorem 6.3.4. *Let M be a Cartan-Hadamard manifold. Suppose that there are positive constants $\alpha > 2$, $\beta < \alpha - 2$, and R_0 such that*

$$-r(x)^{2\beta} \leq \mathrm{Ric}_M(x), \quad K_M(x) \leq -\frac{\alpha(\alpha-1)}{r(x)^2}, \quad \forall r(x) \geq R_0.$$

Then the Dirichlet problem at infinity is solvable.

Proof. We need to verify (6.1.9). We can take

$$\begin{cases} f(t) = t^{1/2-\lambda}, & \text{by LEMMA 6.3.2;} \\ L_k = C_1 k^{\beta}, & \text{by the lower bound of the Ricci curvature;} \\ h_k = C_2 k^{-\alpha}, & \text{by LEMMA 6.3.1.} \end{cases}$$

Now it is straightforward to verify (6.1.9) if $0 < \lambda < (\alpha-2-\beta)/2(\alpha-\beta)$. □

The following corollary is immediate.

Corollary 6.3.5. *Let M be a Cartan-Hadamard manifold. Suppose that the Ricci curvature is bounded from below by a constant. If there are positive R_0 and $c > 2$ such that*

$$K_M(x) \le -\frac{c}{r(x)^2}, \quad \forall r(x) \ge R_0,$$

then the Dirichlet problem at infinity is solvable.

6.4. Radially symmetric manifolds

If a Cartan-Hadamard manifold M is radially symmetric, the problem of solvability of the Dirichlet problem at infinity affords an elegant probabilistic solution by taking advantage of the special structure of Brownian motion. The Riemannian metric has the form $ds^2 = dr^2 + G(r)^2 d\theta^2$. As always we identify the boundary at infinity $\mathbb{S}_\infty(M)$ with the unit sphere $\mathbb{S}^{d-1}$ in T_oM.

We have shown in Example 3.3.3 that the radial process $r_t = r(X_t)$ is the solution of the equation

$$r_t = r_0 + \beta_t + \frac{d-1}{2}\int_0^t \frac{G'(r_s)}{G(r_s)} ds, \tag{6.4.1}$$

where β is a one-dimensional Brownian motion, and the angular process $\theta_t = \theta(X_t)$ is a time-changed Brownian motion on $\mathbb{S}^{d-1}$. Define a new time scale

$$l_t = \int_0^t \frac{ds}{G(r_s)^2}. \tag{6.4.2}$$

Then there is a Brownian motion Z on $\mathbb{S}^{d-1}$ independent of the Brownian motion β such that $\theta_t = Z_{l_t}$. Because Z is recurrent on $\mathbb{S}^{d-1}$, the only way the limiting angle $\theta_e = \lim_{t\uparrow e} \theta_t$ can exist is when $l_e < \infty$. This argument reduces the angular convergence problem to the one-dimensional problem of finding conditions on the function G such that $\mathbb{P}_x\{l_e < \infty\} = 1$.

We now observe that l_e is the lifetime of the radial process after a time change. In fact, let $\tau : [0, l_e) \to [0, e)$ be the inverse function of $l : [0, e) \to [0, l_e)$ and let $Y_t = r_{\tau_t}$, $t \le l_e$. Then it is clear that

$$e(Y) = l_e = \int_0^{e(r)} \frac{ds}{G(r_s)^2}.$$

In terms of Y we have

$$\tau_t = \int_0^t G(Y_s)^2 ds.$$

Replacing t in (6.4.1) by τ_t, we have

$$Y_t = Y_0 + \beta_{\tau_t} + \frac{d-1}{2}\int_0^t G'(Y_s)G(Y_s)\, ds.$$

Of course, $\{\beta_{\tau_t}\}$ is no longer a Brownian motion; it is a local martingale with the quadratic variation τ_t. Hence there is a Brownian motion W such that

$$\beta_{\tau_t} = \int_0^t G(Y_s)\, dW_s.$$

It follows that the time-changed radial process is the solution of

$$dY_t = G(Y_t)\, dW_t + \frac{d-1}{2} G'(Y_t) G(Y_t)\, dt. \tag{6.4.3}$$

It is therefore an L-diffusion on $[0, \infty)$ with the generator

$$L = \frac{1}{2}\left[G(x)^2 \left(\frac{d}{dx}\right)^2 + (d-1)G'(x)G(x)\frac{d}{dx}\right].$$

We have stated in PROPOSITION 4.2.2 the exact condition under which Y has finite lifetime with probability one. It is now a simple matter of expressing this condition explicitly in terms of G. Define

$$J(G) = \int_c^\infty G(r)^{1-d} dr \int_c^r G(s)^{d-3} ds.$$

Proposition 6.4.1. *Let M be a radially symmetric Cartan-Hadamard manifold of dimension $d \geq 2$ with the metric $ds^2 = dr^2 + G(r)^2 d\theta^2$. Then*

$$\mathbb{P}_x \left\{ \lim_{t \uparrow e} X_t = X_e \text{ exists} \right\} = 1$$

if and only if $J(G) < \infty$.

Proof. We first note that $J(G) < \infty$ implies that X is transient; thus the limit $X_t = (r_t, \theta_t) \to (r_e, \theta_e)$ exists with probability one if and only if the limiting angle θ_e exists with probability one. Let Y be the one-dimensional diffusion defined by (6.4.3). Then, as we have argued before, the limiting angle θ_e exists with probability one if and only if $\mathbb{P}_x \{e(Y) < \infty\} = 1$. According to PROPOSITION 4.2.2 this happens if and only if one of the following three cases happens:

(1) $l(0) < +\infty$ and $l(\infty) < \infty$;

(2) $s(0) = -\infty$ and $l(\infty) < \infty$;

(3) $l(0) < +\infty$ and $s(\infty) = \infty$.

In the present case, we have

$$s(r) = \int_c^r G(s)^{1-d} ds,$$

$$l(r) = \int_c^r G(s)^{1-d} ds \int_c^s G(u)^{d-3} du.$$

From $G(r) \sim r$ as $r \downarrow 0$ we have $s(0) = -\infty$ and $l(0) = \infty$; thus only (2) can happen. It follows that $\mathbb{P}_x\{e(Y) < \infty\} = 1$ if and only if $l(\infty) = J(G) < \infty$. □

It requires some work to show that the same condition also guarantees the unique solvability of the Dirichlet problem at infinity.

Theorem 6.4.2. *Let M be a radially symmetric Cartan-Hadamard manifold of dimension $d \geq 2$ with the metric $ds^2 = dr^2 + G(r)^2 d\theta^2$. Then there are two possibilities.*

(i) *If $J(G) < \infty$, then for any $f \in C(\mathbb{S}_\infty(M))$, the function $u(x) = \mathbb{E}_x f(X_e)$ is the unique solution of the Dirichlet problem at infinity with boundary function f.*

(ii) *If $J(G) = \infty$, then every bounded harmonic function on M is a constant.*

Proof. (i) By Proposition 6.4.1, the random variable X_e exists. Using the Markov property, we have $u(x) = \mathbb{E}_x\{u(X_{\tau_B})\}$ with $B = B(x; r)$, from which it is clear that u is harmonic (see Lemma 4.4.3). Thus it is enough to show that u takes the boundary value f.

We have $u(x) = \mathbb{E}_x f(\theta_e)$ with $\theta_e = Z_{l_e}$, where Z is a Brownian motion on $\mathbb{S}^{d-1}$ independent of l_e. Hence we can write

$$u(x) = \mathbb{E}_x \int_{\mathbb{S}^{d-1}} p_{\mathbb{S}^{d-1}}(l_e, \theta(x), \theta) f(\theta) d\theta. \tag{6.4.4}$$

Thus the distribution of θ_e has a density $\mathbb{E}_x p_{\mathbb{S}^{d-1}}(l_e, \theta(x), \theta)$ with respect to the volume measure on $\mathbb{S}^{d-1}$. To show that f is the boundary value of u, we first note that from

$$\lim_{t\downarrow 0} \int_{\mathbb{S}^{d-1}} p_{\mathbb{S}^{d-1}}(t, \theta_0, \theta) f(\theta) d\theta = f(\theta_0)$$

and the fact that the heat kernel $p_{\mathbb{S}^{d-1}}(t, \theta_1, \theta_2)$ is rotationally invariant, we have

$$\lim_{t\downarrow 0,\, \theta_1 \to \theta_0} \int_{\mathbb{S}^{d-1}} p_{\mathbb{S}^{d-1}}(t, \theta_1, \theta) f(\theta) d\theta = f(\theta_0).$$

If we can show that for any $t > 0$,

$$\lim_{r(x)\to\infty} \mathbb{P}_x\{l_e \geq t\} = 0, \tag{6.4.5}$$

then it is a simple matter to show from (6.4.4) that $u(x) \to f(\theta_0)$ as $x \to \theta_0 \in \mathbb{S}_\infty(M)$ (namely, $r(x) \to \infty$ and $\theta(x) \to \theta_0$). Thus it is enough to show (6.4.5).

Recall that $l_e = e(Y)$, the lifetime of Y defined in (6.4.3). We have

$$\sigma_r \stackrel{\text{def}}{=} \inf\{t : Y_t = r\} \uparrow e(Y) \quad \text{as } r \uparrow \infty.$$

Fix an $r_0 > 0$. Since we know that $\mathbb{P}_{r_0}\{e(Y) < \infty\} = 1$, for any $\epsilon > 0$, there is an n such that

$$\mathbb{P}_{r_0}\{e(Y) \geq \sigma_n + t\} \leq \epsilon.$$

If $r \geq n$, then using the Markov property at σ_r we have

$$\epsilon \geq \mathbb{P}_{r_0}\{e(Y) \geq \sigma_n + t\} \geq \mathbb{P}_{r_0}\{e(Y) \geq \sigma_r + t\} = \mathbb{P}_r\{e(Y) \geq t\}.$$

This implies (6.4.5).

(ii) Suppose that $J(G) = \infty$ and let u be a bounded harmonic function on M. Then $\{u(X_t), t < e\}$ is a bounded martingale. Therefore we have

$$u(x) = \mathbb{E}_x H, \qquad H \stackrel{\text{def}}{=} \lim_{t \uparrow e} u(X_t).$$

If Brownian motion X is recurrent, then obviously H is a constant $\mathbb{P}_x$-almost surely and this constant is independent of x. Hence u is a constant. If X is transient, we will use the fact that

$$p_{\mathbb{S}^{d-1}}(t, \theta_1, \theta_2) \to \frac{1}{|\mathbb{S}^{d-1}|}$$

uniformly on $\mathbb{S}^{d-1} \times \mathbb{S}^{d-1}$ as $t \uparrow \infty$. This is a consequence of the L^2-expansion of the heat kernel; see (7.3.1). Now suppose that $x, y \in M$ are such that $r(x) = r(y)$. The radial process r_t has the same law under the probabilities $\mathbb{P}_x$ and $\mathbb{P}_y$. Let $\tau_R = \inf\{t : r_t = R\}$. Then

$$u(x) = \mathbb{E}_x u(X_{\tau_R}) = \mathbb{E}_x \int_{\mathbb{S}^{d-1}} p(l_{\tau_R}, \theta(x), \theta) u(R, \theta) d\theta.$$

It follows that

$$|u(x) - u(y)| \leq \|u\|_\infty \mathbb{E}_x \int_{\mathbb{S}^{d-1}} \left| p_{\mathbb{S}^{d-1}}(l_{\tau_R}, \theta(x), \theta) - p_{\mathbb{S}^{d-1}}(l_{\tau_R}, \theta(y), \theta) \right| d\theta.$$

Letting $R \uparrow \infty$ and using the fact that

$$\mathbb{P}_x\left\{\lim_{R \uparrow \infty} l_{\tau_R} = l_e = \infty\right\} = 1,$$

we have $u(x) = u(y)$. This shows that u is radially symmetric: $u(x) = f(r(x))$ for some function f. Since the limit

$$H = \lim_{t \uparrow e} u(X_t) = \lim_{t \uparrow e} f(r_t)$$

exists and $r_t \to \infty$ as $t \uparrow e$, we see that $\lim_{r \to \infty} f(r) = f(\infty)$ exists and $H = f(\infty)$. It follows that $u(x) = \mathbb{E}_x H = f(\infty)$ is a constant, and the proof is completed. □

To give the above results a geometric flavor, we now convert the condition on $J(G)$ into a condition on the radial sectional curvature

$$\kappa(r) = -\frac{G''(r)}{G(r)}.$$

We first prove the following one-dimensional comparison result.

Lemma 6.4.3. *Suppose that $\kappa_1(r) \le \kappa_2(r) \le 0$. Let G_i be the solution of the Jacobi equation*

$$G_i''(r) + \kappa_i(r)G_i(r) = 0, \quad G_i(0) = 0, \quad G_i'(0) = 1.$$

Then for all $r > 0$,

$$\frac{G_1'(r)}{G_1(r)} \ge \frac{G_2'(r)}{G_2(r)} \quad \text{and} \quad G_1(r) \ge G_2(r).$$

Proof. Let $r_0 = \inf\{r : G_i(r) < 0\}$. Then $r_0 > 0$ and $G_i''(r) \ge 0$ for $r \in [0, r_0]$. This implies $G_i'(r) \ge 0$ and $G_i(r) \ge r$ in the same range, a contradiction unless r_0 is infinite. Hence $G_i(r) \ge 0$. Using this in the Jacobi equation, we have

$$G_i''(r) \ge 0, \quad G_i'(r) \ge 1, \quad G_i(r) \ge r.$$

Let

$$F = \frac{G_1'}{G_1} - \frac{G_2'}{G_2}.$$

Differentiating and using the differential equation for G_i, we have

$$F' = \kappa_1^2 - \kappa_2^2 - \left(\frac{G_1'}{G_1} + \frac{G_2'}{G_2}\right) F.$$

Since $\kappa_1^2 \ge \kappa_2^2$, we obtain

$$\frac{F'}{F} \ge -\frac{G_1'}{G_1} - \frac{G_2'}{G_2}.$$

Integrating from ϵ to r, we have

$$F(r) \ge \frac{G_1(\epsilon)G_2(\epsilon)}{G_1(r)G_2(r)} F(\epsilon). \tag{6.4.6}$$

From the boundary conditions at $r = 0$ we have

$$G_i(\epsilon) = \epsilon + O(\epsilon^2), \quad G_i'(\epsilon) = 1 + O(\epsilon),$$

from which we also have $F(\epsilon) = O(1)$. Hence the limit on the right side of (6.4.6) is zero and we have $F(r) \ge 0$, or

$$\frac{G_1'(r)}{G_1(r)} \ge \frac{G_2'(r)}{G_2(r)}.$$

Integrating from ϵ to r we have

$$\frac{G_1(r)}{G_2(r)} \geq \frac{G_1(\epsilon)}{G_2(\epsilon)}.$$

Since $\lim_{\epsilon \downarrow 0} G_1(\epsilon)/G_2(\epsilon) = 1$, we obtain $G_1(r) \geq G_2(r)$. □

Remark 6.4.4. On a radially symmetric manifold with metric $ds^2 = dr^2 + G(r)^2 d\theta^2$ we have $\Delta r = (d-1)G'(r)/G(r)$. Thus the above lemma is just the Laplacian comparison THEOREM 3.4.2 applied to radially symmetric manifolds.

In the following proposition $c_2 = 1/2$ and $c_d = 1$ for $d \geq 3$.

Proposition 6.4.5. *Suppose that κ is a nonpositive continuous function on $[0, \infty)$. Let G be the solution of the Jacobi equation*

$$G''(r) + \kappa(r)G(r) = 0, \quad G(0) = 0, \quad G'(0) = 1.$$

(i) If there exist $c > c_d$ and r_0 such that

$$\kappa(r) \leq -\frac{c}{r^2 \ln r}, \qquad \forall r \geq r_0,$$

then $J(G) < \infty$.

(ii) If there exists $r_0 > 0$ such that

$$\kappa(r) \geq -\frac{c_d}{r^2 \ln r}, \qquad \forall r \geq r_0,$$

then $J(G) = \infty$.

Proof. Our proof is based on the following probabilistic fact from the last line of the proof of PROPOSITION 6.4.1. Let r_t be the solution of

$$r_t = \beta_t + \frac{d-1}{2} \int_0^t \frac{G'(r_s)}{G(r_s)} ds.$$

Define

$$l_e(G) = \int_0^{e(r)} \frac{ds}{G(r_s)^2}.$$

Then

$$J(G) < \infty \iff \mathbb{P}\{l_e(G) < \infty\} = 1. \tag{6.4.7}$$

(i) Fix an $\alpha \in (c_d, c)$ and let $G_0(r) = r(\ln r)^\alpha$. Then $J(G_0) < \infty$ (for all dimensions) and

$$\frac{G_0''(r)}{G_0(r)} = \frac{\alpha}{r^2 \ln r} + \frac{\alpha(\alpha-1)}{r^2(\ln r)^2}.$$

Define

$$\kappa_1(r) = \max\left\{\kappa(r), -\frac{G_0''(r)}{G_0(r)}\right\}.$$

Then $\kappa_1(r) \geq \kappa(r)$ and $G_0''(r)/G_0(r) = -\kappa_1(r)$ for all sufficiently large r, say $r \geq R$. Let G_1 be the solution of the Jacobi equation for κ_1. On $[R, \infty)$, G_0 and G_1 satisfies the same Jacobi equation; hence

$$G_1(r) = C_1 G_0(r) \int_R^r \frac{ds}{G_0(s)^2} + C_2 G_0(r),$$

where C_1, C_2 are constants determined by the initial conditions at $r = R$. From this we see that as $r \to \infty$,

$$\frac{G_1(r)}{G_0(r)} \to C_3 \stackrel{\text{def}}{=} C_1 \int_R^\infty \frac{ds}{G_0(s)^2} + C_2.$$

We claim that $C_3 > 0$. In fact,

$$G_1(r) = C_3 G_0(r) - C_1 G_0(r) \int_r^\infty \frac{ds}{G_0(s)^2}.$$

From $G_0(r) = r(\ln r)^\alpha$ we see that the last term goes to zero. From this and the fact that $G_1(r) \geq r$ we have $C_3 > 0$. Now $G_1(r)/G_0(r) \to C_3 > 0$ and $J(G_0) < \infty$ imply immediately that $J(G_1) < \infty$.

Now let r^1 be the 1-dimensional diffusion process defined by

$$r_t^1 = \beta_t + \frac{d-1}{2} \int_0^t \frac{G_1'(r_s^1)}{G_1(r_s^1)} ds.$$

From LEMMA 6.4.3 we have $G_1'/G_1 \leq G'/G$ and $G_1 \leq G$. Therefore by the radial process comparison THEOREM 3.5.3, we have $r_t^1 \leq r_t$ for all t. This implies that $e(r^1) \geq e(r)$ and

$$l_e(G_1) = \int_0^{e(r^1)} \frac{ds}{G_1(r_s^1)^2} \geq \int_0^{e(r)} \frac{ds}{G(r_s)^2} = l_e(G).$$

Now we use (6.4.7). From $J(G_1) < \infty$, we have $\mathbb{P}\{l_e(G_1) < \infty\} = 1$; from the above inequality we have $\mathbb{P}\{l_e(G) < \infty\} = 1$, and this implies in turn that $J(G) < \infty$.

(ii) Choose the comparison function $G_0(r) = r(\ln r)^{1/2}(\ln\ln r)^{1/2}$ for $d = 2$ and $G_0(r) = r \ln r$ for $d \geq 3$. We have $J(G_0) = \infty$ (for all dimensions). When $d = 2$,

$$\begin{aligned} \frac{G_0''(r)}{G_0(r)} = &\frac{1}{2r^2 \ln r} + \frac{1}{r^2 \ln r \ln\ln r} \\ &- \frac{1}{4r^2(\ln r)^2} - \frac{1}{r^2(\ln r)^2(\ln\ln r)^2}. \end{aligned}$$

When $d \geq 3$, we have

$$\frac{G_0''(r)}{G_0(r)} = \frac{1}{r^2 \ln r}.$$

In both cases we have

$$\kappa(r) \geq -\frac{G_0''(r)}{G_0(r)}$$

for sufficiently large r. Now the proof can be completed, *mutatis mutandis*, as in (i). □

Combining THEOREM 6.4.2 and PROPOSITION 6.4.5, we obtain the following result.

Theorem 6.4.6. *Suppose that M is a radially symmetric Cartan-Hadamard manifold of dimension d. Let*

$$\kappa(r) = -\frac{G''(r)}{G(r)}$$

be the radial sectional curvature of M. Define the constant $c_2 = 1/2$ for $d = 2$ and $c_d = 1$ for $d \geq 3$.

(i) If there exist $c > c_d$ and r_0 such that

$$\kappa(r) \leq -\frac{c}{r^2 \ln r}, \qquad \forall r \geq r_0,$$

then the Dirichlet problem at infinity is uniquely solvable.

(ii) If there exists a constant r_0 such that

$$\kappa(r) \geq -\frac{c_d}{r^2 \ln r}, \qquad \forall r \geq r_0,$$

then every bounded harmonic function on M is a constant.

6.5. Coupling of Brownian motion

By a coupling of Brownian motion on a Riemannian manifold M we mean a stochastic process $Z = (X, Y)$ on $M \times M$ defined on a probability space $(\Omega, \mathscr{F}, \mathbb{P})$ such that the marginal processes X and Y are Brownian motions on M. The coupling time is the time they first meet:

$$T = \inf\{t : X_t = Y_t\}.$$

We always define $X_t = Y_t$ for $t \geq T$. In this and the next sections we will describe the Kendall-Cranston coupling, a probabilistic tool with various geometric applications. In the last section of the chapter, we use this coupling to prove sharp lower bounds for the first eigenvalue (spectral gap) of the Laplace-Beltrami operator on a compact Riemannian manifold of nonnegative Ricci curvature.

The original idea of coupling two processes of the same distribution comes from the theory of Markov chains. It is natural that we try to minimize the coupling time in some sense. Consider a Brownian motion $X = (X^1, X^2)$ on a euclidean plane starting from $(0, a)$ with a positive a.

In order that another Brownian motion Y starting from $(0, -a)$ to meet with X as soon as possible, they should always head towards each other. This means that we take $Y = (X^1, -X^2)$. With this simple example in mind, we now define, for two distinct points x and y on a general complete Riemannian manifold such that $x \notin C_y$, the mirror map

$$m_{xy} : T_xM \to T_yM$$

as follows: for each $w \in T_xM$, $m_{xy}w$ is obtained by first parallel-translating w along the unique minimal geodesic from x to y and then reflecting the resulting vector with respect to the hyperplane in T_yM perpendicular to the geodesic. It is clear that m_{xy} is an isometry. Let

$$\begin{aligned} D_M &= \{(x, y) \in M \times M : x = y\}, \\ C_M &= \{(x, y) \in M \times M : x \in C_y\}, \\ E_M &= M \backslash (C_M \cup D_M), \\ D_{\mathscr{O}(M)} &= \{(u, v) \in \mathscr{O}(M) \times \mathscr{O}(M) : \pi u = \pi v\}, \\ C_{\mathscr{O}(M)} &= \{(u, v) \in \mathscr{O}(M) \times \mathscr{O}(M) : (\pi u, \pi y) \in C_M\}, \\ E_{\mathscr{O}(M)} &= \mathscr{O}(M) \times \mathscr{O}(M) \backslash \left(C_{\mathscr{O}(M)} \cup D_{\mathscr{O}(M)}\right). \end{aligned}$$

We now define the Kendall coupling. Let W be a d-dimensional euclidean Bronwian motion and consider the following system of equations for a process $\Sigma = (U, V)$ on $E_{\mathscr{O}(M)}$:

$$\text{(6.5.1)} \qquad \begin{cases} dU_t = H_i(U_t) \circ dW_t^i, & U_0 = u_x, \\ dV_t = H_i(V_t) \circ dB_t^i, & V_0 = m_{xy}u_x, \\ dB_t = V_t^{-1} m_{X_tY_t} U_t dW_s, \\ X_t = \pi\, U_t, \\ Y_t = \pi\, V_t. \end{cases}$$

Here U is simply a horizontal Brownian motion driven by W, and X a Brownian motion on M. Because U_t, V_t, and $m_{X_tY_t}$ are isometries, from the third equation B is a euclidean Brownian mtoion; hence from the second equation V is another horizontal Brownian motion, and Y is another Brownian motion on M. It is not difficult to convince ourselves that (6.5.1) is a Stratonovich type stochastic differential equation on $E_{\mathscr{O}(M)}$. Indeed, for $\sigma = (u, v) \in E_{\mathscr{O}(M)}$, the matrix-valued function

$$e^*(\sigma) = \left\{e_{ij}^*(\sigma)\right\} \stackrel{\text{def}}{=} v^{-1} m_{\pi u \pi v} u.$$

This is a smooth function function on $\{\mathscr{O}(M) \times \mathscr{O}(M)\} \backslash C_{\mathscr{O}(M)}$; hence with the aid of the first two equations, we can rewrite the differential of B in the third equation of (6.5.1) as a Stratonovich differential with respect to dW and dt with coefficients are smooth functions of U_t and V_t. Substituting this expression of dB in the second equation, the first two equation become a the Stratonovich type equation for $\Sigma = (U, V)$ on $E_{\mathscr{O}(M)}$ driven by the semimartingale $t \mapsto (W_t, t)$. By general theory (6.5.1) has a unique solution up to $T \wedge T_{C_M}$.

We now describe the generator of the coupled horizontal Brownian motions. In the following, for a function f defined on $\mathscr{O}(M) \times \mathscr{O}(M)$, we will use $H_{i,1}f$ and $H_{2,i}f$ to denote the derivatives of f with respect to the horizontal vector field H_i on the first and the second variable respectively. The horizontal Laplacian on the first and the second variable are

$$\Delta_{\mathscr{O}(M),1} = \sum_{i=1}^{d} H_{i,1}^2, \quad \Delta_{\mathscr{O}(M),2} = \sum_{i=1}^{d} H_{i,2}^2.$$

Recall that whenever $e = \{e_i\} \in \mathbb{R}^d$ we write

$$H_e = \sum_{i=1}^{d} e_i H_i.$$

Define a second order differential operator on $E_{\mathscr{O}(M)}$ by

$$\Delta_{\mathscr{O}(M)}^c = \Delta_{\mathscr{O}(M),1} + \Delta_{\mathscr{O}(M),2} + 2H_{e_i^*,2}H_{i,1}, \tag{6.5.2}$$

where

$$e_i^*(\sigma) = v^{-1} m_{\pi u \pi v} u e_i \in \mathbb{R}^d \qquad \text{for } \sigma = (u, v) E_{\mathscr{O}(M)}.$$

It is a lift of the following operator on E_M:

$$\Delta_M^c = \Delta_{M,1} + \Delta_{M,2} + 2\langle m_{xy} X_i, Y_j \rangle X_{i,1} Y_{j,2},$$

where $\{X_i\}$ and $\{Y_j\}$ are any orthonormal bases at x and y respectively. The last term is independent of the choice of these bases.

Theorem 6.5.1. *The coupled horizontal Brownian motion $\Sigma = (U, V)$ (well defined up to $T \wedge T_{C_M}$) is a diffusion on $E_{\mathscr{O}(M)}$ generated by $\Delta_{\mathscr{O}(M)}^c/2$. The coupled Brownian motions $Z = (X, Y)$ is a diffusion on E_M generated by Δ_M^c.*

Proof. Let f be a smooth function on $\mathscr{O}(M) \times \mathscr{O}(M)$. Applying Itô's formula to $f(\Sigma_t)$ we have

$$d\{f(\Sigma_t)\} = H_{i,1} f(\Sigma) \circ dW_t^i + H_{i,2} f(\Sigma_t) \circ dB_t^i. \tag{6.5.3}$$

Applying Itô's formula again, we have

$$d\{H_{i,1} f(\Sigma_t)\} = H_{j,1} H_{i,1} f(\Sigma_t) \circ dW_t^j + H_{j,2} H_{i,1} f(\Sigma_t) \circ dB_t^j.$$

We have

$$dB_t^i = \langle e_k^*, e_j \rangle (\Sigma_t)\, dW_t^k.$$

Hence, the first term on the right side of (6.5.3) is

$$H_{i,1}f(\Sigma_t) \circ dW_t^i = d\,\{\text{martingale}\} + \frac{1}{2}H_{i,1}^2 f(\Sigma_t)\,dt + \frac{1}{2}H_{e_i^*,2}H_{i,1}f(\Sigma_t)\,dt.$$

Likewise the second term on the right side of (6.5.3) is

$$H_{i,2}f(\Sigma_t) \circ dB_t^i = d\,\{\text{martingale}\} + \frac{1}{2}H_{i,2}^2 f(\Sigma_t)\,dt + \frac{1}{2}H_{e_j^*,2}H_{j,1}f(\Sigma_t)\,dt.$$

It follows that

$$d\,\{f(\Sigma_t)\} = d\,\{\text{martingale}\} + \frac{1}{2}\Delta^c_{\mathscr{O}(M)}f(\Sigma_t)\,dt.$$

This completes the proof. □

6.6. Coupling and index form

We have seen the important role that the radial process of a Brownian motion plays in many applications to geometric problems. It is therefore natural that we study the distance between the coupled Brownian motions $\rho_t = \rho(X_t, Y_t)$. Here and for the rest of this section, $\rho(x, y)$ is the Riemannian distance between two points x and y. Using (6.5.1) on the distance function, we have

$$d\rho_t = dM_t + \frac{1}{2}\Delta^c_M \rho(Z_t)\,dt,$$

where

$$dM_t = \left\{H_{i,1} + H_{e_i^*,2}\right\}\tilde{\rho}(\Sigma_t)dW_t^i.$$

Here $\tilde{\rho}$ is the lift of the distance function: $\tilde{\rho}(\sigma) = \rho(\pi u, \pi v)$. Let us first look at the martingale part. If we let

$$\hat{H}_i(\sigma) = H_{i,1}(\sigma) + H_{e_i^*,2}(\sigma),$$

then the quadratic variation of the martingale is given by

$$d\langle M\rangle_t = \left\{\sum_{i=1}^d \left|\hat{H}_i\tilde{\rho}(\Sigma_t)\right|^2\right\}dt. \tag{6.6.1}$$

Each $\hat{H}_i\tilde{\rho}(\sigma)$ is a well knwon quantity in differential geometry – it is the first variation of the geodesic from $x = \pi u$ to $y = \pi v$ along the Jacobi field J_i along the minimal geodesic $\gamma_{xy} : [0, \rho] \to M$:

$$\nabla_T\nabla_T J_i + R(J_i, T)T = 0, \quad J_i(0) = ue_i, \quad J_i(\rho) = m_{xy}ue_i,$$

where T is the tangent vector field along the geodesic. The sum in (6.6.1) is independent of the choice of the orthonormal basis $\{e_i\}$. We choose e_1 so that ue_1 is tangent to the geodesic. Then $ue_1 = T$ and $m_{xy}ue_1 = -T$.

For $i \geq 2$, ue_i is perpendicular to the geodesic and $m_{xy}ue_i$ is just the parallel translation of ue_i. Using the first variation formula (see Cheeger and Ebin [9], p.5)

$$\hat{H}_i\tilde{\rho}(\sigma) = \langle J_i, T\rangle_y - \langle J_i, T\rangle_x,$$

we have

$$\hat{H}_i\tilde{\rho}(\sigma) = \begin{cases} -2, & \text{if } i = 1, \\ 0, & \text{if } i \geq 2. \end{cases}$$

It follows that there is a Brownian motion β such that

$$d\rho_t = 2d\beta_t + \frac{1}{2}\Delta^c_M \rho(Z_t)\, dt.$$

The following lemma identifies the bounded variation part of ρ_t. Recall that the index form of the Jacobi field J_i is defined by

$$I(J_i, J_i) = \int_{\gamma_{xy}} |\nabla_T J_i|^2 - \langle R(J_i, T)T, J_i\rangle,$$

where the integral is along the minimal geodesic γ_{xy} from x to y.

Lemma 6.6.1. $\Delta^c_M \rho(z) = \sum_{i=2}^d I(J_i, J_i)$.

Proof. Fix a $\sigma \in E_{\mathscr{O}(M)}$ such that $z = \pi\sigma$. We have

$$\Delta^c_M \rho(z) = \Delta^c_{\mathscr{O}(M)}\tilde{\rho}(\sigma),$$

where $\Delta^c_{\mathscr{O}(M)}$ is given by (6.5.2). It is easy to verify that the right side is independent of the choice of the basis $\{e_i\}$, so we can choose it so that ue_1 is tangent to the geodesic γ_{xy}. We have

$$\Delta_{\mathscr{O}(M),2} = \sum_{i,j,k=1}^d \langle e_i^*, e_j\rangle\langle e_i^*, e_k\rangle H_{j,2}H_{k,2} = \sum_{i,k=1}^d \langle e_i^*, e_k\rangle H_{e_i^*,2}H_{k,2}.$$

Note that $e_i^* = e_i^*(\sigma)$ may not be a constant function on $E_{\mathscr{O}(M)}$, so in general

$$\langle e_i^*, e_k\rangle H_{e_i^*,2}H_{k,2} \neq H^2_{e_i^*,2}.$$

Now, from (6.5.2),

$$\Delta^2_{\mathscr{O}(M)}\tilde{\rho}(\sigma) = \sum_{i=1}^d \left\{ H^2_{i,1} + \sum_{k=1}^d \langle e_i^*, e_k\rangle H_{e_i^*,2}H_{k,2} + 2H_{e_i^*,2}H_{i,1} \right\} \tilde{\rho}(\sigma).$$

The term for $i = 1$ is equal to zero; therefore it is enough to show that for each $i \geq 2$,

$$\left\{ H^2_{i,1} + \sum_{j,k=1}^d \langle e_j^*, e_k\rangle H_{e_i^*,2}H_{k,2} + 2H_{e_i^*,2}H_{i,1} \right\} \tilde{\rho}(\sigma) = I(J_i, J_i). \tag{6.6.2}$$

Let γ_1 and γ_2 be the geodesics from x and y with the initial directions ue_i and $m_{xy}ue_i$ respectively. By the second variation formula (see Cheeger and Ebin [**9**], pp.21–22),

$$\left(\frac{d}{dt}\right)^2 \rho(\gamma_1(t), \gamma_2(t)) = I(J_i, J_i), \tag{6.6.3}$$

where the derivative is evaluated at $t = 0$. Let $\{u_t\}$ and $\{v_t\}$ be the horizontal lifts of γ_1 and γ_2 starting from u and v respectively. Then we have

$$\left(\frac{d}{dt}\right)^2 \rho(\gamma_1(t), \gamma_2(t)) = \left(\frac{d}{dt}\right)^2 \tilde{\rho}(u_t, v_t). \tag{6.6.4}$$

By the chain rule, we have

$$\left(\frac{d}{dt}\right)^2 \tilde{\rho}(u_t, v_t) = \left(\frac{d}{dt}\right)^2 \tilde{\rho}(u_t, v) + \left(\frac{d}{dt}\right)^2 \tilde{\rho}(u, v_t) + 2\frac{d}{ds}\frac{d}{dt}\tilde{\rho}(u_s, v_t), \tag{6.6.5}$$

all derivatives being evaluated at $t = s = 0$. Now, by the choices of $\{u_t\}$ and $\{v_t\}$ we have

$$\begin{cases} \left(\dfrac{d}{dt}\right)^2 \tilde{\rho}(u_t, v) = H_{i,1}^2 \tilde{\rho}(\sigma), \\ \left(\dfrac{d}{dt}\right)^2 \tilde{\rho}(u, v_t) = \langle e_i^*, e_j\rangle \langle e_i^*, e_k\rangle H_{j,2} H_{k,2} \tilde{\rho}(\sigma), \\ \dfrac{d}{ds}\dfrac{d}{dt}\tilde{\rho}(u_s, v_t) = H_{e_i^*,2} H_{i,1} \tilde{\rho}(\sigma). \end{cases} \tag{6.6.6}$$

The desired relation (6.6.2) follows from (6.6.3) to (6.6.6). □

To summarize, there is a Brownian motion β such that

$$d\rho_t = 2\beta_t + \frac{1}{2}\left\{\sum_{i=2}^{d} I(J_i, J_i)\right\} dt. \tag{6.6.7}$$

Eor each t, $I(J_i, J_i)$ is the index form of the Jacobi field J_i along the minimal geodesic $\gamma_{X_tY_t}$ with boundary values $J_i(0) = X_i$ and $J_i(\rho_t) = m_{X_tY_t}X_i$, where $\{X_i\}$ is a set of orthonormal vectors in $T_{X_t}M$ perpendicular to $\dot{\gamma}_{X_tY_t}(0)$.

So far we have defined the coupled Brownian motions only up to the stopping time $T \wedge T_{C_M}$. As with the case of the radial process of a Brownian motion, the coupling $Z = (X, Y)$ becomes useful only when we can pass beyond the first hitting time T_{C_M} of the joint cutlocus C_M and extend Z up to T, the coupling time. The following theorem shows that such an extension is indeed possible; cf. THEOREM 3.5.1.

Theorem 6.6.2. (i) *The process $Z = (X, Y)$ defined up to $T_{C_M} \wedge T$ can be extended beyond T_{C_M} up to the coupling time T such that X and Y*

are Brownian motions on M and $Z = (X, Y)$ (or rather, its horizontal lift $\Sigma = (U, V)$) satisfies (6.5.1) when $Z_t \notin C_M$.

(ii) *Let $\rho_t = \rho(X_t, Y_t)$ be the distance between the coupled Brownian motions. There are a Brownian motion β and a nondecreasing process L which increases only when $Z_t \in C_M$, the joint cutlocus of M, such that, before the coupling time of Z,*

$$\rho_t = \rho_o + 2\beta_t + \frac{1}{2}\int_0^t \left\{ \sum_{i=2}^{d} I(J_i, J_i) \right\} ds - L_t.$$

Proof. We sketch the proof. Let X be a Brownian motion on M starting from x. Let U be a horizontal lift of X and W its anti-development. Let y be another point on M different from x such that $y \notin C_x$. For each fixed small positive δ, we define a coupled Brownian motion Y^δ as follows. Let Y^δ be the Kendall coupling before the stopping time

$$\tau_1^\delta = \inf\left\{ t : d_M(Y_t^\delta, C_{X_t}) \le \delta \right\}.$$

Beginning at τ_δ^1, we let Y^δ continue as a Brownian motion independent of X until

$$\tau_2^\delta = \inf\left\{ t > \tau_1^\delta : d_M(Y_t^\delta, C_{X_t}) = 2\delta \right\}.$$

During the time interval $[\tau_1^\delta, \tau_2^\delta]$, the increment of Y^δ is independent of X. A slight extension of the proof of THEOREM 3.5.1 (now with an independently moving initial point instead of a fixed one) shows that there are a Brownian motion β^δ and a nondecreasing process L^δ which increases only when $Y_t^\delta \in C_{X_t}$ such that

$$d\rho(X_t, Y_t^\delta) = 2d\beta_t^\delta + \frac{1}{2}\left\{ \Delta_{M,1} + \Delta_{M,2} \right\} \rho(Z_t^\delta)\, dt - dL_t^\delta. \tag{6.6.8}$$

Note that, because of the independence during the time interval $[\tau_1^\delta, \tau_2^\delta]$, the generator for $Z^\delta = (X, Y^\delta)$, is $\left\{ \Delta_{M,1} + \Delta_{M,2} \right\}/2$. At the end of this time interval $Y_{\tau_2^\delta}^\delta \notin C_{X_{\tau_2^\delta}}$, and we let Y^δ resume the Kendall coupling after time τ_2^δ. This process is to be continued until the coupling time. It is clear from the construction that Y^δ is a Brownian motion on M.

We now examine the situation as $\delta \downarrow 0$. Let

$$A(\delta) = \bigcup_{l=1}^{\infty} [\tau_{2l-1}^\delta, \tau_{2l}^\delta]$$

be the union of the time intervals during which Y^δ runs as an independent Brownian motion. For $t \in A(\delta)$, the process Y_t^δ lies in the 2δ-neighborhood of the cutlocus C_{X_t}. Therefore, from

$$\mathbb{E}\big|A(\delta)\cap[0,t]\big| \le \sup_{x\in M}\int_0^t ds\int_{\rho(y,C_x)\le 2\delta} p_M(s,x,y)\,dy$$

we have

$$\lim_{\delta\downarrow 0}\mathbb{E}\big|A(\delta)\cap[0,t]\big| = 0. \tag{6.6.9}$$

Now, fix a frame v at y and let V^δ be the horizontal lift of Y^δ from v and $\Sigma^\delta = (U, V^\delta)$. The equation for V^δ can be written as

$$dV^\delta = I_{A(\delta)^c}(t)\,H_i(V_t^\delta)\circ e^*(\Sigma_t^\delta)\,dW_t + I_{A(\delta)}(t)\,dV_t^\delta.$$

Since H_i and e^* are smooth in their respective variables, using (6.6.9) it is easy to show that if δ runs through a sequence which goes to zero fast enough, with probability one, V^δ converges to a horizontal Brownian motion V pathwise such that $dV_t = H_i(V_t)\circ e^*(\Sigma_t)dW_t$, where $\Sigma = (U,V)$. This proves the first part of the theorem.

From (6.6.8) for $t\in A(\delta)$ and (6.6.7) for $t\in A(\delta)^c$ we have

$$\begin{aligned}\rho(X_t,Y_t^\delta) = &\int_0^t I_{A(\delta)^c}(s)\left[2d\beta_s + \frac{1}{2}\left\{\sum_{i=2}^d I(J_i,J_i)\right\}ds\right]\\ &+\int_0^t I_{A(\delta)}(s)\left[2d\beta_s^\delta + \frac{1}{2}\left\{\Delta_{M,1}+\Delta_{M,2}\right\}\rho(Z_s^\delta)\,ds\right] - L_t^\delta.\end{aligned}$$

We note that $\rho\Delta_{M,i}\rho$ is uniformly bounded on $M\times M$ (see COROLLARY 3.4.5). Thanks to (6.6.9), as $\delta\downarrow 0$, the second term on the right side tends to zero. It follows that $\lim_{\delta\downarrow 0} L_t^\delta = L_t$ must exists and satisfies the properties stated in the second part of the theorem. □

As before, we define $Y_t = X_t$ for $t\ge T$. The resulting pair $Z=(X,Y)$ of coupled Brownian motion is called the Kendall-Cranston coupling..

6.7. Eigenvalue estimates

In this section we explain the method of Cheng and Wang [**10**] of using the Kendall-Cranston coupling to study the first eigenvalue (spectral gap) of the Laplace-Beltrami operator on a compact Riemannian manifold with nonnegative Ricci curvature. A lower bound for the eigenvalue will be obtained under the assumptions $\mathrm{Ric}_M \ge (d-1)K$ for some $K\ge 0$.

It has long been known in probability theory that fast coupling implies large spectral gap for the generator of the diffusion process. Thus estimating the first eigenvalue from below relies on an appropriate upper bound on the drift part of $\rho_t = \rho(X_t,Y_t)$. From the previous section, we see that the drift part is expressed in terms of the index forms. The following well known fact about index forms gives an effective way of obtaining upper bounds for index forms.

Lemma 6.7.1. (Index lemma) *Let C be a geodesic segment in M from x to y such that there are no points conjugate to x on C. Let V be a vector field along C and J the Jacobi field along C with the same boundary values as V. Then $I(J,J) \le I(V,V)$. In other words, among vector fields with the same boundary conditions, the index form takes the minimum value at the Jacobi field.*

Proof. See Cheeger and Ebin [**9**], LEMMA 1.21. □

Let M be a compact Riemannian manifold such that $\mathrm{Ric}_M(x) \ge (d-1)K$. We first need to find an upper bound for the index form. Suppose that $x, y \in M$ are not on each other's cutlocus, and let $C = \{C_s, 0 \le s \le \rho\}$ be the distance-minimizing geodesic from x to y. Let $\{e_i(0)\}$ be an orthonormal basis on T_xM such that $e_1(0) = \dot{C}(0)$. Let $\{e^i(s), 0 \le s \le \rho\}$ be the parallel translation of $e_i(0)$ along C. For each $i \ge 2$, let J_i be the Jacobi field along C such that $J_i(0) = e_i(0)$ and $J_i(\rho) = e_i(\rho)$. We will use the index LEMMA 6.7.1 to obtain an upper bound for the the sum of the index forms $\sum_{i=2}^d I(J_i, J_i)$.

If M has constant sectional curvature K, then the Jacobi field is

$$V_i(s) = j(s)e_i(s),$$

where j is the solution of the Jacobi equation

$$j''(s) + Kj(s) = 0, \quad j(0) = 1, \quad j(\rho) = 1.$$

Explicitly,

$$j(s) = \cos s\sqrt{K} + \frac{1 - \cos \rho\sqrt{K}}{\sin \rho\sqrt{K}} \sin s\sqrt{K}.$$

For a general M, we define the vector field V_i along C by the above formula. By the index LEMMA 6.7.1, we have $I(J_i, J_i) \le I(V_i, V_i)$. On the other hand, using the definition of the Ricci curvature, we have

$$\begin{aligned}
\sum_{i=2}^d I(V_i, V_i) &= \sum_{i=2}^d \int_C |\nabla_T V_i|^2 - \langle R(V_i, T)T, V_i\rangle \\
&= \int_0^\rho \left[(d-1)|j'(s)|^2 - \sum_{i=2}^d \langle R(V_i, T)T, V_i\rangle \right] ds \\
&= \int_0^\rho \left[(d-1)|j'(s)|^2 - j(s)^2 \mathrm{Ric}(T,T) \right] ds \\
&\le (d-1) \int_0^\rho \left[|j'(s)|^2 - Kj(s)^2 \right] ds.
\end{aligned}$$

The last integral can be computed easily by integrating by parts and using the Jacobi equation for j. We obtain

$$\int_0^t \left[|j'(s)|^2 - Kj(s)^2\right] ds = j'(\rho) - j'(0) = -2\tan\frac{\sqrt{K}\rho}{2}.$$

Compare the above computation with the proof of the Laplacian comparison THEOREM 3.4.2. The following proposition summarizes what we have obtained so far.

Proposition 6.7.2. *Let M be a compact Riemannian manifold such that $\mathrm{Ric}_M(x) \geq (d-1)K$. Then for the distance $\rho_t = \rho(X_t, Y_t)$ of the Kendall-Cranston coupling, we have*

$$d\rho_t = 2d\beta_t - (d-1)\left[\tan\frac{\sqrt{K}\rho_t}{2}\right] dt - dA_t, \tag{6.7.1}$$

where A is a nondecreasing process.

Let $\lambda_1(M)$ be the first nonzero eigenvalue, i.e., the spectral gap, of the Laplace-Beltrami operator Δ_M. We have the following Lichnerowicz theorem.

Theorem 6.7.3. *Let M be a compact Riemannian manifold of dimension d such that $\mathrm{Ric}_M(x) \geq (d-1)K \geq 0$. Then $\lambda_1(M) \geq dK$.*

Proof. By Myers' theorem the diameter of M is $d(M) \leq \pi/\sqrt{K}$. Let

$$\tilde{\rho}_t = \sin\frac{\sqrt{K}}{2}\rho_t, \qquad \rho_t = \rho(X_t, Y_t).$$

Then (6.7.1) and Itô's formula imply that

$$\tilde{\rho}_t \leq \text{martingale} - \frac{dK}{2}\int_0^t \tilde{\rho}_s\, ds.$$

Taking the expectation we have

$$\mathbb{E}\,\tilde{\rho}_t \leq \tilde{\rho}_0 - \frac{dK}{2}\int_0^t \mathbb{E}\,\tilde{\rho}_s ds.$$

Hence by Gronwall's lemma,

$$\mathbb{E}\,\tilde{\rho}_t \leq e^{-dKt/2}\sin\frac{\sqrt{K}\rho(x,y)}{2}. \tag{6.7.2}$$

Now let ϕ_1 be an eigenfunction of $\lambda_1(M)$. Since the marginal processes X and Y are Brownian motions, we have

$$\mathbb{E}\,\phi_1(X_t) = e^{-\lambda_1(M)t/2}\phi_1(x),$$
$$\mathbb{E}\,\phi_1(Y_t) = e^{-\lambda_1(M)t/2}\phi_1(y);$$

hence

$$e^{-\lambda_1(M)t/2}\left\{\phi_1(x) - \phi_1(y)\right\} = \mathbb{E}\left\{\phi_1(X_t) - \phi_1(Y_t)\right\}.$$

We have

$$\left|\phi_1(X_t) - \phi_1(Y_t)\right| \le \|\nabla\phi_1\|_\infty \rho(X_t, Y_t).$$

On the other hand,

$$\rho(X_t, Y_t) = \rho_t \le \frac{\pi}{\sqrt{K}}\,\tilde{\rho}_t.$$

Hence

$$e^{-\lambda_1(M)t/2}\left|\phi_1(x) - \phi_1(y)\right| \le \frac{\pi\|\nabla\phi_1\|_\infty}{\sqrt{K}}\,\mathbb{E}\,\tilde{\rho}_t.$$

Using (6.7.2), we have

$$e^{-\lambda_1(M)t/2}\left|\phi_1(x) - \phi_1(y)\right| \le \frac{\pi\|\nabla\phi_1\|_\infty}{\sqrt{K}}\,\sin\frac{\sqrt{K}\rho(x,y)}{2}\,e^{-dKt/2}.$$

Taking two points $x, y \in M$ such that $\phi_1(x) \ne \phi_1(y)$ and letting $t \to \infty$ in the above inequality, we obtain $\lambda_1(M) \ge dK$. □

The above result says nothing about the case $\mathrm{Ric}_M(x) \ge 0$. In this case we need to choose a different function $\widetilde{\rho}$ of ρ. Let $d(M)$ be the diameter of M. The following result is due to Zhong and Yang [**72**].

Theorem 6.7.4. *Let M be a compact Riemannian manifold with nonnegative Ricci curvature. Then $\lambda_1(M) \ge \pi^2/d(M)^2$.*

Proof. We take

$$\tilde{\rho}_t = \sin\frac{\pi}{2}\frac{\rho_t}{d(M)}.$$

By (6.7.1), the decomposition of ρ_t in this case is simply $d\rho_t = 2d\beta_t - dA_t$; hence

$$\tilde{\rho}_t \le \text{martingale} - \frac{1}{2}\frac{\pi^2}{d(M)^2}\int_0^t \tilde{\rho}_s\,ds.$$

This implies as before,

$$\mathbb{E}\,\rho_t \le e^{-\pi^2 t/2d(M)^2}\sin\frac{\pi}{2}\frac{\rho(x,y)}{d(M)}.$$

Now the proof can be completed in the same way as in Theorem 6.7.3. □

Nothing is gained from the above proof of the Lichnerowicz lower bound $\lambda_1(M) \ge dK$ using coupling of Brownian motion, even for a probabilist, for the geometric proof is rather easy (see Li [**55**]). However, with the Zhong-Yang lower bound $\lambda_1(M) \ge \pi^2/d(M)^2$, the reader will surely draw an entirely different conclusion from comparing the analytic proofs (e.g., Li [**55**], or Schoen and Yau [**64**]) with the well-motivated probabilistic proof presented here.

Chapter 7

Brownian Motion and Index Theorems

This chapter is devoted to the probabilistic proofs of two index theorems, the Gauss-Bonnet-Chern theorem and the Atiyah-Singer index theorem, i.e., the index theorem for the Dirac operator on a twisted spin bundle. Our proof of the Gauss-Bonnet-Chern theorem gives a fairly good overview of the application of stochastic analysis in this area. The proof of the Atiyah-Singer index theorem requires more technical preparations and can be skipped at the reader's discretion.

We will start with a review of some relevant facts on differential forms on manifolds. There are two invariantly defined Laplacians on differential forms: covariant Laplacian (also called rough Laplacian) and the Hodge-de Rham Laplacian. While the former is naturally associated with Brownian motion, the latter is geometrically more useful because it commutes with exterior differentiation. They are related by the Weitzenböck formula. We will review the proof of this formula and show how the heat equation on forms can be solved by using a multiplicative Feynman-Kac functional defined by the curvature tensor. The probabilistic representation of the solution to the heat equation gives rise to a reprensentation of the heat kernel on differential forms in terms of a Brownian bridge and the heat kernel on functions. Based on this representation, we prove Patodi's local Gauss-Bonnet-Chern theorem, which, after integrating over the manifold, gives the usual Gauss-Bonnet-Chern theorem.

The proof of the Atiyah-Singer index theorem for the Dirac operator on a twisted spin bundle follows basically the same idea, although we have to devote a significant number of pages to necessary geometric preliminaries.

After a quick review of Clifford algebra and the spin group, we introduce the Dirac operator on a twisted spin bundle. Unlike the case of the Gauss-Bonnet-Chern theorem, we need to calculate the first order term of the Brownian holonomy, and for this purpose we also need to show that in the normal coordinates a properly scaled Brownian bridge on a Riemannian manifold converges to a euclidean Brownian bridge. These technical details are carried out in the last section.

7.1. Weitzenböck formula

In this section we set up the necessary geometric framework to be used in the rest of the chapter. A good reference for this section is the book *Riemannian Geometry and Geometric Analysis* by Jost [**50**].

Let M be a Riemannian manifold and Δ_M the Laplace-Beltrami operator. By definition, Δ_M is the trace of the Hessian $\nabla^2 f$; namely,

$$\Delta_M f(x) = \sum_{i=1}^{d} \nabla^2 f(X_i, X_i),$$

where $\{X_i\}$ is any orthonormal basis of $T_x M$. Bochner's horizontal Laplacian on the orthonormal frame bundle is defined by

$$\Delta_{\mathscr{O}(M)} = \sum_{i=1}^{d} H_i^2,$$

where $\{H_i\}$ are the fundamental horizontal vector fields on $\mathscr{O}(M)$. It is the lift of the Laplace-Beltrami operator Δ_M in the sense that

$$\Delta_M f(x) = \Delta_{\mathscr{O}(M)} \tilde{f}(u), \tag{7.1.1}$$

where $\tilde{f} = f \circ \pi$ and $\pi u = x$; see PROPOSITION 3.1.2.

The Laplace-Beltrami operator Δ_M on functions can be extended to tensor fields by the same relation:

$$\Delta_M \theta = \sum_{i=1}^{d} \nabla^2 \theta(X_i, X_i).$$

If $\theta \in \Gamma(T^{p,q}M)$ is a (p,q)-tensor field on M, then $\nabla\theta$, the covariant derivative of θ, is a $(p, q+1)$-tensor field. Taking the covariant derivative again, we obtain a $(p, q+2)$-tensor field $\nabla^2\theta$. Thus $\Delta_M\theta$ is obtained from $\nabla^2\theta$ by contracting (with respect to the Riemannian metric) the two new components. The Δ_M thus defined on differential forms is called the covariant (or rough) Laplacian .

On the orthonormal frame bundle $\mathscr{O}(M)$, a (p,q)-tensor field θ is lifted to its scalarization $\tilde{\theta}$ defined by

$$\tilde{\theta}(u) = u^{-1}\theta(\pi u).$$

Here a frame $u : \mathbb{R}^d \to T_x M$ is assumed to be extended canonically to an isometry $u : T^{p,q}\mathbb{R}^d \to T^{p,q}_{\pi u} M$. By definition, $\tilde{\theta}$ is a function on $\mathscr{O}(M)$ taking values in the vector space $T^{p,q}\mathbb{R}^d$ and is $O(d)$-invariant in the sense that $\tilde{\theta}(gu) = g\tilde{\theta}(u)$ for $g \in O(d)$. The horizontal derivatives $H_i\tilde{\theta}$ and Bochner's horizontal Laplacian $\Delta_{\mathscr{O}(M)}\tilde{\theta}$ are well defined, and we have a generalization of (7.1.1) (see PROPOSITION 2.2.1):

$$\Delta_{\mathscr{O}(M)}\tilde{\theta}(u) = u^{-1}\Delta_M \theta(x), \qquad \pi u = x.$$

Let us now turn to the Hodge-de Rham Laplacian on differential forms.

A p-form at $x \in M$ is an alternating multilinear form on the tangent space $T_x M$, or equivalently an alternating $(0,p)$-tensor. We denote the space of p-forms at x by $\wedge^p_x M$ and the vector bundle of p-forms of M by $\wedge^p M$. Thus $\Gamma(\wedge^p M)$ is the space of p-forms on M. If θ_1 and θ_2 are a p-form and a q-form respectively, then the wedge product $\theta_1 \wedge \theta_2$ is a $(p+q)$-form obtained by anti-symmetrizing the tensor product $\theta_1 \otimes \theta_2$. Locally a p-form can be expressed as

$$\theta = \sum_{1 \le i_1, \ldots, i_p \le d} f_{i_1 \cdots i_p} dx^{i_1} \wedge \cdots \wedge dx^{i_p}, \tag{7.1.2}$$

where $f_{i_1 \cdots i_p}$ are smooth functions alternating in its supscripts. The exterior differentiation

$$d : \Gamma(\wedge^p M) \to \Gamma(\wedge^{p+1} M)$$

is uniquely determined by the following conditions:

(i) $df(X) = Xf$ for $f \in C^\infty(M)$ and $X \in \Gamma(TM)$;

(ii) d is an anti-derivation, i.e.,

$$d(\theta_1 \wedge \theta_2) = d\theta_1 \wedge \theta_2 + (-1)^{\deg\theta_1}\theta_1 \wedge d\theta_2;$$

(iii) $d^2 = 0$.

Locally we have

$$d\theta = \sum_{1 \le i_1, \ldots, i_p, i \le d} \partial_{x_i} f_{i_1 \cdots i_p} dx^i \wedge dx^{i_1} \wedge \cdots \wedge dx^{i_p}.$$

Although the covariant Laplacian Δ_M is naturally associated with Brownian motion on a manifold, it does not commute with the exterior differentiation. Geometrically more significant is the Hodge-de Rham Laplacian

$\square_M$, which we now define. For two differential forms α and β of the same degree and with compact support, we define their inner product by

$$(\alpha, \beta) = \int_M \langle \alpha, \beta \rangle_x dx,$$

where $\langle \alpha, \beta \rangle_x$ is the canonical inner product of α_x and β_x on T_xM. Let $\delta : \Gamma(\wedge^p M) \to \Gamma(\wedge^{p-1} M)$ be the formal adjoint of d with respect to this inner product, i.e.,

$$(d\alpha, \beta) = (\alpha, \delta\beta).$$

Then the Hodge-de Rham Laplacian is

$$\square_M = -(d\delta + \delta d).$$

Note that with the sign convention we have chosen $\square_M$ coincides with Δ_M on functions. Using the fact that $d^2 = 0$ (and hence also $\delta^2 = 0$) we verify easily that $d\square_M = \square_M d$. The difference between $\square_M$ and Δ_M is given by the Weitzenböck formula. It turns out that at each point $\square_M - \Delta_M$ is a linear transform on $\wedge_x^* M$ determined by the curvature tensor.

In order to derive the Weitzenböck formula, we need a covariant expression for both the exterior differentiation d and its adjoint δ. For a vector X and a p-form θ, the interior product $i(X)\theta$ is a $(p-1)$-form defined by

$$i(X)\theta(X_1, \ldots, X_{p-1}) = \theta(X, X_1, \ldots, X_{p-1}).$$

Thus $i(X)\theta$ is obtained from $X \otimes \theta$ by contracting the first two components.

Lemma 7.1.1. *Let $\{X_i\}$ be an orthonormal basis of T_xM and $\{X^i\}$ the dual basis for T_x^*M, i.e., $X^i(X_j) = \delta_j^i$. Then*

$$d\theta = X^i \wedge \nabla_{X_i}\theta \tag{7.1.3}$$

and

$$\delta\theta = -i(X_i)\nabla_{X_i}\theta. \tag{7.1.4}$$

Proof. Let $X_i = \partial/\partial x^i$, where $\{x^i\}$ are the normal coordinates at x. Then $X^i = dx^i$. Writing θ in local coordinates as in (7.1.2) and using

$$\nabla_{X_i}(dx^j) = -\Gamma_{ik}^j dx^k,$$

we verify by a straightforward calculation. We leave the details as an exercise. $\square$

We now describe the action of the curvature tensor R on differential forms at a fixed point in the Weitzenböck formula. Let $V = T_xM$ for simplicity, and keep in mind that the following discussion is valid for any finite dimensional inner product space V. Suppose that $T : V \to V$ is a linear transform and $T^* : V^* \to V^*$ its dual. Whenever feasible, we will write T^* simply as T to lessen the notation. The linear map T on V^* can be extended to the full exterior algebra $\wedge^* V = \sum_{p=0}^{n} \oplus \wedge^p V$ in two ways: as a Lie group $GL(d, \mathbb{R}^d)$-action:

$$\wedge^* T(\theta_1 \wedge \theta_2) = T\theta_1 \wedge T\theta_2,$$

or as a Lie algebra $\mathfrak{gl}(d, \mathbb{R}^d)$-action (derivation):

$$D^* T(\theta_1 \wedge \theta_2) = D^* T\theta_1 \wedge \theta_2 + \theta_1 \wedge D^* T\theta_2.$$

It is the second action of T (written as D^*T to distinguish it from the first) that is important to us. The relation between these two actions is as follows. Define the exponential e^T of a linear map T by

$$e^T = \sum_{n=0}^{\infty} \frac{T^n}{n!},$$

where T^n is the nth iteration of T. Then, as linear operators on $\wedge^* V$,

$$\wedge^* e^T = e^{D^* T}. \tag{7.1.5}$$

If no confusion is possible, we write $\wedge^* T$ simply as T.

Let $\operatorname{End}(V)$ denote the space of linear maps from V to itself. By the definition of tensor products we have $\operatorname{End}(V) = V^* \otimes V$. We now define a bilinear map

$$D^* : \operatorname{End}(V) \otimes \operatorname{End}(V) \to \operatorname{End}(\wedge^* V)$$

by

$$D^*(T_1 \otimes T_2) = D^* T_1 \circ D^* T_2.$$

By definition, the curvature tensor $R \in V^* \otimes V^* \otimes V^* \otimes V$. Using the isometry $V^* \to V$ induced by the inner product on the second component, we can write

$$R \in V^* \otimes V \otimes V^* \otimes V = \operatorname{End}(V) \otimes \operatorname{End}(V).$$

Thus the above definition can be applied, and we obtain a linear map

$$D^* R : \wedge^* V \to \wedge^* V.$$

For the proof of the Weitzenböck formula, it is helpful to write D^*R in terms of a local orthonormal basis.

Lemma 7.1.2. *Let $\{X_i\}$ be an orthonormal basis for V and $\{X^i\}$ the dual basis. Suppose that $T \in V^* \otimes V^*$. Then*

$$D^* T = T(X_i, X_j) X^i \wedge i(X_j).$$

Suppose that $R \in V^* \otimes V^* \otimes V^* \otimes V$. Then

$$D^* R = \sum_{i,j=1}^{d} X^i \wedge i(X_j) D^* R(X_i, X_j),$$

where $D^*R(X_i, X_j)$ *is the extension of* $R(X_i, X_j) : V \to V$ *to* $\wedge^* V$ *as a derivation.*

Proof. The first identity follows from the definition of D^*T. For the second identity it is enough to assume that $R = T_1 \otimes T_2$. In this case, by the first identity the right side reduces to

$$T_1(X_i, X_j)X^i \wedge i(X_j)\, D^*T_2 = D^*T_1 \circ D^*T_2,$$

which is equal to $D^*(T_1 \otimes T_2)$. □

Theorem 7.1.3. (Weitzenböck formula) *Let* Δ_M *and* $\Box_M$ *be the covariant Laplacian and the Hodge-de Rham Laplacian on a Riemannian manifold, and* R *the Riemannian curvature tensor. Then*

$$\Box_M = \Delta_M + D^*R.$$

Proof. Both sides are invariantly defined, so we need only to verify it in a specially chosen local orthonormal frame. Let $\{x^i\}$ be the normal coordinates at a fixed point x. Let $X_i = \partial/\partial x^i$ at x and define the vector field X_i in the neighborhood of x by parallel-translating along radial geodesics. Using the same proof as in LEMMA 2.5.5, we can show easily that $\nabla_{X_i} X_j = 0$ at x, i.e., all the Christoffel symbols vanish at x. Since the connection is torsion-free, we also have $[X_i, X_j] = 0$ at x. This gives

$$R(X_i, X_j) = \nabla_{X_i}\nabla_{X_j} - \nabla_{X_i}\nabla_{X_j}.$$

We know that $R(X_i, X_j) \in \mathrm{End}(T_x M)$. Its action $D^*R(X_i, X_j)$ (as a derivation) is given by the same formula, i.e.,

$$D^*R(X_i, X_j) = \nabla_{X_i}\nabla_{X_j} - \nabla_{X_i}\nabla_{X_j}.$$

Now,

$$\Delta_M \theta = \nabla^2\theta(X_i, X_i) = \nabla_{X_i}\nabla_{X_i}\theta - \nabla_{\nabla_{X_i}X_i}\theta = \nabla_{X_i}\nabla_{X_i}\theta.$$

Using (7.1.3), (7.1.4), and $\nabla_{X_j} X^i = 0$ at x, we have

$$-\delta d = i(X_j)\nabla_{X_j}(X^i \wedge \nabla_{X_i}) = \nabla_{X_i}\nabla_{X_i} - X^i \wedge i(X_j)\nabla_{X_j}\nabla_{X_i}.$$

On the other hand, because covariant differentiations commute with contractions,

$$\nabla_X i(Y) = i(Y)\nabla_X + i(\nabla_X Y).$$

Hence, using the fact that $\nabla_{X_i} X_j = 0$ again, we have

$$-d\delta = X^i \wedge \nabla_{X_i}(i(X_j)\nabla_{X_j}) = X^i \wedge i(X_j)\nabla_{X_i}\nabla_{X_j}.$$

Adding the two identities, we obtain

$$\Box_M = \nabla_{X_i}\nabla_{X_i} + X^i \wedge i(X_j)\left\{\nabla_{X_i}\nabla_{X_j} - \nabla_{X_j}\nabla_{X_i}\right\}.$$

The first term on the right side is the covariant Laplacian Δ_M, and the last term is equal to D^*R by LEMMA 7.1.2. □

The Weitzenböck formula takes an especially simple appearance on 1-forms.

Corollary 7.1.4. *If θ is a 1-form on M, then*

$$\Box_M\theta = \Delta_M\theta - \mathrm{Ric}\,\theta,$$

where Ric: $T_x^*M \to T_x^*M$ *is the Ricci transform.*

Proof. Let $\{X_i\}$ be an orthonormal basis for T_xM and $\{X^i\}$ the dual basis for T_x^*M. The curvature tensor is

$$R = \langle R(X_i, X_j)X_k, X_l\rangle X^i \otimes X^j \otimes X^k \otimes X_l.$$

By the definition of the action of (1,3)-tensor on 1-forms we have

$$D^*(X^i \otimes X^j \otimes X^k \otimes X_l)\,\theta = \delta_j^k\,\theta(X_l)\,X^i.$$

Hence we have

$$(D^*R)\,\theta = X^i\langle R(X_i, X_j)\theta, X_j\rangle = -\mathrm{Ric}\,\theta.$$

□

For our purpose it is more convenient to write the Weitzenböck formula on the orthonormal frame bundle $\mathscr{O}(M)$. The curvature form Ω of M is the $\mathfrak{o}(d)$-valued horizontal 2-form on $\mathscr{O}(M)$ defined by

$$\Omega(X, Y) = u^{-1}R(\pi_*X, \pi_*Y)\,u.$$

At a frame $u \in \mathscr{O}(M)$, Ω can be identified with an element in $(\mathbb{R}^d)^* \otimes \mathbb{R}^d \otimes (\mathbb{R}^d)^* \otimes \mathbb{R}^d$ (still denoted by Ω) as follows:

$$\Omega = \sum_{1\le i,j,k,l\le d} \Omega(H_i, H_j)_{kl}e^i \otimes e_j \otimes e^k \otimes e_l,$$

where $\{e_i\}$ is the canonical orthonormal basis for $\mathbb{R}^d$, $\{e^i\}$ its dual basis, and $\{H_i\}$ are the fundamental horizontal vector fields on $\mathscr{O}(M)$. By the procedure described above, $D^*\Omega$ is a linear map on the exterior algebra $\wedge^*\mathbb{R}^d$. Let

$$\Box_{\mathscr{O}(M)} = \Delta_{\mathscr{O}(M)} + D^*\Omega.$$

Then $\Box_{\mathscr{O}(M)}$ is a lift of the Hodge-de Rham Laplacian in the sense that

$$\Box_{\mathscr{O}(M)}\tilde{\theta}(u) = u^{-1}\Box_M\theta(x), \quad \pi u = x.$$

Finally, we express $D^*\Omega$ in component form. Let $\{\Omega_{ijkl}\}$ be the components of the anti-symmetric matrix $\Omega(H_i, H_j)$. The scalarization of a p-form θ can be written as

$$\tilde{\theta} = \theta_{i_1\cdots i_p}e^{i_1} \wedge \cdots \wedge e^{i_p},$$

where $\theta_{i_1\cdots i_p}$ are smooth functions on $\mathscr{O}(M)$ alternating in its indices. We leave it as an exercise to verify that

$$D^*\Omega\,\tilde\theta = \sum_{\alpha,\beta}\sum_{j_\alpha,l_\beta} \Omega_{l_\beta i_\beta j_\alpha i_\alpha}\theta_{i_1\cdots l_\beta\cdots i_p}e^{i_1}\wedge\cdots\wedge e^{j_\alpha}\wedge\cdots\wedge e^{i_p}. \tag{7.1.6}$$

7.2. Heat equation on differential forms

Consider the following initial-value problem for $\theta = \theta(t,x)$:

$$\begin{cases} \dfrac{\partial\theta}{\partial t} = \dfrac{1}{2}\Box_M\theta, & (t,x)\in(0,\infty)\times M; \\ \theta(0,x) = \theta_0(x), & x\in M. \end{cases} \tag{7.2.1}$$

We can rewrite the above equation on $\mathscr{O}(M)$. Let $\tilde\theta$ be the scalarization of θ. Then the above equation is equivalent to

$$\begin{cases} \dfrac{\partial\tilde\theta}{\partial t} = \dfrac{1}{2}\Box_{\mathscr{O}(M)}\tilde\theta, & (t,u)\in(0,\infty)\times \mathscr{O}(M); \\ \tilde\theta(0,u) = \tilde\theta_0(u), & u\in \mathscr{O}(M). \end{cases} \tag{7.2.2}$$

By the Weitzenböck formula

$$\Box_{\mathscr{O}(M)} = \Delta_{\mathscr{O}(M)} + D^*\Omega,$$

where $\Delta_{\mathscr{O}(M)}$ is Bochner's horizontal Laplacian. The difference $\Box_{\mathscr{O}(M)} - \Delta_{\mathscr{O}(M)} = D^*\Omega$ is a fibre-wise linear operator (a "potential"). The solution of the heat equation (7.2.1) can be obtained by using a matrix version of the well-known Feynman-Kac formula. Let M_t be the $\mathrm{End}(\wedge^*\mathbb{R}^d)$-valued multiplicative functional determined by

$$\frac{dM_t}{dt} = \frac{1}{2}M_t\,D^*\Omega(U_t), \qquad M_t = I_n \quad \text{(the identity matrix)}. \tag{7.2.3}$$

Theorem 7.2.1. *The solution of the initial value problem (7.2.2) is given by*

$$\tilde\theta(t,u) = \mathbb{E}_u\left\{M_t\tilde\theta_0(U_t)\right\}.$$

Correspondingly, the solution of (7.2.1) is given by

$$\theta(t,x) = \mathbb{E}_x\left\{M_tU_t^{-1}\theta_0(X_t)\right\},$$

where U is the horizontal lift of a Brownian motion X.

Proof. In view of the equivalence of (7.2.1) and (7.2.2), the second formula in the statement of the theorem is just a rewriting of the first formula.

Supppose that $\tilde{\theta}$ is a solution. Differentiating $M_s\tilde{\theta}(t-s, U_s)$, we have

$$\begin{aligned} d\left\{M_s\tilde{\theta}(t-s,U_s)\right\} &= M_s\, d\tilde{\theta}(t-s,U_s) + \frac{1}{2}M_s D^*\Omega(U_s)\tilde{\theta}(t-s,U_s)\,ds \\ &= M_s H_i U(t-s,U_s)\,dW_s^i \\ &\quad + M_s\left[\frac{\partial}{\partial s} + \frac{1}{2}\Delta_{\mathscr{O}(M)} + \frac{1}{2}D^*\Omega(U_s)\right]\tilde{\theta}(t-s,U_s)\,ds. \end{aligned}$$

The last term vanishes because $\tilde{\theta}$ is a solution of the heat equation. Hence $\left\{M_s\tilde{\theta}(t-s,U_s), 0 \le s \le t\right\}$ is a martingale. Equating the expected values at $s=0$ and $s=t$, we obtain $\tilde{\theta}(t,u) = \mathbb{E}_u\left\{M_t\tilde{\theta}_0(U_t)\right\}$. □

Remark 7.2.2. We often use a fixed frame U_0 to identify the tangent space T_xM with $\mathbb{R}^d$. Under this identification U_0 becomes the identity map and is often dropped from the notation. Thus a more precise writing of the second relation in the above theorem is $\theta(t,x) = \mathbb{E}_x\left\{U_0M_tU_t^{-1}\theta_0(X_t)\right\}$.

We use the above representation to prove a heat semigroup domination inequality. We need an elementary fact about matrix equations.

Lemma 7.2.3. *Suppose that $t \mapsto S_t$ is a continuous function on $[0,T]$ taking values in the space of symmetric $(n\times n)$-matrices such that $vS_tv^* \ge a|v|^2$ (i.e., the smallest eigenvalue of S_t is greater than or equal to a) for all t. Let A_t be the solution of the equation*

$$\frac{dA_t}{dt} + A_tS_t = 0, \qquad A_0 = I.$$

Then $|A|_{2,2} \le e^{-at}$.

Proof. Suppose that v is a column vector and let $f(t) = |v^*A_t|_2^2 = v^*A_tA_t^*v$. Differentiating with respect to t, we have

$$f'(t) = -2v^*A_tS_tA_t^*v \le -2a|v^*A_t|^2 = -2af(t).$$

Hence $f(t) \le e^{-2at}|v|^2$, or equivalently $|v^*A| \le e^{-at}|v|$. The inequality we wanted to prove follows by duality:

$$|Au|_2 = \sup_{|v|_2=1} v^*Au \le |u|_2 \sup_{|v|_2=1}|v^*A|_2 \le e^{-at}|u|_2.$$

□

Theorem 7.2.4. *Suppose that M is a complete Riemannian manifold with* $\mathrm{Ric}_M(x) \ge K$. *Let*

$$P_tf(x) = \mathbb{E}_xf(X_t) = \int_M p_M(t,x,y)f(y)dy$$

be the heat semigroup. Then

$$|\nabla P_t f| \le e^{-Kt/2} P_t |\nabla f|.$$

Proof. Consider $\theta(t,x) = dP_t f(x) = \nabla P_t f(x)$. Define P_t on differential forms as $P_t = e^{\square_M t/2}$. Since d commutes with $\square_M$, we have $\theta(t,x) = P_t(df)(x)$. Thus $\theta(t,x)$ is a solution to the heat equation (7.2.1). Hence

$$\nabla P_t f(x) = \mathbb{E}_x \left\{ M_t U_t^{-1} \nabla f(X_t) \right\}.$$

From LEMMAS 7.1.4 and 7.2.3 we have $|M_t|_{2,2} \le e^{-Kt/2}$. The desired result follows from this and the fact that U_t is an isometry. □

The solution of the heat equation on functions

$$\begin{cases} \dfrac{\partial f}{\partial t} = \dfrac{1}{2}\Delta_M f, & (t,x) \in \mathbb{R}_+ \times M, \\ f(0,x) = f(x), & x \in M, \end{cases}$$

is given by

$$f(t,x) = \int_M p_M(t,x,y) f(y)\, dy.$$

Similarly, in view of THEOREM 7.2.1, let

$$p_M^*(t,x,y) : \wedge_y^* M \to \wedge_x^* M$$

be the heat kernel on differential forms. Then

$$\mathbb{E}_x \left\{ M_t U_t^{-1} \theta(X_t) \right\} = \int_M p_M^*(t,x,y)\, \theta(y)\, dy. \tag{7.2.4}$$

Let $\mathbb{P}_{x,y;t}$ be the law of a Brownian bridge from x to y in time t, i.e., the Wiener mesure on the bridge space $L_{x,y;t}$. From (7.2.4) the heat kernel on differential forms can be written as

$$p_M^*(t,x,y) = p_M(t,x,y)\, \mathbb{E}_{x,y;t} \left\{ M_t U_t^{-1} \right\}.$$

Of particular importance is the value of the heat kernel on the diagonal

$$p_M^*(t,x,x) = p_M(t,x,x)\, \mathbb{E}_{x,x;t} \left\{ M_t U_t^{-1} \right\}, \tag{7.2.5}$$

where, under the probability $\mathbb{P}_{x,x;t}$,

$$U_t^{-1} : \wedge_x^* M \to \wedge_x^* M$$

is the stochastic parallel transport along the reversed Brownian bridge at x (Brownian holonomy). This representation of the heat kernel on the diagonal is the starting point of the probabilistic proof of the Gauss-Bonnet-Chern formula in the next section.

Remark 7.2.5. We have restricted ourselves in this section to the bundle of differential forms $\wedge^* M$, but everything we have said remains true for any vector bundle $\pi : E \to M$ over M equipped with a connection and a second order elliptic operator $\Box_E$ on $\Gamma(E)$ provided that there is a Weitzenböck formula for $\Box_E$, i.e., the difference $\Box_E - \Delta_E$ is a fibre-wise linear transform (a 0th order operator). We will see an example of this situation when we discuss the index theorem for the Dirac operator on a spin manifold.

7.3. Gauss-Bonnet-Chern formula

The Gauss-Bonnet-Chern theorem, one of the most beautiful theorems in differential geometry, is the index theorem for the operator $d + \delta$ on differential forms. It states that for an even dimensional, compact, and oriented Riemannian manifold M, its Euler characteristic $\chi(M)$, a topological invariant, is the integral of a function e defined in terms of the curvature tensor:

$$\chi(M) = \int_M e(x)\, dx.$$

The connection of this formula with Brownian motion is made through the heat kernel on differential forms (7.2.4), in particular its value on the diagonal (7.2.5). We start our probabilistic proof of the Gauss-Bonnet-Chern theorem with the relation between the Euler characteristic and the heat kernel on forms.

Let V be a finite-dimensional inner product space and $T \in \mathrm{End}(V)$ a linear map on V. As we mentioned in the last section, T can be extended to the exterior algebra as a degree-preserving derivation $D^*T : \wedge^* V \to \wedge^* V$. We define the supertrace $\mathrm{Trace}\, T$ by

$$\mathrm{Trace}\, T = \sum_{p=0}^{n} (-1)^p \mathrm{Tr}_{\wedge^p V} D^* T.$$

Let M be a compact, oriented Riemannian manifold. A form θ is called harmonic if $\Box_M \theta = 0$. Let $H^p(M)$ be the space of harmonic p-forms. From the theory of elliptic operators we know that each $H^p(M)$ is finite dimensional. For our purpose we define the Euler characteristc by

$$\chi(M) = \sum_{p=0}^{d} (-1)^p \dim H^p(M).$$

The Hodge-de Rham theory (see Warner [**71**]) shows that $H^p(M)$ is isomorphic to the pth cohomology group of M; hence $\chi(M)$ is in fact a topological invariant independent of the choice of the Riemannian metric on M. Either from geometric or topological consideration, the Poincaré duality gives a natural isomorphism between $H^p(M)$ and $H^{d-p}(M)$. Hence $\chi(M) = 0$ if d

is odd. This fact will also follow from our proof of the Gauss-Bonnet-Chern theorem.

We now connect the Euler characteristic $\chi(M)$ with the heat kernel $p_M^*(t,x,y)$ on forms. L^2-theory of elliptic operators shows that the spectrum of $\Box_M$ on $\Gamma(\wedge^* M)$ is discrete. Let

$$\lambda_0 = 0 \le \lambda_1 \le \lambda_2 \le \cdots$$

be the eigenvalues of $-\Box_M$ counted with multiplicity. There are orthonormal eigenforms $f_n \in \Gamma(\wedge^* M)$ such that $\Box_M f_n = -\lambda_n f_n$ and the heat kernel has the following L^2-expansion:

$$p_M^*(t,x,y) = \sum_{n=0}^{\infty} e^{-\lambda_n t/2} f_n(x) \otimes f_n^*(y). \tag{7.3.1}$$

The diagonal map $p_M^*(t,x,x) : \wedge_x^* M \to \wedge_x^* M$ is a degree-preserving map. Hence the supertrace $\mathrm{Trace} p_M^*(t,x,x)$ is well defined.

Theorem 7.3.1. *For any $t > 0$, we have*

$$\chi(M) = \int_M \mathrm{Trace}\, p_M^*(t,x,x)\, dx.$$

Proof. Let

$$\mu_0 = 0 < \mu_1 < \mu_2 < \cdots$$

be the *distinct* eigenvalues of $-\Box_M$. For each p, let E_i^p be the space of p-forms in the μ_i-eigenspace. If f is a normalized differential form, then it is clear that

$$\mathrm{Trace}(f(x) \otimes f^*(x)) = |f(x)|^2,$$

and hence

$$\int_M \mathrm{Trace}(f(x) \otimes f^*(x)) = 1.$$

It follows that

$$\int_M \mathrm{Trace}\, p_M^*(t,x,x)\, dx = \sum_{i=0}^{\infty} e^{-\mu_i t/2} \sum_{p=0}^{d} (-1)^p \dim E_i^p(M).$$

For $\mu_0 = 0$, the eigenspace $E_0^p(M) = H^p(M)$, the space of harmonic p-forms. Hence

$$\sum_{p=0}^{d} (-1)^p \dim E_0^p(M) = \chi(M).$$

It is therefore enough to show that for any $\mu_i > 0$,

$$\sum_{p=0}^{d} (-1)^p \dim E_i^p(M) = 0. \tag{7.3.2}$$

In order to prove this relation, we first note that if f is a p-eigenform with eigenvalue μ_i, then df is a $(p+1)$-eigenform, and δf is a $(p-1)$-eigenform, both with the same eigenvalue μ_i. This follows from the fact that both d and its adjoint δ commute with $\Box_M = -(d\delta + \delta d)$. Therefore we have

$$dE_i^{p-1} \subseteq E_i^p, \quad \delta E_i^{p+1} \subseteq E_i^p.$$

We claim that for any eigenvalue $\mu_i > 0$,

$$E_i^p = dE_i^{p-1} \oplus \delta E_i^{p+1}. \tag{7.3.3}$$

This is the Hodge decomposition. The orthogonality is obvious, because from $d^2 = 0$ we have

$$(df, \delta g) = (d^2 f, g) = 0.$$

Now suppose that an eigenform h is orthogonal to the subspace on the right side of (7.3.3). Since both dh and δh are eigenforms with the same eigenvalue, we have

$$(dh, dh) = (h, \delta dh) = 0, \quad (\delta h, \delta h) = (h, d\delta h) = 0.$$

Hence

$$-\mu_i(h, h) = (\Box_M h, h) = (dh, dh) + (\delta h, \delta h) = 0.$$

This implies $h = 0$, and (7.3.3) is proved.

From the decomposition (7.3.3) it is clear that the map $d : \delta E_i^{p+1} \to dE_i^p$ is onto; it is also one-to-one, for $d\delta h = 0$ implies $(\delta h, \delta h) = (d\delta h, h) = 0$; hence $\delta h = 0$. If we let $n_p = \dim(dE_i^p)$, then dim $E_i^p = n_{p-1} + n_p$, and we have

$$\sum_{p=0}^{d} (-1)^p \dim E_i^p = \sum_{p=0}^{d} (-1)^p (n_{p-1} + n_p) = 0.$$

□

We will use the representation of the heat kernel

$$p_M^*(t, x, x) = p_M(t, x, x) \, \mathbb{E}_{x,x;t} \left\{ M_t U_t^{-1} \right\}$$

to show that the limit

$$e(x) = \lim_{t \downarrow 0} \text{Trace}\, p_M^*(t, x, x)$$

exists and to express the limit explicitly in terms of the curvature tensor. This is Patodi's local Gauss-Bonnet-Chern theorem. Once this is proved, from THEOREM 7.3.1 we will have immediately the Gauss-Bonnet-Chern formula

$$\chi(M) = \int_M e(x) dx.$$

Note that from THEOREM 5.1.1, the behavior of the heat kernel on the diagonal is $p_M(t, x, x) \sim (2\pi t)^{-d/2}$ as $t \downarrow 0$. Therefore we expect some sort of cancellation in $\mathbb{E}_{x,x;t}\text{Trace}\{M_t U_t^{-1}\}$ so that it has the order of $t^{d/2}$. What we will show is that this "fantastic" cancellation (to quote McKean and Singer [**56**]) occurs at the path leve, namely,

$$\text{Trace}\left\{M_t U_t^{-1}\right\} \sim \text{const.}\, t^{d/2} \quad \text{as } t \downarrow 0.$$

This cancellation is a consequence of the following purely algebraic fact.

Lemma 7.3.2. *Let $T_i \in \text{End}(V), 1 \le i \le k$. If $k < \dim V$, then*

$$\text{Trace}(T_1 \circ \cdots \circ T_k) = 0.$$

If $k = \dim V$, then

$$\phi(T_1 \circ \cdots \circ T_k) = (-1)^k \cdot \textit{coefficient of } x_1 \cdots x_k \textit{ in } \det\left(\sum_{i=1}^k x_i T_i\right).$$

Proof. It is an easy exercise to verify that, for a linear matrix T,

$$\det(I - T) = \text{Trace}\, T.$$

The right side is the supertrace of T. Using this identity and the fact that $\wedge^* e^T = e^{D^*T}$ (see (7.1.5)), we have

$$\det\left(I - e^{x_1 T_1} \cdots e^{x_k T_k}\right) = \text{Trace}\left\{e^{x_1 D^* T_1} \cdots e^{x_k D^* T_k}\right\}.$$

The result follows by comparing the coefficients of $x_1 \cdots x_k$ on both sides. □

An element $S = T_1 \otimes T_2 \in \text{End}(V) \otimes \text{End}(V)$ acts on $\wedge^* V$ by $D^* S = D^* T_1 \circ D^* T_2$, and this definition is extended linearly to all of $\text{End}(V) \otimes \text{End}(V)$. The following corollary is immediate from the above definition.

Corollary 7.3.3. *Let $S_1, \ldots, S_k \in \text{End}(V) \otimes_{\mathbb{R}} \text{End}(V)$ and $T_1, \ldots, T_l \in \text{End}(V)$. If $2k + l < \dim V$ then*

$$\text{Trace}\left(S_1 \circ \cdots \circ S_k \circ T_1 \circ \cdots \circ T_l\right) = 0.$$

The same result holds if we make a permutation on $(S_1, \ldots, S_k, T_1, \ldots, T_l)$.

We now analyze the expression $\text{Trace}\{M_t U_t^{-1}\}$. Recall that U_t^{-1} is the parallel transport along a Brownian bridge. Therefore it, or more precisely, $\wedge^* U_t^{-1}$, is an isometry on $\wedge_x^* M$. For a small time, U_t is close to the identity map I, and we can find a unique $u_t \in \mathfrak{so}(d)$ such that $U_t = \exp u_t$ in $O(d)$. From (7.1.5) the action of U_t (as an isometry) on $\wedge_x M$ is given by

$$\wedge^* U_t = \exp D^* u_t.$$

We now show that $U_t - I$ has the order of t as $t \downarrow 0$, or equivalently $u_t = O(t)$. More precisely we have the following result, which is the only technical part of our probabilistic proof of the Gauss-Bonnet-Chern theorem.

Lemma 7.3.4. *For any positive integer N there is a constant K_N such that*

$$\mathbb{E}_{x,x;t}\,|U_t - I|^N \le K_N t^N.$$

Equivalently, there is a constant C_N such that

$$\mathbb{E}_{x,x;t}|u_t|^N \le C_N t^N.$$

Proof. We assume that $N = 1$. For the general case only trivial modifications are needed. We work under the probability $\mathbb{P}_{x,x;t}$, the law of a Brownian bridge X at x in time t. The horizontal lift U of X is the solution of the following stochastic differential equation:

$$dU_t = H_i(U_t) \circ dW_t^i + h(t-s, U_s)\,ds,$$

where $h(s,u)$ is the horizontal lift of $\nabla \ln p(s,y,x)$; see THEOREM 5.4.4. We use the normal coordinates $\{x^i\}$ at x to explicitly calculate U_t. Let O, O_1 be two neighborhoods of x covered by the coordinates such that the closure of O_1 is contained in O. In these coordinates the coordinates for the frame bundle $\mathscr{O}(M)$ are $u = \left\{x^i, e^i_j\right\}$, where e^i_j are determined by $ue_i = e^i_j(\partial/\partial x^i)$. Let $F : \mathscr{O}(M) \to \mathscr{M}(d,d)$ be a smooth function such that $F(u) = \left\{e^i_j\right\}$ on $\pi^{-1}(O_1)$ and zero on $\pi^{-1}(M\backslash O)$. Then we have $F(U_0) = I$ and $F(U_t) = U_t$. Using Itô's formula on $F(U_s)$, we have

$$\begin{aligned} U_t - I &= \int_0^t \langle \nabla^H F(U_s), dW_s\rangle + \int_0^t \langle \nabla^H F(U_s), h(t-s,U_s)\rangle ds \\ &\quad + \frac{1}{2}\int_0^t \Delta_{O(M)} F(U_s)\,ds \\ &\stackrel{\text{def}}{=} R_1(t) + R_2(t) + R_3(t). \end{aligned} \tag{7.3.4}$$

The key to the proof is the following observation: there is a constant C such that

$$|\nabla^H F(u)| \le C d(x, \pi u). \tag{7.3.5}$$

To see this, we write H_i in the local coordinates on O,

$$H_i = e^j_i \frac{\partial}{\partial x^j} - \Gamma^q_{kl}(y) e^k_i e^l_p \frac{\partial}{\partial e^q_p},$$

where $\Gamma^q_{kl}(y)$ are the Christoffel symbols at $y = \pi u$; see PROPOSITION 2.1.3. By LEMMA 2.5.5 in the normal coordinates all the Christoffel symbols vanish at x. From this the inequality (7.3.5) follows immediately.

We now estimate the three terms in (7.3.4). For $R_1(t)$ we have, by (7.3.5),

$$\mathbb{E}_{x,x;t}|R_1(t)|^2 \le C_1 \int_0^t \mathbb{E}_{x,x;t} d(x, X_s)\,ds.$$

By LEMMA 5.5.4 $\mathbb{E}_{x,x;t}d(x, X_s) \le C_2\sqrt{s}$; hence

$$\mathbb{E}_{x,x;t}|R_1(t)|^2 \le C_3 \int_0^t \sqrt{s}ds \le C_4 t^{3/2}.$$

To estimate $R_2(t)$, we need the estimate on $\nabla \ln p(s, x, y)$ in THEOREM 5.5.3. This estimate combined with (7.3.5) gives

$$|\langle \nabla^H F(U_s), h(t-s, U_s)\rangle| \le C_5 \left\{ \frac{d(x, X_s)^2}{t-s} + \frac{d(x, X_s)}{\sqrt{t-s}} \right\}.$$

Hence, because $\mathbb{E}_{x,x;t}d(x, X_s)^2 \le C_6(t-s)$ (see LEMMA 5.5.4), we have

$$\mathbb{E}_{x,x;t}|R_2(t)| \le C_7 \int_0^t \left[\frac{\mathbb{E}_{x,x;t}d(x, X_s)^2}{t-s} + \frac{\mathbb{E}_{x,x;t}d(x, X_s)}{\sqrt{t-s}} \right] ds \le C_8 t.$$

For $R_3(t)$, because the integrand is obviously uniformly bounded, we have $\mathbb{E}_{x,x;t}|R_3(t)| \le C_9 t$. Combining the estimates for $R_1(t)$, $R_2(t)$, and $R_3(t)$, we obtain $\mathbb{E}_{x,x;t}|U_t - I| \le C_{10}t$, as desired. □

We now come to the proof of the local Gauss-Bonnet-Chern theorem.

Theorem 7.3.5. *The limit*

$$e(x) = \lim_{t\downarrow} \operatorname{Trace} p_M^*(t, x, x)$$

exists and

$$e(x) = \begin{cases} \dfrac{\operatorname{Trace}(D^*\Omega)^l}{(4\pi)^l l!}, & \text{if } d = 2l \text{ is even,} \\ 0, & \text{if } d \text{ is odd.} \end{cases}.$$

Proof. Recall that the heat kernel on differential forms has the representation

$$p_M^*(t, x, x) = p_M(t, x, x)\mathbb{E}_{x,x;t}\left\{M_t U_t^{-1}\right\}.$$

Since

$$p_M(t, x, x) \sim \left(\frac{1}{2\pi t}\right)^{d/2},$$

we need to show that the following limit exists:

$$e(x) = \lim_{t\to 0} \left(\frac{1}{2\pi t}\right)^{n/2} \operatorname{Trace}\left\{M_t U_t^{-1}\right\}. \tag{7.3.6}$$

Note that U_t^{-1} in the above relation is $\wedge^* U_t^{-1}$, the action of U_t^{-1} on $\wedge_x^* M$ as an isometry. Let $l = d/2$ and $l_* = [d/2] + 1$. From $U_t^{-1} = e^{v_t}$ we have $\wedge^* U_t^{-1} = \exp D^* v_t$. Therefore

$$\wedge^* U_t^{-1} = \sum_{i=0}^{[l]} \frac{\{D^* v_t\}^i}{i!} + R(t),$$

where the remainder $R(t)$ satisfies the estimate $\mathbb{E}_{x,x;t}|R(t)| \le C_1 t^{l^*}$ by LEMMA 7.3.4. On the other hand, iterating the equation

$$M_s = I + \frac{1}{2}\int_0^t M_\tau D^*\Omega(U_\tau)\, d\tau,$$

we have

$$M_t = \sum_{i=0}^{[l]} m_i(t) + Q(t), \tag{7.3.7}$$

where $m_0(s) = I$ and

$$m_i(s) = \frac{1}{2^i}\int_0^s m_{i-1}(\tau) D^*\Omega(U_\tau)\, d\tau.$$

The remainder in (7.3.7) satisfies $|Q(t)| \le C_2 t^{l_*}$ for some constant C_2. From the expansions of M_t and U_t^{-1} we have

$$M_t U_t^{-1} = \sum_{i,j \le [l]} \frac{1}{j!} m_i(t)(D^* v_t)^j + S(t),$$

where the remainder satisfies $\mathbb{E}_{x,x;t}|S_t| \le C_3 t^{l^*}$.

We now come to the crucial point. By the definition of $m_i(t)$, it is clearly the limit of a sequence of linear combinations of terms of the form $D^*S_1 \circ \cdots \circ D^*S_i$ with $S_j \in \mathrm{End}(\mathbb{R}^d) \otimes \mathrm{End}(\mathbb{R}^d)$. Hence, by COROLLARY 7.3.3,

$$\mathrm{Trace}\left\{m_i(t)(D^*v_t)^j\right\} = 0 \quad \text{if } 2i + j < d.$$

On the other hand, by PROPOSITION 7.3.4,

$$\mathbb{E}_{x,x;t}|m_i(t)(D^*v_t)^j| \le C_4 t^{i+j}.$$

We conclude from these two facts that the limit (7.3.6) is possibly nonzero only if $i + j \le d/2$ and $2i + j \ge d$, that is, $j = 0$ and $2i = d$. In this case $l = d/2$ must be an integer, and using the fact that

$$\lim_{t\to 0} \frac{m_l(t)}{t^l} = \frac{D^*\Omega)^l}{2^l},$$

we have

$$\begin{aligned} e(x) &= \lim_{t\to 0}\left(\frac{1}{2\pi t}\right)^{d/2} \mathrm{Trace}\left\{M_t U_t^{-1}\right\} \\ &= \lim_{t\to 0}\left(\frac{1}{2\pi t}\right)^{l} \mathrm{Trace}\, m_l(t) \\ &= \frac{\mathrm{Trace}\,(D^*\Omega)^l}{(4\pi)^l l!}. \end{aligned}$$

This completes the proof. □

Using the explicit formula for $D^*\Omega$ in (7.1.6), we can show that, for an even d,

$$e(x) = \frac{1}{(4\pi)^{d/2}(d/2)!} \sum_{I,J} \operatorname{sgn}(I;J)\Omega_{i_1 i_2 j_1 j_2} \cdots \Omega_{i_{d-1} i_d j_{d-1} j_d},$$

where $I = \{i_1, \ldots, i_d\}$ and $J = \{j_1, \ldots, j_d\}$. Let dx be the volume form of M. The d-form $e = e(x)\,dx$ is called the Euler form of the manifold M. Define the Pfaffian $\operatorname{Pf}(\Omega)$ by

$$\operatorname{Pf}(\Omega) = \sum_I \Omega_{i_1 i_2} \wedge \cdots \wedge \Omega_{i_{d-1} i_d}.$$

$\operatorname{Pf}(\Omega)$ is a horizontal d-form on $\mathscr{O}(M)$, and we can verify that the pullback of the Euler form e on M is given by

$$\pi^* e = \frac{\operatorname{Pf}(\Omega)}{(4\pi)^{d/2}\,(d/2)!}.$$

Let us conclude this section by a formal statement of the Gauss-Bonnet-Chern theorem.

Theorem 7.3.6. *Suppose that M is an even dimensional, compact, and oriented Riemannian manifold and $e(x)\,dx$ is its Euler form. Then*

$$\chi(M) = \int_M e(x)\,dx.$$

Proof. Combine THEOREMS 7.3.1 and 7.3.5. □

7.4. Clifford algebra and spin group

In the next few sections we will prove the Atiyah-Singer index theorem for the Diract operator on a twisted spin bundle. Besides the difference in necessary algebraic and geometric backgrounds for the two problems, probabilistically the main new feature for the Dirac operator lies in the fact that, while in the previous case of the Gauss-Bonnet-Chern theorem it is enough to show that the Brownian holonomy satisfies $U_t - I = O(t)$, in the present case we need to compute precisely the limit of $(U_t - I)/t$ as $t \downarrow 0$.

We will start with the general setup of a spin bundle over a spin manifold and the associated Dirac operator, much of which is parallel to the case of exterior vector bundle of a manifold and the operator $d + \delta$. For analytic treatment of Dirac operators we recommend *Dirac Operators in Riemannian Geometry* by Friedrich [**28**] and *Invariance Theory, the Heat Equation, and the Atiyah-Singer Index Theorem* by Gilkey [**31**].

Let $\mathbb{R}^d$ be a euclidean space with even dimension $d = 2l$. We fix an oriented orthonormal basis $\{e_i\}$. The inner product on $\mathbb{R}^d$ is denoted by

$\langle \cdot, \cdot \rangle$. The Clifford algebra $C(\mathbb{R}^d)$ is the algebra generated by $\mathbb{R}^d$ and the relation

$$uv + vu + 2\langle u, v\rangle \cdot 1 = 0.$$

Thus $C(\mathbb{R}^d)$ is a vector space over $\mathbb{R}$ of dimension 2^d with basis

$$e_I = e_{i_1} \cdots e_{i_k}, \quad I = \{i_1 < \ldots < i_k\} \in \mathscr{I},$$

where I runs through the collection $\mathscr{I}$ of ascending subsets of $\{1, 2, \ldots, d\}$.

The spin group $\mathrm{Spin}(d)$ is the set of all $w \in C(\mathbb{R}^d)$ of the form

$$w = v_1 v_2 \cdots v_{2k-1} v_{2k}, \quad v_i \in \mathbb{R}^d, \ |v_i| = 1$$

(product of an even number of unit vectors in $\mathbb{R}^d$). It is the (two-sheeted) universal covering of the special orthogonal group $SO(d)$. The covering map $\sigma : \mathrm{Spin}(d) \to SO(d)$ is given explicitly by

$$\sigma \prod_{j=1}^{l} \left(\cos \frac{\theta_j}{2} + \sin \frac{\theta_j}{2} e_{2j-1} e_{2j} \right)$$

$$= \begin{pmatrix} \cos\theta_1 & \sin\theta_1 & & & & \\ -\sin\theta_1 & \cos\theta_1 & & & & \\ & & \cdot & & & \\ & & & \cdot & & \\ & & & & \cdot & \\ & & & & & \cos\theta_l & \sin\theta_l \\ & & & & & -\sin\theta_l & \cos\theta_l \end{pmatrix}.$$

From now on we assume that $C(\mathbb{R}^d)$ is complexified. A special element in $\mathrm{Spin}(d)$ is

$$\tau = \sqrt{-1}^{\,l} e_1 \cdots e_d.$$

It acts on the Clifford algebra $C(\mathbb{R}^d)$ by left multiplication. Since $\tau^2 = 1$, this action decomposes $C(\mathbb{R}^d)$ into eigenspaces of eigenvalues 1 and -1:

$$C(\mathbb{R}^d) = C(\mathbb{R}^d)^+ \oplus C(\mathbb{R}^d)^-.$$

The spin group $\mathrm{Spin}(d)$ acts on $C(\mathbb{R}^d)$ by left multiplication, which makes it into a left $\mathrm{Spin}(d)$-module. This representation of the spin group is reducible; in fact it is the direct sum of 2^l isomorphic representations:

$$C(\mathbb{R}^d) = 2^l \Delta.$$

Δ is called the spin representation of $\mathrm{Spin}(d)$. By the action of τ, each Δ can be further decomposed into two irreducible, nonisomorphic representations:

$$\Delta = \Delta^+ \oplus \Delta^-. \tag{7.4.1}$$

$\Delta^\pm$ are the half-spin representations.

The Lie algebra of $\mathrm{Spin}(d)$ is just $\mathfrak{so}(d)$, the space of anti-symmetric matrices. The exponential map $\exp : \mathfrak{so}(d) \to \mathrm{Spin}(d)$ can be described as follows. Let $A \in \mathfrak{so}(d)$. Then there is a basis $\{e_i\}$ of $\mathbb{R}^d$ such that A has the following form:

$$A = \begin{pmatrix} 0, & \theta_1 & & & & & \\ -\theta_1 & 0 & & & & & \\ & & \cdot & & & & \\ & & & \cdot & & & \\ & & & & \cdot & & \\ & & & & & 0 & \theta_l \\ & & & & & -\theta_l & 0 \end{pmatrix} \tag{7.4.2}$$

In this basis we have

$$\exp A = \prod_{j=1}^{l} \left(\cos \frac{\theta_j}{2} + \sin \frac{\theta_j}{2} e_{2j-1} e_{2j} \right). \tag{7.4.3}$$

The representation of $\mathrm{Spin}(d)$ on $C(\mathbb{R}^d)$ described above induces in naturally a representation D^* of $\mathfrak{so}(d)$ on $C(\mathbb{R}^d)$, which preserves the decompositions of $C(\mathbb{R}^d)$. This action is described in the following lemma.

Lemma 7.4.1. *The action $D^*A : \Delta^\pm \to \Delta^\pm$ of $A = (a_{ij}) \in \mathfrak{so}(d)$ on $C(\mathbb{R}^d)$ is given by*

$$D^*A = \textit{multiplication on the left by } \frac{1}{4} \sum_{1 \le i,j \le d} a_{ij} e_i e_j.$$

Proof. We may assume that A has the special form in (7.4.2). Recall that the action of $\exp tA$ on $C(\mathbb{R}^d)$ is the multiplication on the left. Replacing A in (7.4.3) by tA and differentiating with respect to t, we have

$$\left. \frac{d \exp tA}{dt} \right|_{t=0} = \frac{1}{2} \sum_{j=1}^{l} \theta_j e_{2j-1} e_{2j}.$$

□

If $A = (a_{ij}) \in \mathfrak{so}(d)$ we define

$$A(e) = \frac{1}{2} \sum_{i,j=1}^{d} a_{ij} e_i \wedge e_j \in \wedge^2 \mathbb{R}^d.$$

The Pfaffian $\mathrm{Pf}(A)$ is defined by the relation

$$A(e)^{\wedge l} = l! \, \mathrm{Pf}(A) \, e_1 \wedge \cdots \wedge e_d.$$

We see that $\mathrm{Pf}(A)$ is a homogeneous polynomial in a_{ij} of degree l, and $\mathrm{Pf}(A)^2 = \det A$.

For any linear transformation $S : \Delta \to \Delta$ which preserves the decomposition (7.4.1) we define the supertrace of S by

$$\operatorname{Trace} S = \operatorname{Tr}_{\Delta^+} S - \operatorname{Tr}_{\Delta^-} S = \operatorname{Tr}_{\Delta}(\tau S).$$

In the following, whenever needed we regard an element $f \in C(\mathbb{R}^d)$ as a linear transformation on $C(\mathbb{R}^d)$ by left multiplication. Thus if f preserves decompostion (7.4.1), the supertrace $\operatorname{Trace} f$ is well defined.

Lemma 7.4.2. *Let $A \in \mathfrak{so}(d)$. Then*

$$\lim_{t\to 0} \frac{\operatorname{Trace}\exp tA}{t^l} = \sqrt{-1}^{-l}\operatorname{Pf}(A).$$

Proof. By the decomposition $C(\mathbb{R}^d) = 2^l\Delta$, we can compute $\operatorname{Trace}\exp tA$ on $C(\mathbb{R}^d)$ and then divide by 2^l. Without loss of generality, we may assume that A has the form (7.4.2). Then we have

$$\exp tA = \prod_{j=1}^{l}\left(\cos\frac{\theta_j t}{2} + \sin\frac{\theta_j t}{2} e_{2j-1}e_{2j}\right). \tag{7.4.4}$$

In order to compute the supertrace, we need to find a suitable basis for $C(\mathbb{R}^d)^{\pm}$. If $I \in \mathscr{I}$, the collection of ascending subsets of $\{1, \dots, d\}$, we let $I^* = (1, \dots, d)\backslash I$, arranged in the ascending order. For an ascending multi-index $I \in \mathscr{I}$ we have $\tau e_I = c(I)e_{I^*}$ for some constant $c(I)$. It is easy to verify that the elements

$$f_I^{\pm} = e_I \pm c(I)e_{I^*}, \quad I \in \mathscr{I},$$

form a basis for $C(\mathbb{R}^d)^{\pm}$ respectively, or more precisely, twice a basis, because $f_I^{\pm}$ is a multiple of $f_{I^*}^{\pm}$. From (7.4.4) and $\tau f_I^+ = f_I^+$ we have

$$(\exp tA)f_I^+ = \prod_{j=1}^{l}\cos\frac{\theta_j t}{2}\cdot f_I^+ + \sqrt{-1}^{-l}\prod_{j=1}^{l}\sin\frac{\theta_j t}{2}\cdot f_I^+ + \sum_{J\neq I} a_J^+ f_J^+.$$

A similar relation holds also for f^- with a minus sign on the second term. Therefore the supertrace of $\exp tA$ on $C(\mathbb{R}^d)$ is

$$\operatorname{Trace}_{C(\mathbb{R}^d)}\exp tA = 2^l\sqrt{-1}^{-l}\prod_{j=1}^{l}\sin\frac{\theta_j t}{2}.$$

Now it is easy to see that

$$\lim_{t\to 0}\frac{\operatorname{Trace}\exp tA}{t^l} = \sqrt{-1}^{-l}\theta_1\theta_2\cdots\theta_l = \sqrt{-1}^{-l}\operatorname{Pf}(A).$$

□

We have the following analogue of Lemma 7.3.2.

Lemma 7.4.3. *Let* $A_1, \ldots, A_l \in \mathfrak{so}(d)$. *If* $k < l$, *then*

$$\text{Trace}\,(D^* A_1 \circ \cdots \circ D^* A_k) = 0.$$

If $k = l$, *then*

$$\begin{aligned}&\text{Trace}\,(D^* A_1 \circ \cdots \circ D^* A_k)\\&\quad = \sqrt{-1}^{\,-l} \cdot \textit{ coeff. of } x_1 x_2 \cdots x_l \textit{ in } \mathrm{Pf}(x_1 A_1 + x_2 A_2 + \cdots + x_l A_l).\end{aligned}$$

Proof. Let $A(x) = x_1 A_1 + x_2 A_2 + \cdots + x_l A_l$. Since the representation D^* of $\mathfrak{so}(d)$ is induced from the representation of $\mathrm{Spin}(d)$ on $C(\mathbb{R}^d)$, as actions on $C(\mathbb{R}^d)$, we have

$$\exp tA(x) = \sum_{n=0}^{\infty} \frac{t^n}{n!}\, D^* A(x)^n.$$

Take the supertrace on both sides and compare coefficients. By Lemma 7.4.2, we have

$$\begin{aligned}&\text{Trace} \sum_{\sigma \in \mathbb{S}_k} D^* A_{\sigma(1)} \circ \cdots \circ D^* A_{\sigma(k)}\\&= \begin{cases} 0, & \text{if } k < l,\\ \text{coefficient of } x_1 x_2 \cdots x_l \text{ in} & \\ l! \sqrt{-1}^{\,-l} \mathrm{Pf}(x_1 A_1 + x_2 A_2 + \cdots + x_l A_l), & \text{if } k = l.\end{cases}\end{aligned}$$

($\mathbb{S}_k$ is the permutation group on $\{1, \ldots, k\}$.) The desired result follows because the supertrace of a product of $D^* A_i$ is independent of the order of the factors. □

The following corollary will be useful later.

Corollary 7.4.4. *We have*

$$\text{Trace}\,(D^* A_1 \circ \cdots \circ D^* A_l)\, e_1 \wedge \cdots \wedge e_d = \sqrt{-1}^{\,-l} A_1(e) \wedge \cdots \wedge A_l(e).$$

Proof. By the definition of Pfaffian we have

$$\begin{aligned}\mathrm{Pf}(A(x))\, e_1 \wedge \cdots \wedge e_d &= \frac{1}{l!}\, [x_1 A_1(e) + \cdots + x_l A_l(e)]^{\wedge l}\\ &= (x_1 \cdots x_l) A_1(e) \wedge \cdots \wedge A_l(e) + \text{other terms.}\end{aligned}$$

The corollary follows by using the lemma on the left side of the above identity. □

We now leave algebra and turn to geometry.

7.5. Spin bundle and the Dirac operator

Let M be a compact, oriented Riemannian manifold of dimension $d = 2l$. The bundle of oriented orthonormal frames $\mathscr{SO}(M)$ is a principal bundle over M with structure group $SO(d)$. We further assume that M is a spin manifold. This means that there exists a principal bundle $\mathscr{SP}(M)$ with structure group $\mathrm{Spin}(d)$ and a two-sheeted covering map

$$\widetilde{\sigma} : \mathscr{SP}(M) \to \mathscr{SO}(M)$$

such that

$$\widetilde{\sigma}\eta u = \sigma(\eta)u, \quad u \in \mathscr{SP}(M),\ \eta \in \mathrm{Spin}(d),$$

and $\sigma : \mathrm{Spin}(d) \to SO(d)$ is the standard covering map. Any such $\mathrm{Spin}(d)$-principal bundle over M is called a spin structure of M. The necessary and sufficient topological conditions for the existence of a spin structure on a Riemannian manifold is known. In the following we assume that M is a spin manifold and fix a spin structure (or a $\mathrm{Spin}(d)$-principal bundle $\mathscr{SP}(M)$) on M.

The spin bundle over M of the spin structure is the associated vector bundle

$$\mathscr{S}(M) = \mathscr{SP}(M) \times_{\mathrm{Spin}(d)} \Delta.$$

From the decomposition (7.4.1) we have a natural (fibrewise) decomposition

$$\mathscr{S}(M) = \mathscr{S}(M)^+ \oplus \mathscr{S}(M)^-, \quad \mathscr{S}(M)^\pm = \mathscr{SP}(M) \times_{\mathrm{Spin}(d)} \Delta^\pm.$$

The (fibre-wise) Clifford multiplication

$$c : T_xM \to \mathrm{Hom}(\mathscr{S}(M)_x, \mathscr{S}(M)_x)$$

is defined as follows. Let $v \in T_xM$ and $\xi \in \mathscr{S}(M)_x$. Take a frame $u : \Delta \to \mathscr{S}(M)_x$. Its projection is $\widetilde{\sigma}u : \mathbb{R}^d \to T_xM$. Then

$$c(v)\xi = u\left((\widetilde{\sigma}u)^{-1}v \cdot u^{-1}\xi\right),$$

where $\cdot$ is the the usual multiplication in $C(\mathbb{R}^d)$. In other words, the Clifford multiplication by $c(v)$ is simply the pushforward (by $u : \Delta \to \mathscr{S}(M)_x$) of the usual left mutiplication by $(\widetilde{\sigma}u)^{-1}v$ on Δ. It is clear that $c(v)$ exchanges $\mathscr{S}(M)_x^\pm$.

Because the principal bundle $\mathscr{SP}(M)$ is a two-sheeted cover of $\mathscr{SO}(M)$, the Levi-Civita connection ∇ on M extends naturally to a connection on the spin bundle $\mathscr{S}(M)$:

$$\nabla : \Gamma(TM) \times \Gamma(\mathscr{S}(M)) \to \Gamma(\mathscr{S}(M)).$$

We describe this connection. Let ω be the connection form of the original Levi-Civita connection ∇. By definition ω is an $\mathfrak{so}(d)$-valued 1-form on $\mathscr{SO}(M)$. Since $\mathscr{SP}(M)$ is a covering of $\mathscr{SO}(M)$, ω can be lifted to a connection form on $\mathscr{SP}(M)$ and defines a connection on the associated bundle $\mathscr{S}(M)$ by the representation of the Lie algebra $\mathfrak{so}(d)$ on Δ. The corresponding curvature tensor of this connection is a map

$$R : \Gamma(TM) \times \Gamma(TM) \times \Gamma(\mathscr{S}(M)) \to \Gamma(\mathscr{S}(M)).$$

The following lemma expresses ∇ and R on the spin bundle $\mathscr{S}(M)$ explicitly in a local moving frame.

Lemma 7.5.1. *Let $\widetilde{\sigma}u = \{X_i\} \in \mathscr{S}\mathscr{O}(M)$ be an oriented local orthonormal frame for the tangent bundle TM and Γ_{ij}^k be the corresponding Christoffel symbols. Let u be the lift of $\widetilde{\sigma}u$ to $\mathscr{S}\mathscr{P}(M)$ and $X_I = u(e_I)$. Then $\{X_I, I \in \mathscr{I}\} \in \mathscr{S}\mathscr{P}(M)$ is a local frame for the spin bundle $\mathscr{S}(M)$. In terms of this frame, the covariant differentiation ∇ on $\mathscr{S}(M)$ is given by*

$$\nabla_{X_i} X_I = \frac{1}{4} \sum_{j,k=1}^{d} \Gamma_{ij}^k c(X_j) c(X_k) X_I.$$

The curvature R on $\mathscr{S}(M)$ is given by

$$R(X_i, X_j) X_I = \frac{1}{4} \sum_{k,l=1}^{d} \langle R(X_i, X_j) X_k, X_l \rangle c(X_k) c(X_l) X_I.$$

Here $c(X)$ denotes the Clifford multiplication.

Proof. The connection form in this case is defined by

$$\omega(X_i) = \left\{ \Gamma_{ij}^k \right\}.$$

LEMMA 7.4.1 describes the representation of $\mathfrak{so}(d)$ on Δ. The action of $\omega(X_i)$ on Δ is $D^*\omega(X_i)$, i.e., the multiplication on the left by the element $\sum_{j,k=1}^{d} \Gamma_{ij}^k e_j e_k / 4$. Hence, from $X_I = u(e_I)$ we have

$$\nabla_{X_i} X_I = u\left(D^*\omega(X_i)\, e_I\right) = \frac{1}{4} \sum_{j,k=1}^{d} \Gamma_{ij}^k u(e_j e_k e_I).$$

This implies the desired expression for the connection because

$$u(e_j e_k e_I) = c(X_j) c(X_k) X_I.$$

The proof for the curvature is similar. □

Remark 7.5.2. $c(X_i)c(X_j)$ corresponds to the action of A_{ij} on Δ, where $A_{ij} \in \mathfrak{so}(n)$ is the matrix such that $a_{ij} = 1, a_{ji} = -1$ and all other entries are 0. More precisely,

$$c(X_i) c(X_j) = 2\, u \circ D^* A_{ij} \circ u^{-1},$$

where $\circ$ denotes "composition of maps".

The setting we have developed so far can be made more interesting by "twisting" the spin bundle $\mathscr{S}(M)$ with a hermitian vector bundle ξ. Namely, let ξ be a hermitian vector bundle on M equipped with a connection ∇^ξ. We denote the corresponding curvature operator R^ξ of ξ by L. The twisted bundle is the product bundle $G = \mathscr{S}(M) \otimes \xi$ with the product connection $\nabla \otimes \nabla^\xi$, which we will simply denote by ∇. We have

$$G = G^+ \oplus G^-, \quad G^\pm = \mathscr{S}(M)^\pm \otimes \xi.$$

The Clifford multiplication exchanges $G_x^\pm$.

The Dirac operator D on the twisted bundle G is defined as a series of compositions:

$$D : \Gamma(G) \xrightarrow{\nabla} \Gamma(T^*M \otimes G) \xrightarrow{\text{dual}} \Gamma(TM \otimes G) \xrightarrow{c} \Gamma(G).$$

If $\{X_i\}$ is an orthonormal basis, then it is immediate from the definition that

$$D = \sum_{i=1}^{d} c(X_i)\nabla_{X_i}. \tag{7.5.1}$$

It is easy to see that D exchanges $\Gamma(G^\pm)$.

At this point it is perhaps helpful to point out that the counterpart of the Dirac operator D for the exterior bundle $\wedge^* M$ is $D = d + \delta$. The decomposition $G = G^+ \oplus G^-$ corresponds to the decomposition of $\wedge^* M$ into the subbundles of even and odd forms. The Hodge-de Rham Laplacian is

$$\square_M = -(d\delta + \delta d) = -D^2.$$

The operator $-D^2$ should play the role of $\square_M$ in the current setting.

Since G is a vector bundle over the Riemannian manifold M with the connection ∇, we have the covariant Laplacian Δ_M defined by

$$\Delta_M f = \sum_{i=1}^{d} \nabla^2 f(X_i, X_i), \quad f \in \Gamma(G),$$

for an orthonormal frame $\{X_i\}$. There exists a Weitzenböck formula relating D^2 and Δ_M.

Theorem 7.5.3. (Lichnerowicz formula) *Let $\{X_i\}$ be an orthonormal basis of T_xM. Then*

$$D^2 = -\Delta_M + \frac{S}{4} + \frac{1}{2}\sum_{j,k=1}^{d} c(X_j)\, c(X_k) \otimes L(X_j, X_k).$$

Here S is the total (scalar) curvature of M.

Proof. We use the notations in LEMMA 7.5.1. Fix a point on the manifold as the origin and let $X = \{X_i\}$ be the moving frame obtained by parallel-translating the frame at the origin along the geodesic rays. This moving frame has the property that all Christoffel symbols vanish at the origin (see the proof of LEMMA2.5.5), and, because the connection is torsin-free, all the brackets $[X_i, X_j]$ vanish at the origin as well. The curvature operator on the twisted bundle G becomes

$$R^G = \nabla_{X_i}\nabla_{X_j} - \nabla_{X_j}\nabla_{X_i}. \tag{7.5.2}$$

These facts will simplify our computation

Using (7.5.1), we have

$$D^2 = c(X_i)c(X_j)\nabla_{X_i}\nabla_{X_j} + c(X_i)c\left(\nabla_{X_i}X_j\right)\nabla_{X_j}.$$

The last term vanishes because the Christoffel symbols vanish at the origin. From the definition of Clifford multiplication and the orthogonality of $\{X_i\}$ we have

$$c(X_i)c(X_j) + c(X_j)c(X_i) + 2\delta_{ij} = 0. \tag{7.5.3}$$

Using this and (7.5.2), we have

$$\begin{aligned} D^2 &= c(X_i)c(X_j)\nabla_{X_i}\nabla_{X_j} \\ &= \frac{1}{2}\left\{c(X_i)c(X_j)\nabla_{X_i}\nabla_{X_j} + c(X_j)c(X_i)\nabla_{X_j}\nabla_{X_i}\right\} \\ &= -\nabla_{X_i}\nabla_{X_i} + \frac{1}{2}c(X_i)c(X_j)\left[\nabla_{X_i}\nabla_{X_j} - \nabla_{X_i}\nabla_{X_j}\right] \\ &= -\Delta_M + \frac{1}{2}c(X_i)c(X_j)R^G(X_i, X_j). \end{aligned}$$

We now compute the curvature R^G on a section of the form $X_I \otimes f$ of the twisted spin bundle G. We have by definition

$$\begin{aligned} &\nabla_{X_i}\nabla_{X_j}(X_I \otimes f) \\ &\quad = \nabla_{X_i}\nabla_{X_j}X_I \otimes f + \nabla_{X_i}X_I \otimes \nabla_{X_j}f \\ &\quad\quad + \nabla_{X_j}X_I \otimes \nabla_{X_i}f + X_I \otimes \nabla_{X_i}\nabla_{X_j}f. \end{aligned}$$

The two terms in the middle vanish by LEMMA 7.5.1. Hence, using (7.5.2), we have

$$R^G(X_I \otimes f) = R(X_i, X_j)X_I \otimes f + X_I \otimes L(X_i, X_j)f,$$

where L is the curvature operator on the factor bundle ξ. The second term on the right side yields the last term in the Lichnerowicz formula. In view of LEMMA 7.5.1, to complete the proof we need to show that

$$\sum_{i,j,k,l=1}^{d} R_{ijkl}\, c(X_i)c(X_j)c(X_k)c(X_l) = 2S,$$

where $R_{ijkl} = \langle R(X_i, X_j)X_k, X_j \rangle$, and $S = \sum_{i,j=1}^d R_{ijji}$ is the scalar curvature. This identity follows from the first Bianchi identity

$$R_{ijkl} + R_{iljk} + R_{iklj} = 0$$

and the commutation relation (7.5.3). We leave this straightforward verification to the reader. □

7.6. Atiyah-Singer index theorem

Let $D^{\pm} = D\big|_{G^{\pm}}$. We define the index of D^+ by

$$\operatorname{Ind}(D^+) = \dim \operatorname{Ker} D^+ - \dim \operatorname{Ker} D^-.$$

The reader can verify that its counterpart for $d + \delta$ on $\wedge^* M$ is exactly the Euler characteristic $\chi(M)$. The Atiyah-Singer index theorem expresses $\operatorname{ind}(D^+)$ as the integral of a d-form on M determined by the curvature.

The supertrace on the twisted bundle $G = \mathscr{S}(M) \otimes \xi$ is defined as follows. If $S : G_x \to G_x$ is a linear map on the fibre G_x which prerserves the decomposition $G_x = G_x^+ \oplus G_x^-$, we define

$$\operatorname{Trace} S = \operatorname{Tr}_{G^+} S - \operatorname{Tr}_{G^-} S.$$

From this definition it is clear that if S has the form $U \otimes V$ with $U : \mathscr{S}(M)_x \to \mathscr{S}(M)_x$ and $V : \xi_x \to \xi_x$, then

$$\operatorname{Trace} S = \operatorname{Trace} U \cdot \operatorname{Tr}_\xi V.$$

Let $p_M^D(t, x, y)$ be the heat kernel for $-D^2$ on the twisted bundle G. We have the following representation for the index of D^+.

Theorem 7.6.1. *For any $t > 0$, we have*

$$\operatorname{Ind}(D^+) = \int_M \operatorname{Trace} p_M^D(t, x, x)\, dx.$$

Proof. See the proof of THEOREM 7.3.1. □

As in the case of the Gauss-Bonnet-Bonnet theorem, we will derive a probabilistic representation of the heat kernel $p_M^D(t, x, x)$ in terms of a Brownian bridge and show that a cancellation takes place after taking its trace. We will then evaluate the limit (the so-called local index)

$$I(x) = \lim_{t \downarrow 0} \operatorname{Trace} p_M^D(t, x, x)$$

explicitly in terms of the curvature.

Let us start with some notations and a precise formulation of the result we will prove. Let M be a Riemannian manifold and $\pi : V \to M$ a vector bundle on M equipped with a connection compatible with the metric. Let R^V be the curvature of this connection. Fix a point $x \in M$ and let $\{x^i\}$ be a local coordinate system in a neighborhood of x such that $X_i = \partial/\partial x^i$ is an orthonormal basis at x. By definition the curvature transform

$$R^V(X_i, X_j) : V_x \to V_x$$

is an anti-symmetric (or anti-hermitian) linear transform on the fibre V_x. It is also anti-symmetric in the indices i and j. Let $\{X^i\}$ be the dual basis. Define the following the $\mathrm{End}(V)$-valued curvature 2-form

$$\Omega^V = \frac{1}{2}\sum_{i,j=1}^{d} R^V(X_i, X_j)X^i \wedge X^j. \tag{7.6.1}$$

If $f(A)$ is a power series in the entries of $A \in \mathfrak{so}(d)$ and is invariant under the group $SO(d)$ in the sense that $f(O^{-1}AO) = f(A)$ for all $O \in SO(d)$, then the form $f(\Omega^V)$ is well-defined and is independent of the choice of the basis. For our purpose the following examples are relevant:

(i) $V = TM$, $R^V = R$, $\Omega^V = \Omega$. The $\hat{A}$-genus of the tangent bundle TM is

$$\hat{A}(TM) = \det\left[\frac{\Omega/4\pi}{\sin\Omega/4\pi}\right]^{1/2}(TM).$$

(ii) $V = \xi$, $R^V = L$, $\Omega^V = \Lambda$. The Chern character of the hermitian bundle ξ is

$$\mathrm{ch}\,\xi = \mathrm{Tr}_\xi \exp\left[\frac{\sqrt{-1}}{2\pi}\Lambda\right].$$

If A and B are two differential forms on M, we use $A \overset{d}{\approx} B$ to denote that A and B have the same d-form components.

Theorem 7.6.2. *The limit*

$$I(x) = \lim_{t\downarrow 0} \mathrm{Trace}\; p_M^D(t, x, x)$$

exists and

$$I(x)\, dx \overset{d}{\approx} \hat{A}(TM) \wedge \mathrm{ch}\,\xi.$$

[dx is the volume form on M.]

The Atiyah-Singer index theorem for the Dirac operator follows from the above local index theorem and THEOREM 7.6.1.

Theorem 7.6.3. *We have*

$$\mathrm{Ind}(D^+) = \int_M \hat{A}(TM) \wedge \mathrm{ch}\,\xi.$$

As before, in order to derive the desired probablistic representation of $p_M^D(t,x,x)$ we need to solve the initial value problem

$$\begin{cases} \dfrac{\partial f}{\partial t} = -\dfrac{1}{2}D^2 f, & (t,x) \in (0,\infty)\times M, \\ u(0,x) = f_0(x), & x \in M. \end{cases} \tag{7.6.2}$$

This can be done by introducing an appropriate multiplicative Feynman-Kac functional using the Lichnerowicz formula.

In the following we always fix a frame U_0 at x and identify T_xM with $\mathbb{R}^d$ using this frame. We use $\{e_i\}$ to denote the usual oriented orthonormal basis of $\mathbb{R}^d$. We use X to denote the coordinate process on the path space $W(M)$ and U the horizontal lift of X starting from U_0. The law of Brownian motion from x and the law of Brownian bridge at x in time t are denoted by $\mathbb{P}_x$ and $\mathbb{P}_{x,x;t}$ respectively. Note that U_t^{-1} is the parallel transport from X_t to $X_0 = x$ along the Brownian path $X[0,t]$.

Proposition 7.6.4. *Let $S(z)$ be the total (scalar) curvature at z and defined*

$$R_t = \exp\left[-\frac{1}{8}\int_0^t S(X_s)\,ds\right]$$

Define the linear map $M_t : G_x \to G_x$ by the following ordinary differential equation:

$$\frac{dM_s}{ds} = -\frac{1}{4}M_s\sum_{j,k=1}^d c(e_j)c(e_k)\otimes U_sL(U_se_j, U_se_k)U_s^{-1}, \quad M_0 = I.$$

Then the solution of the initial value problem (7.6.2) is

$$f(t,x) = \mathbb{E}_x\left\{R_tM_tU_t^{-1}f(X_t)\right\}.$$

Proof. Apply Itô's formula to $R_sM_s\tilde{f}(t-s,U_s)$, where $\tilde{f}$ is the lift of f to the orthonormal frame bundle of the twisted bundle G; see the proof of THEOREM 7.2.1. □

From

$$f(t,x) = \int_M p_M^D(t,x,y)f(y)dy$$

and the above proposition we have the representation we have

$$I(t,x) = p_M(t,x,x)\mathbb{E}_{x,x;t}\left\{R_t\text{Trace}\left(M_tU_t^{-1}\right)\right\}.$$

Since $R_t \to 1$ as $t \to 0$, this factor does not play a role. As before, the equation for M can be iterated to obtain a series for M_t in the form ($l = d/2$)

$$M_t = \sum_{i=0}^{[l]} m_i(t) + Q(t),$$

where

$$\mathbb{E}_{x,x;t}|Q(t)| \le C_1 t^{l+1}$$

and

$$\frac{m_i(t)}{t^l} \to \frac{F^k}{i!} \qquad \text{as } t \to 0$$

with

$$F = -\frac{1}{4}\sum_{j,k=1}^{d} c(e_j)c(e_k) \otimes L_x(e_j, e_k). \tag{7.6.3}$$

Next, we need to compute the stochastic parallel transport $V_t \stackrel{\text{def}}{=} U_t^{-1}$. It moves the two components of the fibre $G_s = \mathscr{S}(M)_x \otimes \xi_x$ separately,

$$V_s = V_s^{\mathscr{S}(M)} \otimes V_s^{\xi},$$

where $V_s^{\mathscr{S}(M)}$ and V^{ξ} are the parallel transports on the respective bundles. Note that under the probability $\mathbb{P}_{x,x;t}$, $V_t^{\mathscr{S}(M)}$ and V_t^{ξ} are orthogonal transforms on the fibres $\mathscr{S}(M)_x$ and ξ_x respectively. In the proof of the Gauss-Bonnet-Chern theorem, we have shown that $V_t^{\mathscr{S}(M)} - I^{\mathscr{S}(M)} = O(t)$, but now we need more precise information on the parallel transport $V_t^{\mathscr{S}(M)}$. In the statement of the following proposition concerning Brownian holonomy in the spin boundle, Y is a euclidean Brownian bridge defined by

$$dY_s = dW_s - \frac{Y_s}{1-s}ds, \quad Y_0 = 0.$$

Proposition 7.6.5. *We have*

$$\lim_{t\to 0}\frac{V_t^{\mathscr{S}(M)_x} - I^{\mathscr{S}(M)_x}}{t} = -\frac{1}{8}\int_0^1 \langle R(e_j,e_k)Y_s, dY_s\rangle\, c(e_j)c(e_k)$$

in $L^N(\Omega, \mathscr{F}_*, \mathbb{P})$ for all $N \ge 1$. Here we identify Y_s with $\sum_{i=1}^d Y_s^i e_i$.

In order not to interrupt our discussion of the local index theorem for the Dirac operator, we will prove this result in the next section, in which the reader will also find a more precise statement.

From $p_M(t,x,x) \sim (2\pi t)^{-l}$ and $R_t \to 1$ as $t \downarrow 0$ we have

$$I(t,x) \sim \left(\frac{1}{2\pi t}\right)^l \mathbb{E}_{x,x;t}\text{Trace}\left\{M_t\left(V_t^{\mathscr{S}(M)} \otimes V_t^{\xi}\right)\right\}.$$

As actions on $\mathscr{S}(M)_x$, $V_t^{\mathscr{S}(M)}$ is close to the identity; therefore there exists a $v_t^{\mathscr{S}(M)} \in \mathfrak{so}(d)$ such that such that $V_t^{\mathscr{S}(M)} = \exp v_t^{\mathscr{S}(M)}$ in Spin(d). Proposition 7.6.5 now implies that

$$C \stackrel{\text{def}}{=} \lim_{t\to 0}\frac{v_t^{\mathscr{S}(M)}}{t} = -\frac{1}{8}\int_0^1 \langle R(e_j,e_k)Y_s, dY_s\rangle c(e_j)c(e_k).$$

As in the proof of the local Gauss-Bonnet-Chern theorem, in the supertrace $\mathrm{Trace}\left\{M_t\left(V_t^{\mathscr{S}(M)}\otimes V_t^{\xi}\right)\right\}$ we replace M_t and $V_t^{\mathscr{S}(M)}$ by their expansions. Thanks to LEMMA 7.4.3 all terms with total exponent less than l vanish. We thus obtain the following formula for the local index

$$I(x)=\left(\frac{1}{2\pi}\right)^l\mathbb{E}\,\mathrm{Trace}\left[\sum_{k+m=l}\frac{F^k}{k!}\frac{C^m\otimes I^{\xi}}{m!}\right].$$

It remains to rewrite this into an elegant form. We multiply $I(x)$ by the volume form $dx=dx^1\wedge\cdots\wedge dx^d$, and use COROLLARY 7.4.4 and REMARK 7.5.2. This gives

$$I(x)\,dx=\mathbb{E}\left[\sum_{k+m=l}\frac{\mathrm{Tr}_{\xi}M^{\wedge k}}{k!}\wedge\frac{J^{\wedge m}}{m!}\right],$$

where

$$M=\frac{\sqrt{-1}}{4\pi}\sum_{j,k-1}^{d}dx^j\wedge dx^k\otimes L(e_j,e_k)=\frac{\sqrt{-1}}{2\pi}\Lambda$$

is the curvature of the hermitian bundle ξ, and, with Ω being the curvature form on $\mathrm{Spin}(M)$ (see (7.6.1)),

$$\begin{aligned}J&=-\frac{\sqrt{-1}}{8\pi}\sum_{j,k=1}^{d}dx^j\wedge dx^k\otimes\int_0^1\langle R(e_j,e_k)Y_s,dY_s\rangle\\&=-\frac{\sqrt{-1}}{4\pi}\int_0^1\langle\Omega Y_s,dY_s\rangle.\end{aligned}$$

Note that M and J commute because they act on ξ and $\mathscr{S}(M)$ separately. Hence we can write

$$I(x)\,dx\overset{d}{\approx}\mathbb{E}\exp J\wedge\mathrm{Tr}_{\xi}\exp M.\tag{7.6.4}$$

The expected value can be evaluated by a matrix version of Lévy's stochastic area formula. Let $Y_s=(Y_s^1,Y_s^2)$ be the standard two-dimensional Brownian bridge from 0 to z. Then Lévy's stochastic area formula is

$$\mathbb{E}\left[\exp\sqrt{-1}\lambda\int_0^1Y_s^1dY_s^2-Y_s^2dY_s^1\right]=\frac{\lambda}{\mathrm{sh}\lambda}\exp\left[\frac{(1-\lambda\coth\lambda)|z|^2}{2}\right];$$

see Ikeda and Watanabe [**48**], p.388. This can be generalized to the following matrix form by diagonalization.

Lemma 7.6.6. *Let Y be the standard Brownian bridge in $\mathbb{R}^d$ from 0 to z, and A an anti-symmetric matrix. Then*

$$\mathbb{E}\left[\exp\sqrt{-1}\int_0^1\langle AY_s,dY_s\rangle\right]=\det\left[\frac{A}{\sin A}\right]^{1/2}\exp\left\langle(I-A\cot A)z,z\right\rangle.$$

Proof. Exercise. □

Finally, using this lemma to evaluate the expectation in (7.6.4), we obtain the following local index formula:

$$\begin{aligned} I(x)\,dx &\overset{d}{\approx} \det\left[\frac{\Omega/4\pi}{\sin(\Omega/4\pi)}\right]^{1/2}(TM)\wedge \mathrm{Tr}_\xi \exp\left(\frac{\sqrt{-1}}{2\pi}\Lambda\right) \\ &= \hat{A}(TM)\wedge \mathrm{ch}(\xi). \end{aligned}$$

Except for the proof of Proposition 7.6.5, which will be carried out in the next section, we have completed the proof of the Atiyah-Singer index theorem for the Dirac operator of a twisted spin bundle.

7.7. Brownian holonomy

We first analyze the Brownian bridge X at x as the time length $t \downarrow 0$. Let $\exp_x : \mathbb{R}^d \to M$ be the exponential map at x. We show that the image $\exp_x^{-1} X$ of the Brownian bridge is asymptotically a euclidean Brownian bridge scaled (in space) by a factor of $\sqrt{t}$.

Strictly speaking, under the probability $\mathbb{P}_{x,x;t}$, the Brownian bridge X will not always stay within the cutlocus of x, so the expression $\exp_x^{-1} X$ is not defined with probability 1. But it can be shown that the probability of the set of loops which do not lie within the cutlocus is exponentially small.

Lemma 7.7.1. *For any positive μ, there is a positive constant λ such that*

$$\mathbb{P}_{x,x;t}\left\{X[0,t] \not\subseteq B(x;\mu)\right\} \le e^{-\lambda/t}.$$

Proof. By symmetry we only need to consider the half interval $[0, t/2]$. Let τ_μ be the first exit time of $B(x;\mu)$. Then by (5.4.1) we have

$$\mathbb{P}_{x,x;t}\left\{\tau_\mu \le t/2\right\} = \frac{\mathbb{E}_x\left\{p_M(t/2, X_{t/2}, x); \tau_\mu \le t/2\right\}}{p_M(t,x,x)}.$$

From Corollary 5.3.5 there is a constant C such that

$$p_M(t/2, X_{t/2}, x) \le \frac{C}{t^{d/2}}, \quad p_M(t,x,x) \ge \frac{C^{-1}}{t^{d/2}}.$$

Hence,

$$\mathbb{P}_{x,x;t} \le C^2\, \mathbb{P}_x\left\{\tau_\mu \le t/2\right\}.$$

The result follows from this inequality and Proposition 5.1.4. □

As $t \downarrow 0$, this small probability is irrelevant for our purpose. This observation justifies our carrying out the following analysis as if X lies in a local coordinate chart with probability 1.

Let $\{x^i\}$ be the normal coordinates based at the fixed point $x \in M$. The Riemannian metric is given by $ds^2 = g_{ij}dx^i dx^j$. Recall that the Laplace-Beltrami operator in local coordinates has the form

$$\Delta_M = \frac{1}{\sqrt{G}}\frac{\partial}{\partial x^i}\left(\sqrt{G}g^{ij}\frac{\partial}{\partial x^j}\right),$$

where $G = \det(g_{ij})$ and $\{g^{ij}\}$ is the inverse of $\{g_{ij}\}$. Let σ be the positive definite matrix square root of $\{g^{ij}\}$,

$$b^i = \frac{1}{2}\sum_{j=1}^{d}\frac{1}{\sqrt{G}}\frac{\partial}{\partial x^j}\left(\sqrt{G}g^{ij}\right) = -\frac{1}{2}g^{jk}\Gamma^i_{jk}.$$

Define

$$C^i(s,z) = g^{ij}(z)\frac{\partial}{\partial z^j}\ln p_M(s,z,x). \tag{7.7.1}$$

In these local coordinates the Brownian bridge X is the solution of the stochastic differential equation

$$dX^t_s = \sigma(X^t_s)\,dW_s + b(X^t_s)\,ds + C(t-s,X^t_s)\,ds,$$

where W is a Brownian motion on $\mathbb{R}^d$. This is just the local version of the equation (5.4.6). Note we have used the notation X^t instead of X to show the dependence on the time length of the Brownian bridge. We want to show that the scaled Brownian bridge

$$Z^t_s = \frac{X^t_{st}}{\sqrt{t}}, \quad 0 \le s \le 1, \tag{7.7.2}$$

converges to a standard euclidean Brownian bridge Y in an appropriate sense. In view of the equation (7.7.4) for Y, this should be clear by a rough calculation using the following three facts:

(1) X^t shrinks to the origin as $t \downarrow 0$;

(2) $\sigma(0) = I$, the identity matrix;

(3) $s\,C(s,z) \to -z$ as $s \downarrow 0$.

To actually talk about convergence, it is more convenient to put Brownian bridges in different time lengths on the same probability space. This can be achieved by replacing the Brownian motion W in the above equation for X^t with the Brownian motion $\{\sqrt{t}W_{s/t}, 0 \le s \le t\}$, where now $\{W_s, 0 \le s \le 1\}$ is a euclidean Brownian motion defined on a fixed probability space $(\Omega, \mathscr{F}_*, \mathbb{P})$. The stochastic differential equation for the scaled Brownian bridge Z^t becomes

$$dZ^t_s = \sigma(\sqrt{t}Z^t_s)\,dW_s + \sqrt{t}b(\sqrt{t}Z^t_s)\,ds + \sqrt{t}C(t(1-s),\sqrt{t}Z^t_s)\,ds. \tag{7.7.3}$$

Let Y be the euclidean Brownian bridge defined by

$$dY_s = dW_s - \frac{Y_s}{1-s}ds, \quad Y_0 = 0. \tag{7.7.4}$$

We will show that $Z^t \to Y$ as $t \downarrow 0$ on the fixed probability space $(\Omega, \mathscr{F}_*, \mathbb{P})$. In the proof of this result, we will need the following two facts about the gradient of the logarithmic heat kernel in (7.7.1):

$$\sqrt{t}\left|C(t(1-s), \sqrt{t}z)\right| \le C_1 \left\{ \frac{|z|}{1-s} + \frac{1}{\sqrt{1-s}} \right\}; \tag{7.7.5}$$

$$\lim_{t\to 0} \sqrt{t}C(t(1-s), \sqrt{t}z) = -\frac{z}{1-s}. \tag{7.7.6}$$

The convergence is uniform for $0 \le s \le 1/2$ and uniformly bounded z. The first fact is a consequence of the estimate on $\nabla \ln p_M(s, x, y)$ in THEOREM 5.5.3. For the proof of the second fact see Bismut [**6**], p. 104 or Norris [**60**].

Lemma 7.7.2. *For any positive integer N,*

$$\lim_{t\to 0} \mathbb{E}|Z^t - Y|_{1,\infty}^N \to 0.$$

Proof. We prove for $N = 2$ and leave the rest to the reader. By the symmetry of a Brownian bridge under time reversal, it is enough to work on the half interval $[0, 1/2]$. Using the inequality (7.7.5) it is easy to show by Doob's martingale inequality that, for any positive integer N, there is a constant C_N such that

$$\mathbb{E}|Z^t|_{1/2,\infty}^N \le C_N, \qquad \mathbb{E}|Y^t|_{1/2,\infty}^N \le C_N. \tag{7.7.7}$$

Since we are restricted to $[0, 1/2]$, $\sqrt{t}C(t(1-s), \sqrt{t}z)$, σ, and b are uniformly bounded. Now using Doob's inequality again, we have

$$\begin{aligned}\mathbb{E}|Z^t - Y|_{s,\infty}^2 \le C_1 \mathbb{E}\int_0^s \left|\sigma(\sqrt{t}Z_\tau^t) - I\right|^2 d\tau + C_1 t \\ + C_1 \mathbb{E}\int_0^s \left|\sqrt{t}C(t(1-\tau), \sqrt{t}Z_\tau^t) + \frac{Y_\tau}{1-\tau}\right|^2 d\tau.\end{aligned}$$

Using the inequality $|\sigma(z) - I| \le \text{const.}|z|$ and (7.7.7), we have

$$\mathbb{E}\int_0^s \left|\sigma(\sqrt{t}Z_\tau^t) - I\right|^2 d\tau \le C_2 t.$$

For the last term on the right side, we have

$$\begin{aligned}&\int_0^s \left|\sqrt{t}C(t(1-\tau), \sqrt{t}Z_\tau^t) + \frac{Y_\tau}{1-\tau}\right|^2 d\tau \le \\ &\quad 2\int_0^s \left|\frac{Z_\tau^t - Y_\tau}{1-\tau}\right|^2 d\tau + 2\int_0^s \left|\sqrt{t}C(t(1-\tau), \sqrt{t}Z_\tau^t) + \frac{Z_\tau^t}{1-\tau}\right|^2 d\tau.\end{aligned}$$

For any positive ϵ we have, for sufficiently small t,

$$\mathbb{E}\int_0^s \left|\sqrt{t}C(t(1-\tau),\sqrt{t}Z_\tau^t)+\frac{Z_\tau^t}{1-\tau}\right|^2 d\tau \le \epsilon.$$

This estimate follows considering two cases: $|z_\tau^t| > \lambda$ and $|z_\tau^t| \le \lambda$. Using (7.7.7), we see that the former case occurs with arbitrarily small probability if λ is a fixed large number, and hence by (7.7.7) again the expected value in this case can be made arbitrarily small; for a fixed λ, using (7.7.6), we see that the expected value in the latter case is arbitrarily small if t is sufficiently small.

Putting these estimates together, we see that for any positive ϵ, the folowing inequality holds for all sufficiently small t and all $s \in [0, 1/2]$:

$$\mathbb{E}|Z^t - Y|_{s,\infty}^2 \le C_3\int_0^s \mathbb{E}|Z^t - Y|_{\tau,\infty}^2 d\tau + C_3(t+\epsilon).$$

From this we conclude by Gronwall's inequality that

$$\lim_{t\to 0}\mathbb{E}|Z^t - Y|_{1/2,\infty}^2 = 0.$$

□

After this long digression on Brownian bridge, we come back to the problem of the asymptotic behavior of the Brownian holonomy on the spin bundle $\mathscr{S}(M)$. Recall that $U_t : T_xM \to T_{X_t}M$ is the stochastic parallel transport along the path $X[0,t]$. In the following theorem, we assume that Brownian bridges at x of different time lengths are defined on the same probability space $(\Omega, \mathscr{F}_*, \mathbb{P})$ in the manner described above.

Proposition 7.7.3. *We have*

$$\lim_{t\to 0}\frac{U_t^{\mathscr{S}(M)_x} - I^{\mathscr{S}(M)_x}}{t} = \frac{1}{8}\int_0^1 \langle R(e_i,e_j)Y_s, dY_s\rangle c(e_i)c(e_j)$$

in $L^N(\Omega, \mathscr{F}_*, \mathbb{P})$ for all positive integers N.

Lemma 7.4.1 describes the action of $\mathfrak{so}(d)$ on the spin bundle $\mathscr{S}(M)$. It is clear from this lemma that it is enough to prove that

$$\lim_{t\to 0}\frac{e_i^j(t) - \delta_i^j}{t} = \frac{1}{2}\int_0^1 \langle R(e_i,e_j)Y_s, dY_s\rangle, \tag{7.7.8}$$

where $U_t = \left\{e_i^j(t)\right\}$. The rest of this section is devoted to the proof of this result.

We will calculate U_t explicitly in the normal coordinates $\left\{x^i\right\}$. The curvature enters into our calculation through the following expansion of the Christoffel symbols in these coordinates.

Lemma 7.7.4. *Let $\{x^i\}$ be the normal coordinates at a point $x \in M$, and denote by $\{X_i\}$ the local orthonormal frame obtained from $\{\partial/\partial x^i\}$ at x by translating along geodesic rays from x. Then the Christoffel symbols have the following expansion at x:*

$$\Gamma_{ij}^k = -\frac{1}{2}\sum_{l=1}^d \langle R(X_i, X_l)X_j, X_k\rangle x^l + O(|x|^2).$$

Proof. We know that the Christoffel symbols vanish at x (see the proof of LEMMA 2.5.5); hence it is enough to prove that, at x,

$$\nabla_{X_l}\Gamma_{ij}^k = -\frac{1}{2}\langle R(X_i, X_l)X_j, X_k\rangle.$$

From the definition of Christoffel symbols and the orthogonality of $\{X_i\}$, we have

$$\Gamma_{ij}^k = \langle \nabla_{X_i}X_j, X_k\rangle.$$

Differentiating along X_l and using $\nabla_{X_l}X_k = 0$ at x, we have, at the same point

$$\nabla_{X_l}\Gamma_{ij}^k = \langle \nabla_{X_l}\nabla_{X_i}X_j, X_k\rangle.$$

Now for $l \neq i$, the unit vector field $(X_l + X_i)/\sqrt{2}$ is parallel along the geodesic ray $\{x_m = 0, m \neq i, l; x_i = x_l\}$. Hence it is the tangent vector field of the geodesic. On the other hand, X_j is parallel along the geodesic; hence $\nabla_{X_l+X_i}X_j = 0$ along the geodsic, which implies trivially that

$$\nabla_{X_l+X_i}\nabla_{X_l+X_i}X_j = 0$$

at x. A similar argument shows also that

$$\nabla_{X_l}\nabla_{X_l}X_j = \nabla_{X_i}\nabla_{X_i}X_j = 0.$$

It follows that

$$\nabla_{X_l}\nabla_{X_i}X_j + \nabla_{X_i}\nabla_{X_l}X_j = 0.$$

Finally, since the bracket $[X_l, X_i] = 0$ at x, by the definition of the curvature tensor we have

$$\begin{aligned}\nabla_{X_l}\Gamma_{ij}^k &= -\frac{1}{2}\langle \nabla_{X_i}\nabla_{X_l}X_j - \nabla_{X_l}\nabla_{X_i}X_j, X_k\rangle \\ &= -\frac{1}{2}\langle R(X_i, X_l)X_j, X_k\rangle.\end{aligned}$$

This completes the proof. □

In the normal coordinates, an orthonormal frame is expressed as

$$u = \left\{x^i, e_i^j\right\}, \qquad ue_i = e_i^j\frac{\partial}{\partial x^j}.$$

From the local equation for horizontal Brownian motion we have (see (3.3.9) in EXAMPLE 3.3.5)

$$de_i^j(s) = -\Gamma_{mk}^j(X_s)\, e_i^k(s) \circ dX_s^m, \quad 0 \le s \le t.$$

Replacing X_s by $\sqrt{t}Z_{st}^t$ (see (7.7.2)), we obtain

$$de_i^j(st) = -t\left[\frac{\Gamma_{mk}^j(\sqrt{t}Z_s^t)\, e_i^k(st)}{\sqrt{t}}\right] \circ dZ_s^{t,m}, \quad 0 \le s \le 1. \tag{7.7.9}$$

As $t \downarrow 0$, the limit of the expression in brackets can be obtained from the local expansion of the Christoffel symbols (LEMMA 7.7.4) and the fact that $Z^t \to Y$ as $t \downarrow 0$ (PROPOSITION 7.7.2), and we have

$$\lim_{t\downarrow 0} \frac{\Gamma_{mk}^j(\sqrt{t}Z_s)\, e_i^k(st)}{\sqrt{t}} = -\frac{1}{2}\langle R(e_i, e_j)Y_s, e_m\rangle.$$

Integrating (7.7.9) from 0 to 1 and taking the limit, we have, at least formally,

$$\begin{aligned} \frac{e_i^j(t) - \delta_i^j}{t} &= -\frac{1}{2}\int_0^1 \left[\frac{\Gamma_{mk}^i(\sqrt{t}Z_s^t)e_j^k(st)}{\sqrt{t}}\right] \circ dZ_s^{t,m} \\ &\to \frac{1}{2}\int_0^1 \langle R(e_i, e_j)Y_s, \circ dY_s\rangle. \end{aligned} \tag{7.7.10}$$

Since the matrix $R(e_i, e_j)$ is anti-symmetric, its diagonal elements are zero; hence the bounded variation part of the Stratonovich integral vanishes, and we have

$$\lim_{t\downarrow 0} \frac{e_i^j(t) - \delta_i^j}{t} = \int_0^1 \langle R(e_i, e_j)Y_s, dY_s\rangle.$$

To justify the passing to the limit in (7.7.10) we consider the half time intervals $[0, 1/2]$ and $[1/2, 1]$ separately. Using the equation (7.7.3) for Z_s^t we can convert the Stratonovich in the equation (7.7.9) for $e_i^j(st)$ into the corresponding Itô integral. Then, integrating from 0 to 1/2 and using (7.7.6) and (7.7.4), we see that

$$\lim_{t\downarrow 0} \frac{e_i^j(t/2) - \delta_i^j}{t} = \frac{1}{2}\int_0^{1/2} \langle R(e_i, e_j)Y_s, dY_s\rangle.$$

Likewise, by symmetry we have

$$\lim_{t\downarrow 0} \frac{e_i^j(t) - e_i^j(t/2)}{t} = \frac{1}{2}\int_{1/2}^1 \langle R(e_i, e_j)Y_s, dY_s\rangle.$$

Adding the two limits, we obtain (7.7.8). We have therefore completed the proof of PROPOSITION 7.7.3.

Chapter 8

Analysis on Path Spaces

Stochastic analysis on path and loop spaces is an active area of current research. In this chapter we concentrate on two topics for the path space over a compact Riemannian manifold: integration by parts and logarithmic Sobolev inequalities. For the quasi-invariance of the Wiener measure we only discuss the euclidean case, for including a proof of this theorem for a general Riemannian manifold would take too much space. Although the integration by parts formula can be derived easily once the quasi-invariance of the Wiener measure under Cameron-Martin shifts is established, fortunately there are other approaches which circumvent the quasi-invariance, thus making the integration by parts formula in path space much more accessible. The reader should consult Malliavin [**57**] for a different approach to the topics in this chapter.

The first two sections of this chapter is devoted to the euclidea theory, where most results can be proved by explicit computations. In SECTION 8.3 we prove several formulas due to Bismut [**6**] involving Brownian motion and the gradient operator on a Riemannian manifold. Driver's integration by parts formula in the path space is proved in SECTION 8.4. In SECTION 8.5, we use the integration by parts formula to extend the Clark-Ocone martingale representation theorem from euclidean Brownian motion to Riemannian Brownian motion. SECTION 8.6 contains a general discussion on logarithmic Sobolev inequalities and hypercontractivity. In the last SECTION 8.7 we prove a logarithmic Sobolev inequality for the path space over a compact Riemannian manifold.

8.1. Quasi-invariance of the Wiener measure

We will restrict ourselves to the unit time interval $I = [0,1]$ and denote by $P_o(\mathbb{R}^n)$ the space of continuous functions from I to $\mathbb{R}^n$ from the origin o. The Wiener measure on $P_o(\mathbb{R}^d)$ is denoted by $\mathbb{P}$.

If an $h \in P_o(\mathbb{R}^n)$ is absolutely continuous and $\dot{h} \in L^2(I;\mathbb{R}^n)$, we define

$$|h|_{\mathscr{H}} = \sqrt{\int_0^1 |\dot{h}_s|^2 ds}\,;$$

otherwise we set $|h|_{\mathscr{H}} = \infty$. The ($\mathbb{R}^n$-valued) Cameron-Martin space is

$$\mathscr{H} = \{h \in P_o(\mathbb{R}^n) : |h|_{\mathscr{H}} < \infty\}.$$

Theorem 8.1.1. *Let $h \in \mathscr{H}$, and let*

$$\xi_h\omega = \omega + h, \qquad \omega \in P_o(\mathbb{R}^n),$$

be a Cameron-Martin shift on the path space. Let $\mathbb{P}$ be the Wiener measure on $P_o(\mathbb{R}^d)$. Then the shifted Wiener measure $\mathbb{P}^h = \mathbb{P} \circ \xi_h^{-1}$ is absolutely continuous with respect to $\mathbb{P}$, and

$$\frac{d\,\mathbb{P}^h}{d\,\mathbb{P}} = \exp\left\{\int_0^1 \left\langle \dot{h}_s, d\omega_s\right\rangle - \frac{1}{2}\int_0^1 |\dot{h}_s|^2 ds\right\}. \tag{8.1.1}$$

When the shifted Wiener measure $\mathbb{P}^h$ is equivalent to $\mathbb{P}$ (meaning they are mutually absolutely continuous), we say that the Wiener measure is quasi-invariant under the Cameron-Martin shift ξ_h. The above theorem is implied by the the following more general Girsanov's theorem

Theorem 8.1.2. *Let $B = \{B_t, t \in I\}$ be an $\mathscr{F}_*$-Brownian motion on a filtered probability space $(\Omega, \mathscr{F}_*, \mathbb{P})$, and V an $\mathbb{R}^n$-valued, $\mathscr{F}_*$-adapted process such that*

$$\int_0^1 |V_s|^2 ds \le C$$

for a fixed (nonrandom) constant C. Define a new probability measure $\mathbb{Q}$ by

$$\frac{d\,\mathbb{Q}}{d\,\mathbb{P}} = \exp\left\{\int_0^1 \langle V_s, dB_s\rangle - \frac{1}{2}\int_0^1 |V_s|^2 ds\right\}.$$

Let $X = \{X_t, t \in I\}$ be defined by

$$X_s = B_s - \int_0^s V_\tau d\tau. \tag{8.1.2}$$

Then X is an $\mathscr{F}_*$-Brownian motion on the probability space $(\Omega, \mathscr{F}_*, \mathbb{Q})$.

Proof. Let $e = \{e_s, s \in I\}$ be the exponential martingale

$$e_s = \exp\left\{\int_0^s \langle V_\tau, dB_\tau\rangle - \frac{1}{2}\int_0^s |V_\tau|^2 d\tau\right\}.$$

Then

$$\left.\frac{d\mathbb{Q}}{d\mathbb{P}}\right|_{\mathscr{F}_s} = \mathbb{E}^{\mathbb{P}}\left\{\left.\frac{d\mathbb{Q}}{d\mathbb{P}}\right|\mathscr{F}_s\right\} = \mathbb{E}^{\mathbb{P}}\left\{e_1|\mathscr{F}_s\right\} = e_s.$$

From this it is easy to check that if $Y_t \in \mathscr{F}_t$ and $s \le t$, then

$$\mathbb{E}^{\mathbb{Q}}\left\{Y_t|\mathscr{F}_s\right\} = e_s^{-1}\mathbb{E}^{\mathbb{P}}\left\{Y_t e_t|\mathscr{F}_s\right\}.$$

This means that a continuous, $\mathscr{F}_*$-adapted process Y is a local martingale under $\mathbb{Q}$ if and only if $eY = \{e_s Y_s, s \in I\}$ is a local martingale under $\mathbb{P}$. Now by Itô's formula we have $de_s = e_s \langle V_s, dB_s\rangle$. Another application of Itô's formula gives

$$d(e_s X_s) = e_s dB_s + e_s X_s \langle V_s, dB_s\rangle .$$

This shows that $eX = \{e_s X_s, s \in I\}$ is a local martingale under $\mathbb{P}$; hence by the above remark X is a local martingale under $\mathbb{Q}$. From (8.1.2) it is clear that X has the same quadratic variations as B. Lévy's criterion now implies that X is a Brownian motion under $\mathbb{Q}$. □

There is a similar result for the loop space

$$L_o(M) = \left\{\omega \in P_o(\mathbb{R}^d) : \omega_1 = o\right\}.$$

Let

$$\mathscr{H}_o = \{h \in \mathscr{H} : h(1) = 0\}.$$

Recall that the Wiener measure $\mathbb{P}_o$ on $L_o(\mathbb{R}^n)$ is the law of a Brownian bridge at o with time length 1. We show that $\mathbb{P}_o$ is quasi-invariant under the Cameron-Martin shift $\xi_h : L_o(\mathbb{R}^n) \to L_o(\mathbb{R}^n)$ for $h \in \mathscr{H}_o$.

On the probability space $(P_o(\mathbb{R}^n), \mathscr{B}_*, \mathbb{P})$ consider the stochastic differential equation for a Brownian bridge

$$dX_s = dW_s - \frac{X_s}{1-s}\,ds, \qquad X_0 = o,$$

where W is the coordinate process on $W(\mathbb{R}^d)$. The assignment $JW = X$ defines a measurable map $J : P_o(\mathbb{R}^n) \to L_o(\mathbb{R}^n)$. The map J can also be viewed as an $L_o(\mathbb{R}^n)$-valued random variable. Suppose that $h \in \mathscr{H}_o$. A simple computation shows that

$$d\{X_s + h_s\} = d\{W_s + k_s\} - \frac{X_s + h_s}{1-s}ds,$$

where

$$k_s = h_s + \int_0^s \frac{h_\tau}{1-\tau}d\tau.$$

This shows that through the map J, the shift $X \mapsto X + h$ in the loop space $L_o(\mathbb{R}^n)$ is equivalent to the shift $W \mapsto W + k$ in the path space $P_o(\mathbb{R}^n)$. The following lemma shows that the latter is a Cameron-Martin shift.

Lemma 8.1.3. *Let $h \in \mathscr{H}_o$ and*

$$k_s = h_s + \int_0^s \frac{h_\tau}{1-\tau} d\tau.$$

Then $k \in \mathscr{H}$ and $|k|_{\mathscr{H}} = |h|_{\mathscr{H}}$.

Proof. We have, for any $t \in (0,1)$,

$$\int_0^t |\dot{k}_s|^2 ds = \int_0^t |\dot{h}_s|^2 ds + 2\int_0^2 \frac{h_s \cdot \dot{h}_s}{1-s} ds + \int_0^t \left| \frac{h_s}{1-s} \right|^2 ds.$$

Integrating by parts in the last integral, we obtain

$$\int_0^t |\dot{k}_s|^2 ds = \int_0^t |\dot{h}_s|^2 ds + \frac{|h_t|^2}{1-t}.$$

As $t \to 1$, the last term tends to zero because

$$\frac{|h_t|^2}{1-t} = \frac{1}{1-t} \left| \int_t^1 \dot{h}_s ds \right|^2 \le \int_t^1 |\dot{h}_s|^2 ds \to 0.$$

This completes the proof. □

The lemma implies that $k \in \mathscr{H}$. Define the exponential martingale

$$e_s = \exp\left\{ \int_0^s \left\langle \dot{k}_\tau, dW_\tau \right\rangle - \frac{1}{2} \int_0^s |\dot{k}_\tau|^2 d\tau \right\}.$$

Let $\mathbb{P}^k$ be the probability measure on $P_o(\mathbb{R}^n)$ defined by

$$\frac{d\,\mathbb{P}^k}{d\,\mathbb{P}} = e_1. \tag{8.1.3}$$

By the Cameron-Martin-Maruyama theorem, $\mathbb{P}^k$ is the law of $W + k$. Since it is absolutely continuous with respect to $\mathbb{P}$, the random variable $J(W+k)$ is well-defined, and under the probability $\mathbb{P}^k$,

$$J(W + k) = X + h. \tag{8.1.4}$$

Let $\mathbb{P}_o^h$ be the law of the shifted Brownian bridge $X + h$. Then for any $C \in \mathscr{B}(L_o(\mathbb{R}^n))$,

$$\begin{aligned} \mathbb{P}_o^h(C) &= \mathbb{P}_o(C - h) = \mathbb{P}(J^{-1}C - k) \\ &= \mathbb{P}^k(J^{-1}C) = \mathbb{P}(e_1; J^{-1}C) \\ &= \mathbb{P}_o(e_1 \circ J^{-1}; C), \end{aligned}$$

where we have used (8.1.4) and (8.1.3) in the second and the fourth steps, respectively. Now it is clear that $\mathbb{P}_o^h$ and $\mathbb{P}_o$ are mutually equivalent on $L_o(\mathbb{R}^n)$, and

$$\frac{d\,\mathbb{P}_o^h}{d\,\mathbb{P}_o} = e_1 \circ J^{-1}. \tag{8.1.5}$$

Finally, it is easy to verify that

$$\int_0^1 \left\langle \dot{k}_s, dW_s \right\rangle = \int_0^1 \left\langle \dot{h}_s, dX_s \right\rangle.$$

This together with LEMMA 8.1.3 implies that

$$e_1 \circ J^{-1} = \exp\left\{ \int_0^1 \left\langle \dot{h}_s, dX_s \right\rangle - \frac{1}{2}\int_0^1 |\dot{h}_s|^2 ds \right\}.$$

We have proved the following result.

Theorem 8.1.4. *Let $h \in \mathscr{H}_o$ and $\xi_h \gamma = \gamma + h$ for $\gamma \in L_o(\mathbb{R}^n)$. Let $\mathbb{P}_o$ be the Wiener measure on the loop space $L_o(\mathbb{R}^n)$. Then the shifted Wiener measure $\mathbb{P}_o^h = \mathbb{P}_o \circ (\xi^h)^{-1}$ on the loop space $L_o(\mathbb{R}^n)$ is absolutely continuous with respect to $\mathbb{P}_o$, and*

$$\frac{d\,\mathbb{P}_o^h}{d\,\mathbb{P}_o} = \exp\left\{ \int_0^1 \left\langle \dot{h}_s, dX_s \right\rangle - \frac{1}{2}\int_0^1 |\dot{h}_s|^2 ds \right\}.$$

The converse of THEOREM 8.1.1 also holds.

Theorem 8.1.5. *Let $h \in P_o(\mathbb{R}^n)$, and let*

$$\xi_h \omega = \omega + h, \qquad \omega \in P_o(\mathbb{R}^n)$$

be the shift on the path space by h. Let $\mathbb{P}$ be the Wiener measure on $P_o(\mathbb{R}^d)$. If the shifted Wiener measure $\mathbb{P}^h = \mathbb{P} \circ (\xi_h)^{-1}$ is absolutely continuous with respect to $\mathbb{P}$, then $h \in \mathscr{H}$.

Proof. We show that if $h \notin \mathscr{H}$, then the measures $\mathbb{P}$ and $\mathbb{P}^h$ are mutually singluar, i.e., there is a set A such that $\mathbb{P}A = 1$ and $\mathbb{P}^h A = 0$. Let

$$\langle f, g \rangle_{\mathscr{H}} = \int_0^1 \dot{f}_s dg_s,$$

whenever the integral is well defined. If $f \in \mathscr{H}$ such that $\dot{f}$ is a step function on $[0, 1]$:

$$\dot{f} = \sum_{i=0}^{l-1} f_i I_{[s_i, s_{i+1})},$$

where $f_i \in \mathbb{R}^n$ and $0 = s_0 < s_1 < \cdots < s_l = 1$, then

$$\langle f, h \rangle_{\mathscr{H}} = \sum_{i=0}^{l-1} f_i \left(h_{s_{i+1}} - h_{s_i} \right)$$

is well defined. It is an easy exercise to show that if there is a constant C such that

$$\langle f, h \rangle_{\mathscr{H}} \le C |f|_{\mathscr{H}}$$

for all step functions $\dot{f}$, then h is absolutely continuous and $\dot{h}$ is square-integrable, namely, $h \in \mathscr{H}$.

Suppose that $h \notin \mathscr{H}$. Then there is a sequence $\{f_l\}$ such that

$$|f_l|_{\mathscr{H}} = 1 \quad \text{and} \quad \langle h, f_l \rangle_{\mathscr{H}} \geq 2l.$$

Let W be the coordinate process on $P_o(\mathbb{R}^d)$. Then it is a Brownian motion under $\mathbb{P}$ and the stochastic integral

$$\langle f_l, W \rangle_{\mathscr{H}} = \int_0^1 \left\langle \dot{f}_{l,s}, dW_s \right\rangle$$

is well defined. Let

$$A_l = \{\langle f_l, W \rangle_{\mathscr{H}} \leq l\}$$

and $A = \limsup_{l \to \infty} A_l$. Since $|f_l|_{\mathscr{H}} = 1$, the random variable $\langle f_l, W \rangle_{\mathscr{H}}$ is standard Gaussian under $\mathbb{P}$; hence

$$\mathbb{P} A_l \geq 1 - e^{-l^2/2}.$$

This shows that $\mathbb{P} A = 1$. On the other hand,

$$\mathbb{P}^h A_l = \mathbb{P}\left\{\langle f_l, W + h \rangle_{\mathscr{H}} \leq l\right\} \leq \mathbb{P}\left\{\langle f_l, W \rangle_{\mathscr{H}} \leq -l\right\}.$$

Hence $\mathbb{P}^h A_l \leq e^{-l^2/2}$ and $\mathbb{P}^h A = 0$. Therefore $\mathbb{P}$ and $\mathbb{P}^h$ are mutually singular. □

8.2. Flat path space

Before we study the general path space $P_o(M)$ over a Riemannian manifold, it is helpful to have a complete understanding of the flat path space $P_o(\mathbb{R}^n)$. We first introduce the basic directional derivative operator D_h along a direction $h \in \mathscr{H}$ and the gradient operator D on the flat path space $P_o(\mathbb{R}^n)$. They are first defined on the set of cylinder functions and then extended to closed operators on appropriate L^p-spaces. The extensions are accomplished by establishing integration by parts formulas for these operators. Along the way we will see that a natural set of directions h along which the D_h are well-defined as closed operators is the Cameron-Martin space $\mathscr{H}$. Thus at this preliminary stage, we can roughly regard $\mathscr{H}$ as the natural tangent space of the infinite-dimensional manifold $P_o(\mathbb{R}^n)$. A parallel theory can be developed for the loop space $L_o(\mathbb{R}^n)$. It should be pointed out that, from the point of view of Gaussian measures, owing to the linear structure of the base space $\mathbb{R}^n$, there are no differences between the flat path and loop spaces.

By analogy with finite dimensional space, each element $h \in P_o(\mathbb{R}^n)$ represents a direction along which one can differentiate a nice function F on $P_o(\mathbb{R}^n)$. Naturally the directional derivative of F along h should be defined by the formula

$$D_h F(\omega) = \lim_{t \to 0} \frac{F(\omega + th) - F(\omega)}{t} \tag{8.2.1}$$

if the limit exists in some sense. The preliminary class of functions on $P_o(\mathbb{R}^n)$ for which the above definition of D_hF makes immediate sense is that of cylinder functions.

Definition 8.2.1. *Let X be a Banach space. A function $F : P_o(\mathbb{R}^n) \to X$ is called an X-valued cylinder function if it has the form*

$$F(\omega) = f(\omega_{s_1}, \cdots, \omega_{s_l}), \tag{8.2.2}$$

where $0 < s_1 < \cdots < s_l \le 1$ and f is an X-valued smooth function on $(\mathbb{R}^n)^l$ such that all its derivatives have at most polynomial growth. The set of X-valued cylinder functions is denoted by $\mathcal{C}(X)$. We denote $\mathcal{C}(\mathbb{R}^1)$ simply by $\mathcal{C}$.

Let $\mathbb{P}$ be the Wiener measure on $P_o(\mathbb{R}^n)$ and $L^p(\mathbb{P}) = L^p(P_o(\mathbb{R}^n), \mathscr{B}, \mathbb{P})$ be the Banach space of $\mathbb{R}$-valued L^p-integrable functions on $P_o(\mathbb{R}^n)$. For a Banach space X, denote by $L^p(\mathbb{P}; X)$ the space of X-valued functions F such that $|F|_X \in L^p(\mathbb{P})$. Then $L^p(\mathbb{P}; X)$ is a Banach space with the norm

$$\|F\|_p = \left\{ \int_{P_o(\mathbb{R}^n)} |F|_X^p \, d\,\mathbb{P} \right\}^{1/p}.$$

It is easy to verify that the set $\mathcal{C}(X)$ is dense in $L^p(\mathbb{P}; X)$ for all $p \in [1, \infty)$. Typically $X = \mathbb{R}^1$, $\mathbb{R}^n$, or $\mathscr{H}$.

If $F \in \mathcal{C}$ is given by (8.2.2), then it is clear that the limit (8.2.1) exists everywhere and

$$D_hF = \sum_{i=1}^{l} \langle \nabla^i F, h_{s_i} \rangle_{\mathbb{R}^n}, \tag{8.2.3}$$

where

$$\nabla^i F(\omega) = \nabla^i f(\omega_{s_1}, \cdots, \omega_{s_l}).$$

Here $\nabla^i f$ denotes the gradient of f with respect to the ith variable.

From functional analysis we know that the directional derivative operator D_h is useful only if it is closable in some $L^p(\mathbb{P})$. The directional derivative D_h operator is closable if it has an integration by parts formula. We will see later that this requirement forces us to restrict ourselves to the directions $h \in \mathscr{H}$, the Cameron-Martin space $\mathscr{H}$, because only for these directions does D_h have an integration by parts formula. The fact that the restriction $h \in \mathscr{H}$ is appropriate can also be seen from the preliminary definition (8.2.1) of D_h. The limit there should be understood at least in measure with respect to $\mathbb{P}$. Since in general F is only defined $\mathbb{P}$-a.e., the function $\omega \mapsto F(\omega + th)$ is not well defined unless the shifted measure $\mathbb{P}^{th}$ (the law of $\omega + th$) is absolutely continuous with respect to $\mathbb{P}$. This means that $h \in \mathscr{H}$ by THEOREM 8.1.5.

It is natural to define the gradient DF of a function $F \in \mathcal{C}$ to be an $\mathscr{H}$-valued function on $P_o(\mathbb{R}^n)$ such that

$$\langle DF, h\rangle_{\mathscr{H}} = D_h F.$$

By a simple calculation we verify that

$$(DF)_s = \sum_{i=1}^{l} \min(s, s_i)\nabla^i F \tag{8.2.4}$$

and

$$|DF|^2_{\mathscr{H}} = \sum_{i=1}^{l} (s_i - s_{i-1}) \left| \sum_{j=i}^{l} \nabla^j F \right|^2 .$$

We can now proceed as follows. First we define D_h on the set $\mathcal{C}$ of cylinder functions by (8.2.3). We then exhibit an integration by parts formula for D_h defined this way. This will gives us a formal adjoint D_h^* on the set of cylinder functions. Using a standard argument from functional analysis, we show that D_h is closable in $L^p(\mathbb{P})$ for all $p > 1$ and $\mathcal{C}$ is a core. The closability of the gradient operator D follows immediately. The same procedure will be followed later when we discuss D_h and D on the path space over a Riemannian manifold.

There are two methods for the integration by parts formula. The first method uses the quasi-invariance of the Wiener measure (Cameron-Martin-Maruyama THEOREM 8.1.1). The second method is to show the formula for cylinder functions with one time point dependence $F(\omega) = f(\omega_s)$, and then to apply induction on the number of time points. We will use the first method for the flat path space in this section and the second method when we discuss general path spaces in SECTION 8.4.

In the following we will use the notation

$$(F, G) = \int_{P_o(\mathbb{R}^n)} FG \, d\mathbb{P}.$$

Theorem 8.2.2. *Let $F, G \in \mathcal{C}$ and $h \in \mathscr{H}$. Then*

$$(D_h F, G) = (F, D_h^* G), \tag{8.2.5}$$

where

$$D_h^* = -D_h + \int_0^1 \langle \dot{h}_s, dW_s\rangle.$$

Proof. Let $\xi_{th}\omega = \omega + th$ be the flow generated by the vector field D_h on the path space $P_o(\mathbb{R}^n)$ and $\mathbb{P}^{th} = \mathbb{P} \circ \xi_{th}^{-1}$ the shifted measure. Then $\mathbb{P}^{th}$ and

$\mathbb{P}$ are mutually absolutely continuous (THEOREM 8.1.1). We have

$$\begin{aligned}\int_{P_o(\mathbb{R}^n)} (F \circ \xi_{th})\, G\, d\,\mathbb{P} &= \int_{P_o(\mathbb{R}^n)} (G \circ \xi_{-th})\, F\, d\,\mathbb{P}^{th} \\ &= \int_{P_o(\mathbb{R}^n)} (G \circ \xi_{-th})\, F \frac{d\,\mathbb{P}^{th}}{d\,\mathbb{P}}\, d\,\mathbb{P}.\end{aligned}$$

We differentiate with respect to t and set $t = 0$. Using the formula for the Radon-Nikodým derivative (8.1.1) in the Cameron-Martin-Maruyama THEOREM 8.1.1, we have at $t = 0$

$$\frac{d}{dt}\left\{\frac{d\,\mathbb{P}^{th}}{d\,\mathbb{P}}\right\} = \int_0^1 \langle \dot{h}_s, dW_s\rangle.$$

The formula follows immediately. □

Since D_h is a derivation, we have

$$D_h^* G = -D_h G + (D_h^* 1) G.$$

Thus the above theorem says that

$$D_h^* 1 = \int_0^1 \left\langle \dot{h}_s, dW_s \right\rangle.$$

To understand the integration by parts formula better, let's look at its finite dimensional analog and find out the proper replacement for the stochastic integral in D_h^*. Let $h \in \mathbb{R}^N$ and consider the differential operator

$$D_h = \sum_{i=1}^N h^i \frac{d}{dx^i}.$$

Let μ be the Gaussian measure on $\mathbb{R}^N$, i.e.,

$$\frac{d\mu}{dx} = \left(\frac{1}{2\pi}\right)^{N/2} e^{-|x|^2/2}.$$

[dx is the Lebesgue measure.] For smooth functions F, G on $\mathbb{R}^N$ with compact support, we have by the usual integration by parts for the Lebesgue measure,

$$(D_h F, G) = (F, D_h^* G), \tag{8.2.6}$$

where $D_h^* = -D_h + \langle h, x\rangle$ at $x \in \mathbb{R}^N$.

An integration by parts formula for the gradient operator D can be obtained as follows. Fix an orthonormal basis $\{h^j\}$ for $\mathscr{H}$. Denote by $\mathcal{C}_0(\mathscr{H})$ the set of $\mathscr{H}$-valued functions G of the form $G = \sum_j G_j h^j$, where each G_j is in $\mathcal{C}$ and almost all of them are equal to zero. It is easy to check that $\mathcal{C}_0(\mathscr{H})$ is dense in $L^p(\mathbb{P}; \mathscr{H})$ for all $p \in [1, \infty)$.

Since $DF = \sum_j (D_{h^j} F) h^j$ in $L^2(\mathbb{P}; \mathscr{H})$, we have

$$(DF, G) = \sum_j (D_{h^j} F, G_j) = \sum_j (F, D_{h^j}^* G_j).$$

By the assumption that $G \in \mathcal{C}_o(\mathscr{H})$, the sums are finite. Let

$$D^* G = \sum_j D_{h^j}^* G_j = -\sum_j D_{h^j} G_j + \sum_j G_j \int_0^1 \left\langle \dot{h}_s^j, dW_s \right\rangle.$$

We rewrite this formula in a more compact form. If

$$J = \sum_{j,k} J_{jk} h^j \otimes h^k$$

is in $\mathscr{H} \otimes \mathscr{H}$, we naturally define

$$\operatorname{Trace} J = \sum_j J_{jj}.$$

For $G = \sum_k G_k h^j \in \mathcal{C}_0(\mathscr{H})$ we define its gradient to be

$$DG = \sum_{j,k} (D_{h^j} G_k) h^j \otimes h^k.$$

Then it is clear that

$$\operatorname{Trace} DG = \sum_j D_{h^j} G_j.$$

For $G = \sum_j G_j h^j$ we define

$$\int_0^1 \left\langle \dot{G}_s, dW_s \right\rangle = \sum_j G_j \int_0^1 \left\langle \dot{h}_s^j, dW_s \right\rangle.$$

Note that in general $\dot{G}_s$ is not $\mathscr{B}_*$-adapted and the above integral is interpreted as the term-by-term integration with respect to a specific basis for $\mathscr{H}$, but it is independent of the choice of the basis. Using the notations introduced above, we can write

$$D^* G = -\operatorname{Trace} DG + \int_0^1 \left\langle \dot{G}_s, dW_s \right\rangle. \tag{8.2.7}$$

We have proved the following integration by parts formula for the gradient operator D.

Theorem 8.2.3. *Let $F \in \mathcal{C}$ and $G \in \mathcal{C}_0(\mathscr{H})$. Then*

$$(DF, G) = (F, D^* G),$$

where $D^* G$ is given by (8.2.7).

Proof. See above. □

One use of the integration by parts formula is to show that both D_h and D are closable. The argument is standard in functional analysis. Let's first review some basic facts.

Definition 8.2.4. *Let B_1, B_2 be Banach spaces and $A : \mathrm{Dom}(A) \to B_2$ a densely defined linear operator. The graph of A is*

$$G(A) = \{(x, Ax) \in B_1 \times B_2 : x \in \mathrm{Dom}(A)\}.$$

A is closed if $G(A)$ is closed. A is closable if the closure $\overline{G(A)}$ is the graph of a linear operator $\overline{A}$, which is called the closure of A. If A is closed, any set $C \subseteq \mathrm{Dom}(A)$ such that $\overline{G(A|_C)} = G(A)$ is called a core of A. Here $A|_C$ is the restriction of A to C.

Definition 8.2.5. *Let $A : \mathrm{Dom}(A) \to B_2$ be a densely defined linear operator from B_1 to B_2. Let $\mathrm{Dom}(A^*)$ be the set of elements $y^* \in B_2^*$ such that there exists a constant C with the property that*

$$|y^*(Ax)| \le C|x|_{B_1}$$

for all $x \in \mathrm{Dom}(A)$. For each element $y^* \in \mathrm{Dom}(A^*)$ there is a unique element $A^*y^* \in B_1^*$ such that

$$A^*y^*(x) = y^*(Ax)$$

for all $x \in \mathrm{Dom}(A)$. The linear operator $A^* : \mathrm{Dom}(A^*) \to B_1^*$ is called the dual operator of A. When $B_1 = B_2 = H$, a Hilbert space, the linear operator A^* on H is called the adjoint of A. In this case we say that A is symmetric if $A \subseteq A^*$ (i.e., $\mathrm{Dom}(A) \subseteq \mathrm{Dom}(A^*)$, and $A^*|_{\mathrm{Dom}(A)} = A$), and selfadjoint if $A = A^*$.

The next lemma contains a basic criterion for closability.

Lemma 8.2.6. *$A : \mathrm{Dom}(A) \to B_2$ is closable if and only if it has the following property: If $x_n \to 0$ in B_1 and Ax_n converges in B_2, then $Ax_n \to 0$.*

Proof. Exercise. □

We now use the integration by parts formula to show that D_h is closable on $L^p(\mathbb{P})$ for any $p > 1$. Recall that by definition $\mathcal{C} \subseteq L^p(\mathbb{P})$ for all $p \in [1, \infty)$.

Theorem 8.2.7. *Let $h \in \mathscr{H}$ and $p \in (1, \infty)$. Then $D_h : \mathcal{C} \to L^p(\mathbb{P})$ is closable. Furthermore, $\mathcal{C} \subseteq \mathrm{Dom}(D_h^*)$ and, for $G \in \mathcal{C}$,*

$$D_h^* G = -D_h G + (D_h^* 1)\, G,$$

where

$$D_h^* 1 = \int_0^1 \left\langle \dot{h}_s, dW_s \right\rangle.$$

Proof. By Lemma 8.2.6, for the closability of D_h it is enough to show that if $\{F_n\} \subset \mathcal{C}$ such that $F_n \to 0$ and $D_h F_n \to K$ in $L^p(\mathbb{P})$, then $K = 0$. The fact that $\mathcal{C}$ is a core of the closure of D_h is clear.

Recall Doob's inequality: for any $r > 1$,

$$\mathbb{E}\left|\int_0^1 \left\langle \dot{h}_s, dW_s \right\rangle\right|^r \leq C(r)|h|_{\mathscr{H}}^r \tag{8.2.8}$$

for some constant $C(r)$ depending on r. Let $q = p/(p-1)$ be the index dual to p. Suppose that G is a real-valued cylinder function on $P_o(\mathbb{R}^n)$. Then $D_h G \in L^q(\mathbb{P})$. By (8.2.8) and Hölder's inequality we have $(D_h^* 1)G \in L^q(\mathbb{P})$; hence $D_h^* G \in L^q(\mathbb{P})$. We have the integraion by parts formula $(D_h F_n, G) = (F_n, D_h^* G)$. Letting $n \to \infty$, we see that $(K, G) = 0$ for all $G \in \mathcal{C}$. Since $\mathcal{C}$ is dense in $L^q(\mathbb{P}) = L^p(\mathbb{P})^*$, we have immediately $K = 0$.

If $G \in \mathcal{C}$, then by Hölder's inequality we have

$$(D_h F, G) = (F, D_h^* G) \leq \|D_h^* G\|_{L^q} \|F\|_{L^p}$$

for all $F \in \mathcal{C}$. Hence $G \in \mathrm{Dom}(D_h^*)$ by definition. This shows that $\mathcal{C} \subseteq \mathrm{Dom}(D_h^*)$ and $D_h^* G$ is given by the indicated formula. □

Similarly we have the following integration by parts formula for the gradient operator D.

Theorem 8.2.8. *Let $1 < p < \infty$. Then the gradient operator $D : \mathcal{C} \to L^p(\mathbb{P}; \mathscr{H})$ is closable. We have $\mathcal{C}_0(\mathscr{H}) \subseteq \mathrm{Dom}(D^*)$ and*

$$D^* G = -\mathrm{Trace}\, DG + \int_0^1 \left\langle \dot{G}_s, dW_s \right\rangle.$$

Proof. Use Theorem 8.2.3 and the same argument as in the proof of the preceding theorem. Note that the assumption $G \in \mathcal{C}_0(\mathscr{H})$ implies that $D^* G \in L^q(\mathbb{P})$ for all $q \in (1, \infty)$. □

We have shown that the gradient operator $D : \mathrm{Dom}(D) \to L^2(\mathbb{P}; \mathscr{H})$ is a closed operator and the set of cylinder functions $\mathcal{C}$ is a core. Let $\mathrm{Dom}(\mathcal{E}) = \mathrm{Dom}(D)$ and define the positive symmetric quadratic form

$$\mathcal{E} : \mathrm{Dom}(\mathcal{E}) \times \mathrm{Dom}(\mathcal{E}) \to \mathbb{R}$$

by

$$\mathcal{E}(F, F) = (DF, DF)_{L^2(\mathbb{P}; \mathscr{H})} = \mathbb{E}|DF|_{\mathscr{H}}^2.$$

Then $(\mathcal{E}, D(\mathcal{E}))$ is a closed quadratic form, i.e., $\mathrm{Dom}(\mathcal{E})$ is complete with respect to the inner product

$$\mathcal{E}_1(F, F) = \mathcal{E}(F, F) + (F, F).$$

By the general theory of closed symmetric forms (Fukushima[**29**], 17-19), there exists a non-positive self-adjoint operator L such that $\mathrm{Dom}(\mathcal{E}) = \mathrm{Dom}(\sqrt{-L})$ and

$$\mathcal{E}(F,F) = (\sqrt{-L}F, \sqrt{-L}F). \tag{8.2.9}$$

L is called the Ornstein-Uhlenbeck operator on the path space $P_o(\mathbb{R}^n)$.

Proposition 8.2.9. *$L = -D^*D$ and $\mathcal{C} \subseteq \mathrm{Dom}(L)$. If $F \in \mathcal{C}$ is given by $F(\omega) = f(\omega_{s_1}, \cdots, \omega_{s_l})$, then*

$$LF = \min\{s_i, s_j\} \nabla^i_{e_k} \nabla^j_{e_k} F(\omega) - \langle W_{s_i}, \nabla^i F\rangle_{\mathbb{R}^n}. \tag{8.2.10}$$

[Summation over repeated indices!] Here $\nabla^i_{e_k} F$ denotes the derivative along the kth coordinate unit vector $e_k \in \mathbb{R}^n$ with respect to ith variable of F.

Proof. Assume that $F \in \mathrm{Dom}(D^*D)$ and $G \in \mathrm{Dom}\left(\sqrt{-L}\right)$. Then they are both in $\mathrm{Dom}(D)$, and we have

$$\left(\sqrt{-L}F, \sqrt{-L}G\right) = \mathcal{E}(F,G) = (DF, DG) = (D^*DF, G).$$

Hence $\sqrt{-L}F \in \mathrm{Dom}\left(\sqrt{-L}\right)$ and $-LF = D^*DF$, that is, $-D^*D \subseteq L$.

If $F \in \mathrm{Dom}(L)$, then $F \in \mathrm{Dom}\left(\sqrt{-L}\right) = \mathrm{Dom}(D)$. For any $G \in \mathrm{Dom}(D)$, we have

$$(DF, DG) = \mathcal{E}(F,G) = \left(\sqrt{-L}F, \sqrt{-L}G\right) = (-LF, G).$$

Thus $DF \in \mathrm{Dom}(D^*)$ and $D^*DF = -LF$. Therefore $L \subseteq -D^*D$. It follows that $L = -D^*D$.

Now we prove the formula for L on cylinder functions. Suppose that $F, G \in \mathcal{C}$ and they depend only on the time points $s_1, \ldots, s_l$. By definition

$$(DF)_s = \sum_i \min\{s, s_i\} \nabla^i F$$

and

$$(DF, DG)_{L^2(\mathbb{P};\mathscr{H})} = \int_0^1 (D_s F, D_s G)_{L^2(\mathbb{P};\mathbb{R}^n)}\, ds,$$

where $D_s F = d(DF)_s/ds$. Hence

$$\begin{aligned}
(DF, DG) &= \min\{s_i, s_j\} (\nabla^i F, \nabla^j G) \\
&= \min\{s_i, s_j\} (\nabla^i_{e_k} F, \nabla^j_{e_k} G) \\
&= \sum_i (\nabla^i_{e_k} F, D_{l_{ik}} G),
\end{aligned}$$

where $l_{ik} \in \mathscr{H}$ is given by

$$l_i(s) = \min\{s_i, s\}\, e_k.$$

Note that both $\nabla^i_{e_k} F$ and $\nabla^j_{e_k} G$ are in $\mathcal{C}$. Integrating by parts in the the last expression, we obtain

$$(DF, DG) = \left(D^*_{l_{ik}} \nabla^i_{e_k} F, G\right).$$

This implies, formally,

$$LF = -D^*_{l_{ik}} \nabla^i_{e_k} F = D_{l_{ik}} \nabla^i_{e_k} F - \left(\int_0^1 \left\langle \dot{l}_{ik}, dW_s \right\rangle, \nabla^i_{e_k} F\right).$$

Using

$$\int_0^1 \left\langle \dot{l}_{ik}, dW_s \right\rangle = \langle W_{s_i}, e_k \rangle$$

in the second term on the right side, we can write

$$LF = \min\{s_i, s_j\} \nabla^i_{e_k} \nabla^j_{e_k} F - \left\langle W_{s_i}, \nabla^i F \right\rangle_{\mathbb{R}^n}.$$

The right side is clearly in $L^2(\mathbb{P})$, and we have just shown that, with LF given as above, $(DF, DG) = (-LF, G)$ for all $G \in \mathcal{C}$. It follows that $DF \in \mathrm{Dom}(D^*)$, $F \in \mathrm{Dom}(L)$, and LF is given by (8.2.10). □

From now on, we fix an orthonormal basis $h^i, i = 0, 1, \ldots$, for $\mathscr{H}$ such that each $h^i \in C^1([0,1], \mathbb{R}^n)$. For example, in the case $n = 1$ we may take

$$h^0_s = s, \qquad h^i_s = \frac{\sqrt{2}}{i\pi}(1 - \cos i\pi s), \quad i \geq 1.$$

For an $h \in \mathscr{H}$ we use the notation

$$\langle h, W \rangle_{\mathscr{H}} = \int_0^1 \left\langle \dot{h}_s, dW_s \right\rangle_{\mathbb{R}^n}.$$

The following result will be useful later.

Proposition 8.2.10. *Let $F : P_o(\mathbb{R}^n) \to \mathbb{R}$ have the form*

$$F = f(\langle h^0, W \rangle_{\mathscr{H}}, \ldots, \langle h^l, W \rangle_{\mathscr{H}}), \tag{8.2.11}$$

where $f : \mathbb{R}^{l+1} \to \mathbb{R}$ is a Schwartz test function. Then $F \in \mathrm{Dom}(L)$. The set of such functions forms a core for both D and L, and

$$DF = \sum_{i=0}^{l} F_{x_i}(W) h^i,$$

$$LF = \sum_{i=0}^{l} \left\{F_{x_i x_i}(W) - \left\langle h^i, W \right\rangle_{\mathscr{H}} F_{x_i}\right\},$$

where

$$F_{x_i}(W) = f_{x_i}(\langle h^0, W \rangle_{\mathscr{H}}, \ldots, \langle h^l, W \rangle_{\mathscr{H}})$$

and

$$F_{x_i x_i}(W) = f_{x_i x_i}(\langle h^0, W \rangle_{\mathscr{H}}, \ldots, \langle h^l, W \rangle_{\mathscr{H}}).$$

Proof. Replacing each $\langle h, W\rangle_{\mathscr{H}}$ in (8.2.11) by the Riemann sum

$$\sum_{j=1}^{N} \dot{h}_{(j-1)/N}(W_{j/N} - W_{(j-1)/N}),$$

the resulting function F_N is a cylinder function. The formulas for DF and LF can be obtained by taking the limits in the explicit formulas of DF_N and LF_N as $N \to \infty$. □

We now describe the spectrum of L. This means finding a set of eigenfunctions of L which span the whole $L^2(P_o(\mathbb{R}^d), \mathscr{B}, \mathbb{P})$. Let us first consider the case of the Gaussian measure μ on $\mathbb{R}^1$. Let

$$D = \frac{d}{dx}, \quad D^* = -\frac{d}{dx} + x.$$

Then D^* is the formal adjoint of D with respect to the standard Gaussian measure μ. Hence,

$$L = -D^*D = \frac{d^2}{dx^2} - x\frac{d}{dx}.$$

For the time being we may assume that L is defined on the space of Schwartz test functions, i.e., the set of smooth functions on $\mathbb{R}^1$ whose derivatives of all orders have at most polynomial growth. Let H_N be the Nth Hermite polynomial:

$$H_N(x) = \frac{(-1)^N}{\sqrt{N!}} e^{x^2/2} \frac{d^N}{dx^N} e^{-x^2/2}.$$

Then the following identities hold:

$$H_N' = \sqrt{N} H_{N-1}, \quad H_N'' - xH_N' + NH_N = 0. \tag{8.2.12}$$

The following result is well known.

Proposition 8.2.11. *$\{H_N, N \in \mathbb{Z}_+\}$ is the orthonormal basis for $L^2(\mathbb{R}, \mu)$ obtained from $\{x^N\}$ by the Gram-Schmidt procedure, and $LH_N = -NH_N$. The operator L defined on the space of Schwartz test functions is essentially self-adjoint. Denote its unique self-adjoint extension still by L. Then*

$$\mathrm{Dom}(L) = \left\{ f \in L^2(\mu; \mathbb{R}^1) : \sum_{N=0}^{\infty} N^2 \langle f, H_N\rangle^2 < \infty \right\}.$$

Hence the spectrum of $-L$ is

$$\mathrm{Spec}(-L) = \{0, 1, 2, \ldots\}.$$

Proof. See Courant and Hilbert [**13**], Chapter II. □

Returning to the path space, for simplicity we will assume in the following that the base space has dimension $d = 1$. Let $\tilde{\mu}$ be the standard Gaussian measure on the product space $\mathbb{R}^{\mathbb{Z}_+}$. Then $T\omega = \left\{\langle h^i, \omega\rangle_{\mathscr{H}}\right\}$ defines a measure-preserving map

$$T : (P_o(\mathbb{R}), \mathbb{P}) \to (\mathbb{R}^{\mathbb{Z}_+}, \tilde{\mathbb{P}})$$

between the two measure spaces. With PROPOSITION 8.2.11 in mind, the following construction is in order. Let $\mathscr{I}$ denote the set of indices $I = \{n_i\}$ such that $n_i \in \mathbb{Z}_+$ and almost all of them are equal to zero. Denote $|I| = n_1 + n_2 + \cdots$. For $I \in \mathscr{I}$ define

$$H_I = \prod_i H_{n_i}(\langle h^i, W\rangle_{\mathscr{H}}).$$

Each H_I is a function on the path space $P_o(\mathbb{R}^n)$. The fact that the Hermite polynomials $\{H_N, N \in \mathbb{Z}_+\}$ form an orthonormal basis for $L^2(\mathbb{R}, \mu)$ (PROPOSITION 8.2.11) implies immediately that $\{H_I, I \in \mathscr{I}\}$ is an orthonormal basis for $L^2(P_o(\mathbb{R}), \mathbb{P})$. Moreover, from PROPOSITION 8.2.10 and (8.2.12) we have $H_I \in \mathrm{Dom}(L)$ and $LH_I = -|I|H_I$. Thus the eigenspace of L for the eigenvalue N is

$$C_N = \text{the linear span of } \{H_I : |I| = N\},$$

and

$$L^2(P_o(\mathbb{R}), \mathbb{P}) = C_0 \oplus C_1 \oplus C_2 \oplus \cdots.$$

This is the Wiener chaos decomposition. The following theorem completely describes the spectrum of the Ornstein-Uhlenbeck operator L on the flat path space.

Theorem 8.2.12. *We have*

$$\mathrm{Spec}(-L) = \mathbb{Z}_+,$$

and C_N is the eigenspace for the eigenvalue N. Let $P_N : L^2(P_o(\mathbb{R}), \mathbb{P}) \to C_N$ be the orthogonal projection to C_N. Then

$$\mathrm{Dom}(L) = \left\{F \in L^2(P_o(\mathbb{R}), \mathbb{P}) : \sum_{N=0}^{\infty} N^2 \|P_N F\|^2 < \infty\right\}$$

and

$$LF = -\sum_{N=0}^{\infty} N P_N F.$$

Note that all eigenspaces are infinite dimensional except for $C_0 = \mathbb{R}$.

Proof. The key point here is that $\mathrm{Dom}(L)$ is exactly as given in the statement of the theorem. This is the result of three facts: (1) L is already known to be self-adjoint (hence closed) because it comes from a closed quadratic form; (2) $\{H_I, I \in \mathscr{I}\}$ is a basis for $L^2(P_o(\mathbb{R}), \mathbb{P})$; and (3) each $H_I \in \mathrm{Dom}(L)$. We leave the details as an exercise. □

Remark 8.2.13. In THEOREM 8.2.9 we have found an explicit formula for LF when F is in the space $\mathcal{C}$ of cylinder functions. It can be shown that $L|\mathcal{C}$ is essentially self-adjoint.

For a self-adjoint operator A on an L^2-space with nonnegative spectrum such that 0 is an eigenvalue whose eigenspace consists of constant functions, the spectral gap of A is defined by

$$SG(A) = \inf\{\mathrm{Spec}(-L)\setminus\{0\}\}. \tag{8.2.13}$$

The above theorem shows that for the Ornstein-Uhlenbeck operator on the flat path space we have $SG(-L) = 1$. For a general path space, although we do not have an explicit description of the spectrum of the generalized Ornstein-Uhlenbeck operator, we will show that it has a positive spectral gap which can be bounded from below in terms of the Ricci curvature of the base manifold; see THEOREM 8.7.3.

Let $\mathscr{P}_t = e^{tL/2}$ in the sense of spectral theory. The Ornstein-Uhlenbeck semigroup $\{\mathscr{P}_t\}$ is the strongly continuous L^2-semigroup generated by the Ornstein-Uhlenbeck operator L. Clearly $\mathscr{P}_t F = e^{-Nt/2}F$ for $F \in C_N$. From the definition it is clear that each $\mathscr{P}_t$ is a conservative L^2-contraction, i.e., $\mathscr{P}_t 1 = 1$ and $\|\mathscr{P}_t F\|_2 \le \|F\|_2$.

Proposition 8.2.14. *The Ornstein-Unlenbeck semigroup $\{\mathscr{P}_t\}$ is positive, i.e., $\mathscr{P}_t F \ge 0$ if $F \ge 0$.*

Proof. The closed quadratic form

$$\mathcal{E}(F, F) = (DF, DF), \quad F \in \mathrm{Dom}(D)$$

comes from the gradient operator D; hence it is a Dirichlet form and $L = -D^*D$ is the self-adjoint operator associated with $\mathcal{E}$. The positivity of the semigroup $\mathscr{P}_t = e^{tL/2}$ follows from the general theory of Dirichlet forms, see Fukushima[**29**], 22-24. □

Using the positivity of $\mathscr{P}_t$, we show that it can be extended to a contractive semigroup on $L^p(\mathbb{P})$ for all $p \in [1, \infty]$.

Proposition 8.2.15. *For each $p \in [1, \infty]$, the Ornstein-Uhlenbeck semigroup $\{\mathscr{P}_t\}$ is a positive, conservative, and contractive L^p-semigroup.*

Proof. Let ϕ be a nonnegative convex function on $\mathbb{R}$ with continuous derivative. Then

$$\phi'(a)(b - a) + \phi(a) \le \phi(b).$$

Let $1 < p < \infty$ and $F \in L^p(\mathbb{P}) \cap L^2(\mathbb{P})$. Put $b = F$ in the above inequality and apply $\mathscr{P}_t$ to both sides. By the positivity and $\mathscr{P}_t 1 = 1$ we have

$$\phi(a) + \phi'(a)(\mathscr{P}_t F - a) \leq \mathscr{P}_t \phi(F).$$

For each fixed a this inequality holds for $\mathbb{P}$-almost all ω. By Fubini's theorem, for $\mathbb{P}$-almost all ω, it holds for almost all a (with respect to the Lebesgue measure, say), hence for all a by continuity. Now we can let $a = \mathscr{P}_t F$ and obtain $\mathscr{P}_t \phi(F) \geq \phi(\mathscr{P}_t F)$, $\mathbb{P}$-almost surely. Choosing $\phi(s) = |s|^p$, we have $|\mathscr{P}_t F|^p \leq \mathscr{P}_t |F|^p$. Therefore

$$\|\mathscr{P}_t F\|_p^p = (|\mathscr{P}_t F|^p, 1) \leq (\mathscr{P}_t |F|^p, 1) = (|F|^p, \mathscr{P}_t 1) = \|F\|^p.$$

Since $L^p(\mathbb{P}) \cap L^2(\mathbb{P})$ is dense in $L^p(\mathbb{P})$, we see that $\mathscr{P}_t$ can be extended uniquely to a positive, conservative, and contractive L^p-semigroup. The case $p = 1$ or ∞ can be handled similarly and more simply. □

8.3. Gradient formulas

In preparation for the integration by parts formula in the next section, we will prove several formulas involving at the same time the gradient operator ∇ and Brownian motion. One of them (Theorem 8.3.3) will serve as the starting point for proving an integration by parts formula in the path space. We start with a review of a formula of this type we have already discussed in Section 7.2. Let M be a compact Riemannian manifold and

$$P_s f(x) = \int_M p_M(s, x, y) f(y) dy$$

be the heat semigroup on M. Let X be the coordinate process on $P_o(M)$ and $\mathbb{P}_x$ the law of Brownian motion on M starting from x. Fix an orthonormal frame at x and let U be the horizontal lift of X to the orthonormal frame bundle $\mathscr{O}(M)$ starting from this frame. We denote the anti-development of U by W. Define a multiplicative functional M by

$$\frac{dM_s}{ds} + \frac{1}{2} M_s \operatorname{Ric}_{U_s} = 0, \qquad M_0 = I. \tag{8.3.1}$$

Here $\operatorname{Ric}_u : \mathbb{R}^n \to \mathbb{R}^n$ is the Ricci curvature transform at $u \in \mathscr{O}(M)$.

Theorem 8.3.1. *For $f \in C^\infty(M)$ we have*

$$\nabla P_T f(x) = \mathbb{E}_x \left\{ M_T U_T^{-1} \nabla f(X_T) \right\}.$$

Proof. This is a special case of Theorem 7.2.1 applied to the 1-form $\theta(t, x) = \nabla P_t f(x)$, which is a solution of the heat equation for the Hodge-de Rham Laplacian $\square_M$. See also Theorem 7.2.4. □

This result tells us how to pass the gradient operator ∇ through the heat semigroup. The next result we will discuss shows how to remove the gradient ∇ under the expectation from the function. It is a special case of the integration by parts formula in the path space $P_o(M)$.

Let f be a smooth function on M and

$$\Phi(s,u) = P_s f(\pi u)$$

the lift of $P_s f$ to $\mathscr{O}(M)$. Then J satisfies the following heat equation:

$$\frac{\partial}{\partial s}\Phi(T-s,u) + \frac{1}{2}\Delta_{\mathscr{O}(M)}\Phi(T-s,u) = 0. \tag{8.3.2}$$

Applying Itô's formula to $\Phi(T-s,U_s)$ and using the heat equation (8.3.2), we have

$$d\{\Phi(T-s,U_s)\} = \langle \nabla^H \Phi(T-s,U_s), dW_s\rangle, \tag{8.3.3}$$

which shows that the process $\{\Phi(T-s,U_s), 0 \le s \le T\}$ is a martingale.

Lemma 8.3.2. *The process $\{M_s \nabla^H \Phi(T-s,U_s), 0 \le s \le T\}$ is a martingale.*

Proof. $\nabla^H J(T-s,u)$ is the scalarization of $d(P_{T-s}f) = P_{T-s}(df)$, where on the right side $P_t = e^{t\Box_M/2}$ is the semigroup generated by the Hodge-de Rham Laplacian $\Box_M$; hence it satisfies the heat equation for this Laplacian. By the Weitzenböck formula $\Box_M = \Delta_M - \mathrm{Ric}$ (applied to 1-forms, see COROLLARY 7.1.4),

$$\frac{d}{ds}\nabla^H \Phi(T-s,u) + \frac{1}{2}\left\{\Delta_{\mathscr{O}(M)} - \mathrm{Ric}\right\}\nabla^H\Phi(T-s,u) = 0.$$

Now, using the above relation and the definition 8.3.1 of M_s, we find that

$$d\left\{M_s \nabla^H \Phi(T-s,U_s)\right\} = \langle M_s \nabla^H \nabla^H \Phi(T-s,U_s), dW_s\rangle.$$

This proves the lemma. □

Theorem 8.3.3. *Suppose that X is a Brownian motion on M, U a horizontal lift of X, and W the corresponding anti-development of X. Let $\{h_s\}$ be an adapted process with sample paths in $\mathscr{H}$ such that $\mathbb{E}|h|^2_{\mathscr{H}} < \infty$. Then for $f \in C^\infty(M)$ we have*

$$\mathbb{E}\,\langle \nabla f(X_T), U_T h_T\rangle = \mathbb{E}\left\{f(X_T)\int_0^T \left\langle \dot h_s + \frac{1}{2}\mathrm{Ric}_{U_s} h_s, dW_s\right\rangle\right\}.$$

Proof. Let $N_s = M_s \nabla^H \Phi(T-s,U_s)$. According to LEMMA 8.3.2 N is a martingale. From (8.3.3) we have

$$f(X_T) - \mathbb{E}f(X_T) = \int_0^T \langle \nabla^H \Phi(T-s,U_s), dW_s\rangle. \tag{8.3.4}$$

This is a special case of the martingale representation theorem, see SECTION 8.5. Suppose that $\{g_s\}$ is an $\mathbb{R}^n$-valued, adapted process with sample paths in $\mathscr{H}$ such that $\mathbb{E}|g|^2_{\mathscr{H}} < \infty$. We have the following sequence of equalities:

$$\begin{aligned}
&\mathbb{E}\left[f(X_T)\int_0^T \left\langle M_s^\dagger \dot{g}_s, dW_s\right\rangle\right] \\
&\quad = \mathbb{E}\left[\{f(X_T) - \mathbb{E}f(X_T)\}\int_0^T \left\langle M_s^\dagger \dot{g}_s, dW_s\right\rangle\right] \\
&\quad = \mathbb{E}\left[\int_0^T \langle \nabla^H \Phi(T-s, U_s), dW_s\rangle \int_0^T \left\langle M_s^\dagger \dot{g}_s, dW_s\right\rangle\right] \\
&\quad = \mathbb{E}\int_0^T \left\langle \nabla^H \Phi(T-s, U_s), M_s^\dagger \dot{g}_s\right\rangle ds \\
&\quad = \mathbb{E}\int_0^T \langle M_s \nabla^H \Phi(T-s, U_s), \dot{g}_s\rangle\, ds \\
&\quad = \mathbb{E}\int_0^T \langle N_s, \dot{g}_s\rangle\, ds \\
&\quad = \mathbb{E}\left\langle M_T \nabla^H \Phi(0, U_T), g_T\right\rangle + \mathbb{E}\int_0^T \langle g_s, dN_s\rangle,
\end{aligned}$$

where $M_s^\dagger$ denotes the transpose of M_s. Here in the last step we have integrated by parts. The last term vanishes because N is a martingale; hence we obtain

$$\mathbb{E}\left\{f(X_T)\int_0^T \left\langle M_s^\dagger \dot{g}_s, dW_s\right\rangle\right\} = \mathbb{E}\left\langle \nabla^H \Phi(0, U_T), M_T^\dagger g_T\right\rangle. \tag{8.3.5}$$

To convert this equality in the form in the theorem, we let $h_s = M_s^\dagger g_s$. It satisfies the equation

$$M_s^\dagger \dot{g}_s = \dot{h}_s + \frac{1}{2}\mathrm{Ric}_{U_s} h_s.$$

On the other hand, $\nabla^H \Phi(0, u) = u^{-1}\nabla f(\pi u)$. Hence the right side of (8.3.5) becomes $\mathbb{E}\langle \nabla f(X_T), U_T h_T\rangle$. The proof is completed. □

Combining THEOREMS 8.3.1 and 8.3.3 gives the celebrated Bismut's celebrated formula for $\nabla p_M(T, x, y)$.

Theorem 8.3.4. (Bismut's formula) *Let M be a compact Riemannian manifold and $x, y \in M$. Let $\mathbb{P}_{x,y;T}$ be the law of a Brownian bridge on M from x to y with time length T. Let X be the coordinate process on the path space $P_o(M)$, U a horizontal lift of X, and W the corresponding anti-development of U. Define an $\mathscr{M}(n,n)$-valued process $\{M_s\}$ by*

$$\frac{dM_s}{ds} + \frac{1}{2}M_s \mathrm{Ric}_{U_s} = 0, \qquad M_0 = I.$$

Then

$$\nabla_x \ln p_M(T,x,y) = \frac{1}{T}\mathbb{E}_{x,y;T}\int_0^T M_s\,dW_s. \tag{8.3.6}$$

Proof. Fix an element $e \in \mathbb{R}^n$ and let $h_s = (s/T)M_s^\dagger e$. The factor s/T is chosen so that $h_0 = 0$ and $h_T = M_T^\dagger e$. We first use THEOREM 8.3.1 to pass the gradient through the heat kernel:

$$\langle \nabla P_T f(x), e\rangle = \mathbb{E}_x \left\langle M_T U_T^{-1}\nabla f(X_T), e\right\rangle = \mathbb{E}_x \left\langle U_T^{-1}\nabla f(X_T), h_T\right\rangle.$$

Because U_T is orthogonal, this is equivalent to

$$\langle \nabla P_T f(x), e\rangle = \mathbb{E}_x \left\langle \nabla f(X_T), U_T h_T\right\rangle.$$

Next we apply THEOREM 8.3.3 to remove the gradient from the function. From the definition of M_s we have

$$\dot{h}_s + \frac{1}{2}\mathrm{Ric}_{U_s} h_s = \frac{M_s^\dagger e}{T}.$$

Therefore THEOREM 8.3.3 gives

$$\langle \nabla P_T f(x), e\rangle = \frac{1}{T}\mathbb{E}_x \left\{ f(X_T)\int_0^T \left\langle M_s^\dagger e, dW_s\right\rangle\right\},$$

or equivalently

$$\nabla P_T f(x) = \frac{1}{T}\mathbb{E}_x \left\{ f(X_T)\int_0^T M_s\,dW_s\right\}.$$

This equality can be rewritten as

$$\begin{aligned}\int_M & \nabla_x p_M(T,x,y) f(y)dy \\ &= \int_M \left[\frac{1}{T}\mathbb{E}_{x,y;T}\int_0^T M_s\,dW_s\right] p_M(T,x,y) f(y) dy.\end{aligned}$$

This gives the desired formula (8.3.6) for almost all y. In this formula, the left side is obviously continuous. In order to show that the formula holds for all y, it is sufficient to verify that the right side is continuous in y.

For any $t < T$, we have

$$\mathbb{E}_{x,y;T}\int_0^t M_s\,dW_s = \frac{\mathbb{E}_x\left[p_M(T-t, X_t, y)\int_0^t M_s\,dW_s\right]}{p_M(T,x,y)},$$

which is obviously continuous in y. On the other hand, from (5.4.7) we have

$$\left|\mathbb{E}_{x,y;T}\int_t^T M_s\,dW_s\right| \le \text{const.}\,\mathbb{E}_{x,y;T}\int_t^T |\nabla p_M(T-s, X_s, y)|\,ds.$$

Using the bound of the gradient of the logarithmic heat kernel

$$|\nabla \ln p_M(T-s, X_t, y)| \le \text{const.} \left[\frac{d(X_s, y)}{T-s} + \frac{1}{\sqrt{T-s}}\right]$$

in THEOREM 5.5.3 and the inequality

$$\mathbb{E}_{x,y;T}\, d(X_s, y) \le \text{const.}\sqrt{T-s}$$

in LEMMA (5.4.2), we see that

$$\left|\mathbb{E}_{x,y;T} \int_0^t M_s\, dW_s\right| \le \text{const. } \sqrt{T-t}.$$

It follows that

$$\mathbb{E}_{x,y;T} \int_0^t M_s\, dW_s \to \mathbb{E}_{x,y;T} \int_0^T M_s\, dW_s$$

uniformly in $y \in M$ as $t \uparrow T$, and the right side is a continuous function of y. □

There is also a more direct proof of this formula. By LEMMA 8.3.2 $\{M_s U_s^{-1} \nabla p_M(T-s, X_s, y),\ 0 \le s \le T\}$ is a martingale under $\mathbb{P}_x$. This implies that $\{M_s U_s^{-1} \nabla \ln p_M(T-s, X_s, y),\ 0 \le s \le T\}$ is a martingale under $\mathbb{P}_{x,;T}$. There is some checking to do at $s = T$, where we define the value of the martingale as the limit from the left; see the end the above proof. Hence for any $0 \le s < T$

$$\nabla \ln p_M(T, x, y) = \mathbb{E}_{x,y;T}\left\{M_s U_s^{-1} \nabla p_M(T-s, X_s, y)\right\}.$$

Integrating from 0 to T, we have

$$T\nabla \ln p_M(T, x, y) = \mathbb{E}_{x,y;T} \int_0^T \left\{M_s U_s^{-1} \nabla p_M(T-s, X_s, y)\right\} ds.$$

This is equivalent to (8.3.6) because from (5.4.7), the anti-development W of the Brownian bridge X is

$$W_s = b_s + \int_0^s U_\tau^{-1} \nabla \ln p_M(T-\tau, X_\tau, y)\, d\tau,$$

where b is a euclidean Brownian motion under $\mathbb{P}_{x,y;T}$.

8.4. Integration by parts in path space

In this section M is a compact Riemannian manifold. We denote a general element of the path space $P_o(M)$ by γ, while reserving the letter ω for a general element in the flat path space $P_o(\mathbb{R}^n)$. The coordinate process of $P_o(M)$ is denoted by X. In the filtered probability space $(P_o(M), \mathscr{B}(P_o(M))_*, \mathbb{P})$, where $\mathbb{P}$ is the Wiener measure, X is a Brownian motion on M starting from o. Let U be the horizontal lift of X to the orthonormal frame bundle $\mathscr{O}(M)$

starting from a fixed orthonormal frame at o and W the corresponding anti-development of X.

Following the euclidean theory in SECTIONS 8.1 and 8.2, we study the gradient and Ornstein-Uhlenbeck operators on the path space $P_o(M)$. The first thing we need to do is to find a proper definition of the directional derivative operator D_h for an $h \in \mathscr{H}$. A vector at a point (path) $\gamma \in P_o(M)$ is a vector field along the path γ in the ordinary differential geometric sense. Likewise a vector field V on $P_o(M)$ is an assignment of a vector field $V(\gamma)$ for each $\gamma \in P_o(M)$. As with the flat path space (see SECTION 8.2), we need to restrict the type of vector fields which are allowed to enter our discussion. In differential geometry, we require that vector fields possess a certain smoothness according to the nature of the problems under consideration. In the present setting, we introduce a special class of vector fields on $P_o(M)$, the Cameron-Martin vector fields. Although the Cameron-Martin vector fields do not form the most general class of vector fields we can work with, they are adequate for most applications we have in mind.

The discussion of the flat path space in SECTION 8.2 should serve as a guide for the definition of the directional derivative operator D_h on a general path space we will give here. On the flat manifold $\mathbb{R}^n$, a vector at a point defines a vector field on the whole manifold by parallel translation. On a general manifold, parallel transport is path dependent. Suppose that $\gamma \in P_o(M)$ is a smooth path and $U(\gamma)$ its horizontal lift starting from an orthonormal frame at o. With T_oM identified with $\mathbb{R}^n$ via this frame, $U(\gamma)$ also serves as the parallel transport along the path γ. Let V be a vector field along γ. Then $h_s = U(\gamma)_s^{-1} V_s$ defines a continuous function $h : [0,1] \to \mathbb{R}^n$. Conversely, each such function h determines a vector field V along a smooth path γ by $V_s = U(\gamma)_s h_s$ or simply $V = U(\gamma)h$. Since h is $\mathbb{R}^n$-valued, it is more convenient to work with h than with V. In the setting of the path space $(P_o(M), \mathscr{B}_*, \mathbb{P}_o)$, the stochastic parallel transport $U(\gamma)$ along a Brownian path γ is well defined amost surely, and $D_h(\gamma) = U(\gamma)h$ is a vector field on $P_o(M)$, $\mathbb{P}$-almost surely defined. The vector fields $\{D_h, h \in \mathscr{H}\}$ form the set of basic directional derivative operators we will work with. Sometimes it may be necessary to allow h to be dependent on the path γ. In this case, we usually assume h is an $\mathbb{R}^n$-valued, adapted process on $(P_o(M), \mathscr{B}_*, \mathbb{P})$ with sample paths in $\mathscr{H}$ such that $\mathbb{E}|h|^2_{\mathscr{H}} < \infty$.

By analogy with the flat case, $D_h F$ should be given by

$$D_h F(\gamma) = \lim_{t \to 0} \frac{F(\zeta_{th}\gamma) - F(\gamma)}{t}, \tag{8.4.1}$$

where $\{\zeta_{th}, t \in \mathbb{R}\}$ should be the flow generated by the vector field D_h, i.e., it is the solution of the ordinary differential equation on $P_o(M)$:

$$\frac{d(\zeta_{th}\gamma)}{dt} = D_h(\zeta_{th}\gamma). \tag{8.4.2}$$

This flow is the substitute of the flow $\xi_{th}\omega = \omega + th$ in the flat path space.

The space $\mathcal{C}$ of cylinder functions on $P_o(M)$ is defined in the same way as before. They are functions of the form

$$F(\gamma) = f(\gamma_{s_1}, \cdots, \gamma_{s_l}), \tag{8.4.3}$$

where $f : M^l \to \mathbb{R}$ is a smooth function and $0 < s_1 < \cdots < s_l \le 1$. By applying (8.4.1) to a cylinder function we find the following definition.

Definition 8.4.1. *Let F be a cylinder function as in (8.4.3). The directional derivative of F along the Cameron-Martin vector field D_h is*

$$D_h F(\gamma) = \sum_{i=1}^{l} \langle \nabla^i F(\gamma), U(\gamma)_{s_i} h(\gamma)_{s_i} \rangle_{T_{\gamma_{s_i}} M},$$

where

$$\nabla^i F(\gamma) = \nabla^i f(\gamma_{s_1}, \cdots, \gamma_{s_i}, \cdots, \gamma_{s_l}) \in T_{\gamma_{s_i}} M$$

is the gradient with respect to the i variable of f.

The gradient $DF : P_o(M) \to \mathscr{H}$ is determined by $\langle DF, h \rangle_{\mathscr{H}} = D_h F$, and we have

$$DF(\gamma)_s = \sum_{i=1}^{l} \min\{s, s_i\}\, U(\gamma)_{s_i}^{-1} \nabla^i F(\gamma).$$

The norm of the gradient is given by

$$|DF(\gamma)|_{\mathscr{H}}^2 = \sum_{i=1}^{l} (s_i - s_{i-1}) \left| \sum_{j=i}^{l} U(\gamma)_{s_j}^{-1} \nabla^j F(\gamma) \right|^2.$$

We now prove the integration by parts formula for D_h by induction on the number of time dependences of a cylinder function. We start with THEOREM 8.3.3:

$$\mathbb{E}\, \langle \nabla f(X_s), U_s h_s \rangle = \mathbb{E} \left\{ f(X_s) \int_0^1 \left\langle \dot{h}_s + \frac{1}{2} \mathrm{Ric}_{U_s} h_s, dW_s \right\rangle \right\}. \tag{8.4.4}$$

This is the integration by parts formula of D_h in the path space for the special cylinder function $F(\gamma) = f(\gamma_s)$. Indeed, it is easy to check that in this case

$$D_h F(\gamma) = \left\langle U(\gamma)_s^{-1} \nabla f(\gamma_s), h_s \right\rangle_{\mathbb{R}^n},$$

and the left side of (8.4.4) is simply $(D_h F, 1)$.

Theorem 8.4.2. (Driver's integration by parts formula) *Let F, G be two cylinder functions and $h \in \mathscr{H}$. Define*

$$D_h^* G = -D_h G + G \int_0^1 \left\langle \dot{h}_s + \frac{1}{2}\mathrm{Ric}_{U_s} h_s,\, dW_s \right\rangle.$$

Then we have

$$(D_h F, G) = (F, D_h^* G). \tag{8.4.5}$$

Proof. Because

$$D_h^* G = -D_h G + (D_h^* 1) G,$$

it is enough to show (8.4.5) for $G = 1$:

$$D_h^* 1 = \int_0^1 \left\langle \dot{h}_s + \frac{1}{2}\mathrm{Ric}_{U_s} h_s, dW_s \right\rangle. \tag{8.4.6}$$

Let F be of the form (8.4.3). We argue by induction on l, the number of time dependences in F. Recall that $\mathbb{P}_x$ is the law of a Brownian motion on M starting from x, and $\mathbb{E}_x$ the attendant expectation operator. Define a new function of $l-1$ variables by

$$g(x_1, \cdots, x_{l-1}) = \mathbb{E}_{x_{l-1}} f(x_1, \cdots, x_{l-1}, X_{s_l - s_{l-1}}).$$

Note that the last variable x_{l-1} appears twice on the right side, once as the starting point of the Brownian motion X, and once in the variables of the function f. Let G be the cylinder function

$$G(\gamma) = g(\gamma_{s_1}, \cdots, \gamma_{s_{l-1}}).$$

[Of course this G has nothing to do with the one in the statement of the theorem!] For $i = 1, \ldots, l-2$,

$$\nabla^i g(x_1, \cdots, x_{l-1}) = \mathbb{E}_{x_{l-1}} \nabla^i f(x_1, \cdots, x_{l-1}, X_{s_l - s_{l-1}}).$$

Evaluating this identity at X and applying $U_{s_i}^{-1}$ to both sides, we obtain an identity in $\mathbb{R}^n$. Take the expected value and using the Markov property at time s_{l-1} on the right side, we have, for $i = 1, \ldots, l-2$,

$$\mathbb{E}\, U_{s_i}^{-1} \nabla^i G = \mathbb{E}\, U_{s_i}^{-1} \nabla^i F. \tag{8.4.7}$$

We now calculate the gradient of $g(x_1, \ldots, x_{l-1})$ with respect to x_{l-1}. This variable appears twice; therefore the gradient has two terms:

$$\begin{aligned}
&\nabla^{l-1} g(x_1, \cdots, x_{l-1}) \\
&\quad = \mathbb{E}_{x_{l-1}} \nabla^{l-1} f(x_1, \cdots, x_{l-1}, X_{s_l - s_{l-1}}) \\
&\qquad + \mathbb{E}_{x_{l-1}} \left\{ M_{s_l - s_{l-1}} U_{s_l - s_{l-1}}^{-1} \nabla^l f(x_1, \cdots, x_{l-1}, X_{s_l - s_{l-1}}) \right\}.
\end{aligned}$$

The second term corresponds to the gradient with respect to the starting point of the Brownian motion and is given by THEOREM 8.3.3. Again we evaluate at X and apply $U_{s_i}^{-1}$ to both sides. Taking the expected value and using the Markov propety on the first term of the right side, we have

$$\mathbb{E}\,U_{s_{l-1}}^{-1}\nabla^{l-1}G = \mathbb{E}\,U_{s_{l-1}}^{-1}\nabla^{l-1}F + \mathbb{E}\,\mathbb{E}_{X_{s_{l-1}}} M_{s_l - s_{l-1}} U_{s_l - s_{l-1}}^{-1} \nabla^l F. \tag{8.4.8}$$

We now come to the inductive step. By definition,

$$\mathbb{E} D_h F = \sum_{i=1}^{l} \mathbb{E}\left\langle \nabla^i F, U_{s_i} h_{s_i}\right\rangle. \tag{8.4.9}$$

We rewrite the right side in terms of G. If $i \le l-2$, then by (8.4.7) we have

$$\mathbb{E}\left\langle \nabla^i F, U_{s_i} h_{s_i}\right\rangle = \mathbb{E}\left\langle \nabla^i G, U_{s_i} h_{s_i}\right\rangle. \tag{8.4.10}$$

For $i = l-1$, by (8.4.8) we have

$$\begin{aligned}\mathbb{E}\left\langle \nabla^{l-1} F, U_{s_{l-1}} h_{s_{l-1}}\right\rangle &= \mathbb{E}\left\langle \nabla^{l-1} G, U_{s_{l-1}} h_{s_{l-1}}\right\rangle \\ &\quad - \mathbb{E}\,\mathbb{E}_{X_{s_{l-1}}}\left\langle M_{s_l - s_{l-1}} U_{s_l - s_{l-1}}^{-1} \nabla^l F, h_{s_{l-1}}\right\rangle.\end{aligned} \tag{8.4.11}$$

For $i = l$, we write

$$\mathbb{E}\left\langle \nabla^l F, U_{s_l} h_{s_l}\right\rangle = \mathbb{E}\left\langle \nabla^l F, U_{s_l}(h_{s_l} - h_{s_{l-1}})\right\rangle + \mathbb{E}\left\langle \nabla^l F, U_{s_l} h_{s_{l-1}}\right\rangle. \tag{8.4.12}$$

Combining (8.4.9)-(8.4.12), we have

$$\begin{aligned}\mathbb{E}\, D_h F &= \sum_{i=1}^{l-1} \mathbb{E}\left\langle \nabla^i G, U_{s_i} h_{s_i}\right\rangle + \mathbb{E}\left\langle \nabla^l F, U_{s_l}(h_{s_l} - h_{s_{l-1}})\right\rangle \\ &\quad + \mathbb{E}\left\langle \nabla^l F, U_{s_l} h_{s_{l-1}})\right\rangle - \mathbb{E}\,\mathbb{E}_{X_{s_{l-1}}}\left\langle M_{s_l - s_{l-1}} \nabla^l F, h_{s_{l-1}}\right\rangle.\end{aligned} \tag{8.4.13}$$

The first term on the right side is equal to $\mathbb{E} D_h G$; hence by the induction hypothesis we have

$$\sum_{i=1}^{l-1} \mathbb{E}\left\langle \nabla^i G, U_{s_i} h_{s_i}\right\rangle = \mathbb{E}\left[G \int_0^{s_{l-1}} \left\langle \dot h_s + \frac{1}{2}\mathrm{Ric}_{U_s} h_s, dW_s\right\rangle\right]. \tag{8.4.14}$$

For the second term on the right side of (8.4.13) we first use the Markov property at s_{l-1}. In the inner expectation of the resulting expression we use the integration by parts formula for one time dependence (8.4.4). Then we use the Markov property again. These steps give

$$\begin{aligned}&\mathbb{E}\left\langle \nabla^l F, U_{s_l}(h_{s_l} - h_{s_{l-1}})\right\rangle \\ &\quad = \mathbb{E}\left[F \int_{s_{l-1}}^{s_l} \left\langle \dot h_s + \frac{1}{2}\mathrm{Ric}_{U_s}(h_s - h_{s_{l-1}}), dW_s\right\rangle\right].\end{aligned} \tag{8.4.15}$$

For the last two terms on the right side of (8.4.13), using the Markov property at time s_{l-1} we have

$$\begin{aligned}
&\mathbb{E}\left\langle \nabla^l F, U_{s_l} h_{s_{l-1}})\right\rangle - \mathbb{E}\,\mathbb{E}_{X_{s_{l-1}}}\left\langle M_{s_l - s_{l-1}} U^{-1}_{s_l - s_{l-1}} \nabla^l F, h_{s_{l-1}}\right\rangle \\
&= \mathbb{E}\,\mathbb{E}_{X_{s_{l-1}}}\left\langle \nabla^l F, U_{s_l - s_{l-1}} h_{s_{l-1}}\right\rangle \\
&\qquad - \mathbb{E}\,\mathbb{E}_{X_{s_{l-1}}}\left\langle \nabla^l F, U_{s_l - s_{l-1}} M^{\dagger}_{s_l - s_{l-1}} h_{s_{l-1}}\right\rangle \\
&= \mathbb{E}\,\mathbb{E}_{X_{s_{l-1}}}\left\langle \nabla^l F, U_{s_l - s_{l-1}} k_{s_l - s_{l-1}}\right\rangle,
\end{aligned}$$

where

$$k_s = h_{s_{l-1}} - M^{\dagger}_{s+s_{l-1}} h_{s_{l-1}}, \quad 0 \le s \le s_l - s_{l-1}.$$

A simple calculation shows that

$$\dot{k}_s + \frac{1}{2}\mathrm{Ric}_{U_s} k_s = \frac{1}{2}\mathrm{Ric}_{U_s} h_{s_{l-1}}.$$

Hence, applying (8.4.4), we have

$$\begin{aligned}
&\mathbb{E}\left\langle \nabla^l F, U_{s_l} h_{s_{l-1}}\right\rangle - \mathbb{E}\,\mathbb{E}_{X_{s_{l-1}}}\left\langle M_{s_l - s_{l-1}} U^{-1}_{s_l - s_{l-1}} \nabla^l F, h_{s_{l-1}}\right\rangle \\
&= \mathbb{E}\left[F \cdot \frac{1}{2}\int_{s_{l-1}}^{s_l} \langle \mathrm{Ric}_{U_s} h_{s_{l-1}}, dW_s\rangle\right].
\end{aligned} \tag{8.4.16}$$

Finally, from (8.4.13) – (8.4.16) we have

$$\begin{aligned}
\mathbb{E} D_h F &= \mathbb{E}\left[G \int_0^{s_{l-1}} \left\langle \dot{h}_s + \frac{1}{2}\mathrm{Ric}_{U_s} h_s, dW_s\right\rangle\right] \\
&\quad + \mathbb{E}\left[F \int_{s_{l-1}}^{s_l} \left\langle \dot{h}_s + \frac{1}{2}\mathrm{Ric}_{U_s}(h_s - h_{s_{l-1}}), dW_s\right\rangle\right] \\
&\quad + \mathbb{E}\left[F \cdot \frac{1}{2}\int_{s_{l-1}}^{s_l} \langle \mathrm{Ric}_{U_s} h_{s_{l-1}}, dW_s\rangle\right] \\
&= \mathbb{E}\left[F \int_0^1 \left\langle \dot{h}_s + \frac{1}{2}\mathrm{Ric}_{U_s} h_s, dW_s\right\rangle\right].
\end{aligned}$$

This completes the proof of (8.4.6). □

The following consequences of the integration by parts can be proved in the same way as their counterparts in the flat case (see SECTION 8.2).

Theorem 8.4.3. *Suppose that $h \in \mathscr{H}$ and $1 < p < \infty$. Then $D_h : \mathcal{C} \to L^p(P_o(M), \mathbb{P})$ is closable. Furthermore $\mathcal{C} \subseteq \mathrm{Dom}(D_h^*)$ and, for $G \in \mathcal{C}$,*

$$D_h^* G = -D_h G + (D_h^* 1) G,$$

where

$$D_h^* 1 = \int_0^1 \left\langle \dot{h}_s + \frac{1}{2}\mathrm{Ric}_{U_s} h_s, dW_s \right\rangle.$$

Theorem 8.4.4. *Suppose that* $1 < p < \infty$. *Then* $D : \mathcal{C} \to L^p(P_o(M), \mathbb{P}; \mathscr{H})$ *is closable. We have* $\mathcal{C}_0(\mathscr{H}) \subseteq \mathrm{Dom}(D^*)$ *and*

$$D^* G = -\mathrm{Trace}\, DG + \int_0^1 \left\langle \dot{G}_s + \frac{1}{2}\mathrm{Ric}_{U_s} G_s,\, dW_s \right\rangle.$$

8.5. Martingale representation theorem

In preparation for the discussion of logarithmic Sobolev inequalities on the path space in SECTION 8.7, we generalize the classical Clark-Ocone martingale representation theorem for euclidean Brownian motion to Riemannian Brownian motion. Let's first review the case of the flat path space $(P_o(\mathbb{R}^n), \mathscr{B}(P_o(\mathbb{R}^n))_*, \mathbb{P})$. Let W be the coordinate process on this path space. The martingale representation theorem asserts that every martingale adapted to the filtration $\mathscr{B}_* = \mathscr{B}(P_o(\mathbb{R}^n))_*$ is a stochastic integral with respect to the Brownian motion W. Suppose that $F \in L^2(P_o(\mathbb{R}^n), \mathbb{P})$. Then $M_s = \mathbb{E}\{F|\mathscr{B}_s\}, 0 \le s \le 1$, is a square-integrable martingale adapted to the filtration $\mathscr{B}_*$. There is a unique $\mathscr{B}_*$-adapted, $\mathbb{R}^n$-valued process $\{H_s\}$ such that

$$F = \mathbb{E}F + \int_0^1 \langle H_s, dW_s \rangle. \tag{8.5.1}$$

The Clark-Ocone formula identifies the integrand process H_s explicitly: if $F \in \mathrm{Dom}(D)$, then

$$H_s = \mathbb{E}\{D_s F|\mathscr{B}_s\}, \qquad D_s F = \frac{d(DF)_s}{ds}.$$

On the general path space $(P_o(M), \mathscr{B}(P_o(M))_*, \mathbb{P})$ of a compact Riemannian manifold M. The coordinate process X and its anti-development W generate the same filtration (after proper completions):

$$\mathscr{B}_* = \mathscr{F}_*^X = \mathscr{F}_*^W.$$

Now suppose that $F \in L^2(P_o(M), \mathscr{B}_1, \mathbb{P})$. Then F is also measurable with respect to $\mathscr{B}_1^W$, the terminal σ-field of the euclidean Brownian motion W. Thus the representation (8.5.1) applies and we have

$$F = \mathbb{E}F + \int_0^1 \left\langle \mathbb{E}\left\{\tilde{D}_s(F \circ J)|\mathscr{B}_s\right\}, dW_s \right\rangle,$$

where $J : W \mapsto X$ is the Itô map and $\tilde{D}(F \circ J)$ is the gradient of F in the path space $P_o(\mathbb{R}^n)$. But this is not what we wanted; we need an explicit representation of the integrand in terms of DF, the gradient of F on $P_o(M)$.

Theorem 8.5.1. *Suppose that* $F \in \mathrm{Dom}(D)$. *Then*

$$F = \mathbb{E}F + \int_0^1 \langle H_s, dW_s \rangle ,$$

where

$$H_s = \mathbb{E}\left[D_s F + \frac{1}{2} M_s^{-1} \int_s^1 M_\tau \mathrm{Ric}_{U_\tau}(D_\tau F)\, d\tau \middle| \mathscr{B}_s \right] \tag{8.5.2}$$

and M *is the solution of the equation*

$$\frac{dM_s}{ds} + \frac{1}{2} M_s \mathrm{Ric}_{U_s} = 0, \qquad M_0 = I.$$

Proof. From the general martingale representation theorem we know that H_s exists and is unique, so the proof is a matter of identifying it. Suppose that $\{j_s, 0 \le s \le 1\}$ is an arbitrary $\mathbb{R}^n$-valued, $\mathscr{B}_*$-adapted process defined on $P_o(M)$ such that

$$\mathbb{E} \int_0^1 |j_s|^2 ds < \infty.$$

Define

$$h_s = \int_0^s j_\tau d\tau.$$

Recall the integration by parts formula in the form

$$D_h^* = -D_h + D_h^* 1,$$

where

$$D_h^* 1 = \int_0^1 \left\langle \dot{h}_s + \frac{1}{2} \mathrm{Ric}_{U_s} h_s, dW_s \right\rangle .$$

We compute $\mathbb{E} D_h F$ in two ways. On the one hand, noting that $\dot{h}_s = j_s \in \mathscr{B}_s$, we have

$$\mathbb{E} D_h F = \mathbb{E} \langle DF, h \rangle_{\mathscr{H}} = \mathbb{E} \int_0^1 \langle D_s F, j_s \rangle \, ds.$$

On the other hand, by the integration by parts formula (THEOREM 8.4.2) we have

$$\mathbb{E} D_h F = \mathbb{E}(F D_h^* 1) = \mathbb{E} \left[\int_0^1 \langle H_s, dW_s \rangle \int_0^1 \left\langle \dot{h}_s + \frac{1}{2} \mathrm{Ric}_{U_s} h_s, dW_s \right\rangle \right] .$$

Hence

$$\mathbb{E} \int_0^1 \langle D_s F, j_s \rangle \, ds = \mathbb{E} \int_0^1 \left\langle H_s, \dot{h}_s + \frac{1}{2} \mathrm{Ric}_{U_s} h_s \right\rangle ds. \tag{8.5.3}$$

The next step is to extract a formula for H_s from the above relation. Set

$$k_s = \dot{h}_s + \frac{1}{2} \mathrm{Ric}_{U_s} h_s.$$

Let $\{M_s\}$ be defined as in the statement of the theorem. Then we can solve for h_s in terms of K_s and M_s. The result is

$$h_s = M_s^\dagger \int_0^s M_\tau^{\dagger -1} k_\tau d\tau.$$

Differentiating with respect to s, we obtain

$$j_s = k_s + \frac{1}{2}\mathrm{Ric}_{U_s} M_s^\dagger \int_0^s M_\tau^{\dagger -1} k_\tau d\tau.$$

Using this expression for j_s in the integral on the left side of (8.5.3) and changing the order of integration, we have

$$\int_0^1 \langle D_s F, j_s \rangle \, ds = \int_0^1 \left\langle D_s F + \frac{1}{2} M_s^{-1} \int_s^1 M_\tau \mathrm{Ric}_{U_\tau}(D_\tau F) d\tau, k_s \right\rangle ds.$$

The expected value of this expression is the left side of (8.5.3). On the right-hand side of the same equation we use the definition of k_s. After these manipulations, (8.5.3) becomes

$$\mathbb{E} \int_0^1 \left\langle D_s F + \frac{1}{2} M_s^{-1} \int_s^1 M_\tau \mathrm{Ric}_{U_\tau}(D_\tau F) \, d\tau, k_s \right\rangle ds = \mathbb{E} \int_0^1 \langle H_s, k_s \rangle \, ds.$$

In this relation $\{k_s\}$ can be an arbitrary $\mathscr{B}_*$-adapted process. The formula for H_s in the theorem follows immediately. □

8.6. Logarithmic Sobolev inequality and hypercontractivity

Various forms of Sobolev inequalities play an important role in analysis and partial differential equations. On $\mathbb{R}^N$ (with the usual Lebesgue measure), the L^2-Sobolev inequality takes the form

$$\|f\|_{2N/N-2} \le C_N \|\nabla f\|_2$$

if both f and ∇f are square integrable. Note that

$$\frac{2N}{N-2} = 2 + \frac{4}{N-2}.$$

Thus this Sobolev inequality shows that the square integrability of both f and its gradient ∇f improves the integrability power of f by $4/(N-2)$, which is the main reason for the importance of this inequality. The improvement of integrability disappears as $N \to \infty$. If we replace the Lebesgue measure by the standard Gaussian measure μ, then the following logarithmic Sobolev inequality holds.

Theorem 8.6.1. *(Gross' logarithmic Sobolev inequality) Let μ be the standard Gaussian measure on $\mathbb{R}^N$ and ∇ the usual gradient operator. Suppose* *that f is a smooth function on $\mathbb{R}^N$ such that both f and ∇f are square integrable with respect to μ. Then*

$$\int_{\mathbb{R}^N} |f|^2 \ln |f|^2 d\mu \le 2 \int_{\mathbb{R}^N} |\nabla f|^2 d\mu + \|f\|_2^2 \ln \|f\|_2^2. \tag{8.6.1}$$

Here $\|f\|_2$ is the norm of f in $L^2(\mathbb{R}^N, \mu)$.

Proof. For a positive s let μ_s be the Gaussian measure

$$\mu_s(dx) = \left(\frac{1}{2\pi s}\right)^{n/2} e^{-|x|^2/2s} dx,$$

where dx denotes the Lebesgue measure. Then $\mu = \mu_1$. Let $h = f^2$ and

$$P_s h(x) = \int_{\mathbb{R}^l} h(x - y)\mu_s(dy).$$

Consider the function $H_s = P_s\phi(P_{1-s}h)$, where $\phi(t) = 2^{-1}t \ln t$. Differentiating with respect to s and noting that the Laplacian operator Δ commutes with $P_s = e^{s\Delta/2}$, we have

$$\begin{aligned}
\frac{dH_s}{ds} &= \frac{1}{2} P_s \Delta\phi(P_{1-s}h) - \frac{1}{2} P_s \left\{\phi'(P_{1-s}h)\Delta P_{1-s}h\right\} \\
&= \frac{1}{2} P_s \left\{\phi'(P_{1-s}h)\Delta P_{1-s}h + \phi''(P_{1-s}h) \left|\nabla P_{1-s}h\right|^2\right\} \\
&\quad - \frac{1}{2} P_s \left\{\phi'(P_{1-s}h)\Delta P_{1-s}h\right\} \\
&= \frac{1}{2} P_s \left\{\phi''(P_{1-s}h) \left|\nabla P_{1-s}h\right|^2\right\} \\
&\le \frac{1}{4} P_s \left\{\frac{(P_{1-s}|\nabla h|)^2}{P_{1-s}h}\right\} \\
&\le P_s \left\{P_{1-s}|\nabla f|^2\right\} \\
&= P_1 |\nabla f|^2.
\end{aligned}$$

Here we have used the fact that $|\nabla P_{1-s}h| \le P_{1-s}|\nabla h|$ (from the translation invariance of the euclidean heat kernel) in the fourth step and the inequality

$$(P_{s-r}|\nabla h|)^2 \le 4P_{s-r}h \cdot P_{s-r}|\nabla f|^2$$

in the fifth step, the latter being a consequence of the Cauchy-Schwarz inequality. Now, integrating from 0 to 1, we obtain the desired result immediately. □

A striking feature of the logarithmic Sobolev inequality (8.6.1) is that the coefficient before the integrated gradient is independent of the dimension N. Therefore it can be transferred immediately to the path space $P_o(\mathbb{R}^n)$. For this reason, logarithmic Soboleve inequalities are important tools in infinite dimensional analysis.

Theorem 8.6.2. *For all $F \in \mathrm{Dom}(D)$ we have*

$$\int_{P_o(\mathbb{R}^n)} |F|^2 \ln |F| d\,\mathbb{P} \le \int_{P_o(\mathbb{R}^n)} |DF|^2_{\mathscr{H}} d\,\mathbb{P} + \|F\|_2^2 \ln \|F\|_2. \tag{8.6.2}$$

Proof. Let $\{h^i\}$ be an orthonormal basis for the Cameron-Martin space $\mathscr{H}$. Let F have the form

$$F = f(\langle h^0, W\rangle_{\mathscr{H}}, \dots, \langle h^l, W\rangle_{\mathscr{H}}),$$

where $f : \mathbb{R}^{l+1} \to \mathbb{R}$ is a Schwartz test function. The set of such functions is a core for D, so it is enough to show (8.6.2) for such an F. We have $DF = \sum_{i=0}^l F_{x_i} h^i$ (see PROPOSITION 8.2.10). Hence

$$|DF|_{\mathscr{H}} = |\nabla f(\langle h^0, W\rangle_{\mathscr{H}}, \cdots, \langle h^l, W\rangle_{\mathscr{H}})|.$$

On the other hand, the distribution of $\{\langle h^0, W\rangle_{\mathscr{H}}, \cdots, \langle h^l, W\rangle_{\mathscr{H}}\}$ is the standard Gaussian measure on $\mathbb{R}^{l+1}$. Therefore (8.6.2) reduces to (8.6.1). □

The logarithmic Sobolev inequality for D is intimately related to the hypercontractivity of the associated semigroup $\mathscr{P}_t = e^{tL/2}$, where $L = -D^*D$: it is an infinitesimal formulation of hypercontractivity. This equivalence holds in the general setting of Dirichlet forms. Let X be a complete metric space and $\mathbb{P}$ a Borel probability measure on X. Suppose that $\mathcal{E}$ is a Dirichlet form on $L^2(X, \mathbb{P})$ and let L be the associated nonpositive self-adjoint operator and $\mathscr{P}_t = e^{tL/2}$ the corresponding strongly continuous, positive, and contractive L^2-semigroup.

Definition 8.6.3. *A semigroup $\{\mathscr{P}_t\}$ is said to be hypercontractive if there exist a $t_0 > 0$ and a pair of indices $1 < p_0 < q_0$ such that $\|\mathscr{P}_{t_0}\|_{q_0,p_0} = 1$.*

Remark 8.6.4. It turns out that under suitable conditions, which are satisfied in the current setting, $\|\mathscr{P}_{t_0}\|_{q_0,p_0} = 1$ for one triple (p_0, q_0, t_0) implies $\|\mathscr{P}_t\|_{q,p} = 1$ for all $1 < p < q$ such that

$$e^{t/C} \ge \frac{q-1}{p-1},$$

where

$$\frac{1}{C} = \frac{4}{t_0}\left(\frac{1}{p_0} - \frac{1}{q_0}\right).$$

See Deuschel and Stroock [**15**], 244-247. □

We will prove the equivalence of logarithmic Sobolev inequality and hypercontractivity.

Lemma 8.6.5. *Let $q > 1$. If $G \in \mathrm{Dom}(L)$ is nonnegative and uniformly bounded, then*

$$(G^{q-1}, LG) \le -\frac{4(q-1)}{q^2}\,\mathcal{E}(G^{q/2}, G^{q/2}).$$

Proof. We have

$$(G^{q-1}, LG) = \lim_{t\to 0}\frac{2}{t}(G^{q-1}, \mathscr{P}_t G - G).$$

Now,

$$\begin{aligned}
&(G^{q-1}, \mathscr{P}_t G - G)\\
&= -\frac{1}{2}\int_X \mathscr{P}_t\left[\{G^{q-1} - G(y)^{q-1}\}\{G - G(y)\}\right]\mathbb{P}(dy) - (G^q, 1 - \mathscr{P}_t 1)\\
&\le -\frac{2(q-1)}{q^2}\int_X \mathscr{P}_t\left(G^{q/2} - G(y)^{q/2}\right)^2 \mathbb{P}(dy) - (G^q, 1 - \mathscr{P}_t 1)\\
&= \frac{4(q-1)}{q^2}(G^{q/2}, \mathscr{P}_t G^{q/2} - G^{q/2}) - \left(1 - \frac{2}{q}\right)^2 (G^q, 1 - \mathscr{P}_t 1)\\
&\le \frac{4(q-1)}{q^2}(G^{q/2}, \mathscr{P}_t G^{q/2} - G^{q/2}).
\end{aligned}$$

Here in the second step we have used the elementary inequality

$$(a^{q-1} - b^{q-1})(a - b) \ge \frac{4(q-1)}{q^2}(a^{q/2} - b^{q/2})^2. \tag{8.6.3}$$

The equalities in the first and the third steps can be easily verified by the symmetry of $\mathscr{P}_t$. The last step holds because $\mathscr{P}_t 1 \le 1$. □

We have the following Gross' equivalence between logarithmic Sobolev inequality and hypercontractivity

Theorem 8.6.6. *Let $\mathcal{E}$ be a Dirichlet form on a probabililty space $(X, \mathscr{B}, \mathbb{P})$ and C a positive constant. Let $\{\mathscr{P}_t\}$ be the associated semigroup. The following statements are equivalent.*

(I) Hypercontractivity for $\{\mathscr{P}_t\}$: $\|\mathscr{P}_t\|_{q,p} = 1$ for all (t, p, q) such that $t > 0, 1 < p < q$ and

$$e^{t/C} \ge \frac{q-1}{p-1}.$$

(II) The logarithmic Sobolev inequality for $\mathcal{E}$:

$$\mathbb{E}\left(F^2 \ln F^2\right) \le 2C\mathcal{E}(F, F) + \mathbb{E}F^2 \ln \mathbb{E}F^2.$$

Proof. We will sketch the proof.

(I) $\Rightarrow$ (II). Let $q = q(t)$ be a smooth, strictly increasing function such that $q(0) = p$ and $q' = q'(t) > 0$. Set

$$f = f(t) = \|\mathscr{P}_t F\|_q^q = ((\mathscr{P}_t F)^q, 1), \quad \theta = \theta(t) = \|\mathscr{P}_t F\|_q = f^{1/q}.$$

Differentiating f, we have

$$f' = q'\left((\mathscr{P}_t F)^q, \ln \mathscr{P}_t F\right) + \frac{q}{2}\left((\mathscr{P}_t F)^{q-1}, L\mathscr{P}_t F\right). \tag{8.6.4}$$

Now differentiating $\ln \theta = q^{-1} \ln f$ and multiplying the result by $q^2 f/q'$, we have

$$\frac{q^2 f}{q'} \cdot \frac{\theta'}{\theta} = \frac{qf'}{q'} - f \ln f. \tag{8.6.5}$$

Choose $q = q(t)$ such that $(q-1)/q' = C$, or

$$e^{t/C} = \frac{q-1}{p-1}.$$

Then the hypercontractivity implies $\theta(t) \leq \theta(0)$ for all t; hence $\theta'(0) \leq 0$. By (8.6.5) this implies

$$\frac{q(0)}{q'(0)} \cdot f'(0) \leq f(0) \ln f(0). \tag{8.6.6}$$

Taking $p = 2$, we have $q(0) = 2$, $q'(0) = C^{-1}$, $f(0) = \mathbb{E}F^2$, and, from (8.6.4),

$$f'(0) = (2C)^{-1}\mathbb{E}(F^2 \ln F^2) - \mathcal{E}(F, F).$$

Thus (8.6.6) reduces to the logarithmic Sobolev inequality.

(II) $\Rightarrow$ (I). Using Lemma 8.6.5 on the last term of (8.6.4) and multiplying by $q/2q'$, we obtain

$$\frac{1}{2}\frac{qf'}{q'} \leq (G^2, \ln G) - \frac{q-1}{q'}\mathcal{E}(G, G), \tag{8.6.7}$$

where $G = (\mathscr{P}_t F)^{q/2}$. Substituting (8.6.7) into (8.6.5), we obtain

$$\frac{1}{2}\frac{q^2 f}{q'} \cdot \frac{\theta'}{\theta} \leq (G^2, \ln G) - \frac{q-1}{q'}\mathcal{E}(G, G) - \|G\|_2^2 \ln \|G\|_2. \tag{8.6.8}$$

With the choice of $q = q(t)$ as before, the logarithmic Sobolev inequality applied to G implies that the right side of (8.6.8) is nonpositive; hence $\theta'(t) \leq 0$ for all t. This shows that $\theta(t) \leq \theta(0)$, which is the hypercontractivity of $\{\mathscr{P}_t\}$. □

Applying Theorem 8.6.6 to the Ornstein-Uhlenbeck semigroup on the flat path space $P_o(\mathbb{R}^n)$, we have Nelson's hypercontractivity theorem.

Theorem 8.6.7. *The Ornstein-Uhlenbeck semigroup $\{\mathscr{P}_t\}$ on the flat path space is hypercontractive. More precisely, $\|\mathscr{P}_t\|_{q,p} = 1$ for all (t, p, q) such that $t > 0, 1 < p < q$ and*

$$e^t \geq \frac{q-1}{p-1}.$$

In general a logarithmic Sobolev inequality for a Dirichlet form implies the existence of a spectral gap for the corresponding self-adjoint operator. This is the content of the following theorem.

Theorem 8.6.8. *Let $\mathcal{E}$ be a Dirichlet form on a probability space $(X, \mathscr{B}, \mathbb{P})$ such that $\mathcal{E}(1,1) = 0$. Suppose that it satisfies a logarithmic Sobolev inequality*

$$\mathbb{E}\left(|F|^2 \ln |F|^2\right) \leq 2C\mathcal{E}(F, F) + \mathbb{E}|F|^2 \ln \mathbb{E}|F|^2.$$

Then we have the Poincaré inequality:

$$\mathbb{E}|F - \mathbb{E}F|^2 \leq C\mathcal{E}(F, F). \tag{8.6.9}$$

Proof. Without loss of generality, we assume that $\mathbb{E}F = 0$. Applying the logarithmic Sobolev inequality to the function $G = 1 + tF$ and expanding in powers of t, we find that

$$\begin{aligned}
\mathbb{E}\left(G^2 \ln G^2\right) &= 3\mathbb{E}F^2 \cdot t^2 + O(t^3), \\
\mathcal{E}(G, G) &= \mathcal{E}(F, F) \cdot t^2 + O(t^3), \\
\mathbb{E}\, G^2 \ln \mathbb{E}\, G^2 &= \mathbb{E}F^2 \cdot t^2 + O(t^3).
\end{aligned}$$

Comparing the coefficients of t^2 in the logarithmic Sobolev inequality for G gives the desired inequality. □

If the Poincaré inequality (8.6.9) holds, then in particular $\mathcal{E}(F, F) = 0$ implies that F is a constant. Thus 0 is an eigenvalue, and the eigenspace consists of constant functions. A simple argument by the spectral theory of self-adjoint operators shows that the spectral gap (defined in (8.2.13)) $SG(-L) \geq 1/C$. This implies in turn that the corresponding semigroup approaches the equilibrium at an exponential rate:

$$\|\mathscr{P}_t F - \mathbb{E}F\|_2 \leq e^{-t/2C}\|F\|_2.$$

8.7. Logarithmic Sobolev inequality on path space

We now turn to the logarithmic Sobolev inequality for a general path space $P_o(M)$.

Theorem 8.7.1. *Let M be a compact Riemannian manifold and let K be the upper bound (in absolute value) of its Ricci curvature. Then we have*

$$\mathbb{E}(G^2 \ln G^2) \le 2C(K)\, \mathbb{E}|DG|^2_{\mathscr{H}} + \mathbb{E}\, G^2 \ln \mathbb{E}\, G^2,$$

where

$$C(K) = 1 + \sqrt{e^K - 1 - K} + \frac{1}{4}(e^K - 1 - K) \le e^K. \tag{8.7.1}$$

Proof. Let $F = G^2$ and consider the martingale $N_s = \mathbb{E}\{F|\mathscr{B}_s\}$. We have

$$N_s = \mathbb{E}F + \int_0^s \langle H_\tau, dW_\tau \rangle,$$

where the adapted process $\{H_s\}$ is given in (8.5.2). Now apply Itô's formula to $N_s \ln N_s$. We have

$$\mathbb{E}N_1 \ln N_1 - \mathbb{E}N_0 \ln N_0 = \frac{1}{2}\mathbb{E}\int_0^1 N_s^{-1}|H_s|^2 ds. \tag{8.7.2}$$

It is easy to see that

$$\text{the left side of (8.7.2)} = \mathbb{E}(G^2 \ln G^2) - \mathbb{E}\, G^2 \ln \mathbb{E}\, G^2. \tag{8.7.3}$$

On the other hand,

$$DF = D(G^2) = 2GDG.$$

Using this relation in the explicit formula (8.5.2) for H_s in THEOREM 8.5.1, we have

$$H_s = 2\mathbb{E}\left[G\left(D_s G + \frac{1}{2} M_s^{-1} \int_s^1 M_\tau \mathrm{Ric}_{U_\tau}(D_\tau G) d\tau \right) \middle| \mathscr{B}_s \right]. \tag{8.7.4}$$

From the equation for $\{M_s\}$ we have

$$M_s^{-1} M_\tau = \frac{1}{2} \int_s^\tau M_s^{-1} M_t \mathrm{Ric}_{U_t}\, dt.$$

Hence by Gronwall's lemma and the bound on the Ricci curvature,

$$\|M_s^{-1} M_\tau\| \le e^{K(\tau - s)/2}.$$

Using the Cauchy-Schwarz inequality in (8.7.4) and the above inequality, we can write

$$|H_s|_2^2 \le 4\mathbb{E}\left\{G^2|\mathscr{B}_s\right\} \mathbb{E}\left[\left(|D_s G| + \frac{1}{2} K \int_s^1 e^{K(\tau - s)/2} |D_\tau G| d\tau \right)^2 \middle| \mathscr{B}_s \right].$$

Note that $N_s = \mathbb{E}\left\{G^2|\mathscr{B}_s\right\}$. Now we have

$$\begin{aligned} &\text{the right side of (8.7.2)} \\ &\le 2\mathbb{E}\int_0^1 \left\{ |D_s G| + \frac{1}{2} K \int_s^1 e^{K(\tau - s)/2} |D_\tau G| d\tau \right\}^2 ds. \end{aligned}$$

It remains to estimate the last expression in terms of $|DG|_{\mathscr{H}}$. By the Cauchy-Schwarz inequality,

$$\begin{aligned}\left\{\int_s^1 e^{K(\tau-s)/2}|D_\tau G|d\tau\right\}^2 &\le \int_s^1 e^{K(\tau-s)}d\tau\cdot\int_s^1 |D_\tau G|^2 d\tau\\ &\le \frac{1}{K}\left\{e^{K(1-s)}-1\right\}|DG|^2_{\mathscr{H}},\end{aligned}$$

and

$$\begin{aligned}&\int_0^1\left\{|D_sG|+\frac{1}{2}K\int_s^1 e^{K(\tau-s)/2}|D_\tau G|d\tau\right\}^2 ds\\ &\le \int_0^1\left\{|D_sG|+\frac{1}{2}\sqrt{K}\sqrt{e^{K(1-s)}-1}|DG|_{\mathscr{H}}\right\}^2 ds\\ &= \int_0^1 |D_sG|^2ds+|DG|_{\mathscr{H}}\sqrt{K}\int_0^1\sqrt{e^{K(1-s)}-1}|D_sG|ds\\ &\quad+\frac{1}{4}|DG|^2_{\mathscr{H}}K\int_0^1\left\{e^{K(1-s)}-1\right\}ds\\ &\le C(K)|DG|^2_{\mathscr{H}}.\end{aligned}$$

It follows that

$$\text{the right side of (8.7.2)} \le 2C(K)\mathbb{E}|DG|^2_{\mathscr{H}}. \tag{8.7.5}$$

The desired inequality now follows from (8.7.2), (8.7.3), and (8.7.5). □

If the manifold M is Ricci flat, we have $C(K) = 1$, and the above logarithmic Sobolev inequality reduces to the usual one for the flat path space.

The hypercontractivity of the Ornstein-Uhlenbeck semigroup $\{\mathscr{P}_t\}$ on the path space $P_o(M)$ follows directly from THEOREM 8.7.1 and the general equivalence THEOREM 8.6.6.

Theorem 8.7.2. *Let M be a compact Riemannian manifold whose Ricci curvature is bounded (in absolute value) by K. Then the Orstein-Uhlenbeck semigroup $\{\mathscr{P}_t\}$ on the path space $P_o(M)$ is hypercontractive. More precisely, $\|\mathscr{P}_t\|_{q,p} = 1$ for all (t,p,q) such that $t > 0, 1 < p < q$ and*

$$e^{t/C(K)} \ge \frac{q-1}{p-1},$$

where $C(K)$ is given in (8.7.1).

The existence of a spectral gap for the Ornstein-Uhlenbeck operator on the path space follows from the logarithmic Sobolev inequality and THEOREM 8.6.8.

Theorem 8.7.3. (Fang's spectral gap theorem) *Let M be a compact Riemannian manifold whose Ricci curvature is bounded (in absolute value) by K. Let L be the Ornstein-Unlenbeck operator on the path space $P_o(M)$. Then*

$$SG(-L) \geq \frac{1}{C(K)},$$

where $C(K)$ is given in (8.7.1). *Equivalently, we have the Poincaré inequality:*

$$\mathbb{E}|F - \mathbb{E}F|^2 \leq C(K)\, |DF|_{\mathscr{H}},$$

Notes and Comments

CHAPTER 1. The material in this chapter is standard and can be found in many books on stochastic differential equations, for example, Elworthy [**21**] and Ikeda and Watanabe [**48**]. Our point of view is very close to that of [**21**]. LEMMA 1.3.3 is from Phillips and Sarason [**62**]. Consult Stroock and Varadhan [**68**] for the general theory of martingale problems. The long chapter in Hackenbroch and Thalmaier [**36**] on stochastic analysis on manifolds can also be consulted by those who read German.

CHAPATER 2. Basic differential geometry of frame bundles can be found in Bishop and Crittenden [**4**] and Kobayashi and Nomizu [**53**]. The textbooks Jost [**50**] and Do Carmo [**17**] are also recommended. The fact that the lifetimes of a semimartingale X and its horizontal lift are equal (THEOREM 2.3.5) was implicit in Schwartz [**65**] and was pointed out to me by Bang-he Li. Consult Emery [**24**], [**25**] for more information on martingales on manifolds. SECTIONS 2.5 and 2.6 are largely adapted from [**24**]. THEOREM 2.6.4 seems to be new.

CHAPTER 3. Ikeda and Watanabe [**48**] and Elworthy [**23**] should be consulted for Brownian motion on manifolds. Basic properties of the Laplace-Beltrami operator can be found in Jost [**50**] and Do Carmo [**17**]. LEMMA 3.3.4 is due to Knight, see Ikeda and Watanabe [**48**], pp. 86–89. For a good discussion on cutlocus, consult Cheeger and Ebin [**9**]. That book is also a good reference for various comparison theorems in Riemannian geometry. THEOREM 3.5.1 on the decomposition of the radial process beyond the cutlocus is due to Kendall [**52**]. The basic exit time estimate THEOREM 3.6.1 can be found in Hsu and March [**37**], but the proof given here is new.

CHAPTER 4. Chavel [**8**] contains a wealth of information on heat kernels. The section on the C_0-property is taken from Hsu [**38**]. THEOREM

4.2.3 first appeared in Varopoulos [**70**]. SECTION 4.4 draws partly from Ichihara [**47**]. The survey paper by Grigor'yan [**33**] should be consulted for stochastic completeness, recurrence and transience, and other related topics. The comparison theorems for the heat kernels in SECTION 4.5 are drawn from Debiard, Gaveau, and Mazet [**16**].

CHAPTER 5. The proof of short-time expansion of the heat kernel by the method of parametrix can be found in Berger, Gauduchon, and Mazet [**3**] and Chavel [**8**]. The proof of Varadhan's asymptotic relation in SECTION 5.2 is taken from Hsu [**40**]. SECTION 5.3 is taken from Azencott et al. [**2**] and Hsu [**39**]. The global logarithmic gradient and Hessian estimates in SECTION 5.5 were proved by Sheu [**66**], and the method presented here is taken from Hsu [**46**].

CHAPTER 6. A discussion of the Dirichlet problem at infinity from the geometric point of view is contained in Schoen and Yau [**64**]. The baisc probabilistic method explained here was established in Hsu and March [**37**]. The idea of studying the number of steps between two geodesic spheres is due to Leclercq [**54**]. SECTION 6.4 is taken from March [**58**] with some improvements. The discussion of coupling of Brownian motion follows Cranston [**14**] closely, see also Kendall [**51**]. The results on index forms used in this chapter can be found in Cheeger and Ebin [**9**]. The probabilistic approach to eigenvalue estimates in SECTION 6.7 is due to Chen and Wang [**10**]. Li [**55**] contains a nice analytic presentation of the Zhong-Yang eigenvalue lower bound discussed in this chapter.

CHAPTER 7. The discussion on th Weitzenböck formula in SECTION 7.1 follows that of Jost [**50**], which should also be consulted as a general textbook on differential geometry. For the heat equation on differential forms in SECTION 7.2 we closely follow Ikeda and Watanabe [**48**]. THEOREM 7.2.4 first appeared in Donnelly and Li [**18**]. The basic relation between the Euler characteristic and the trace of the heat kernel (THEOREM 7.3.1) is due to McKean and Singer [**56**]. The algebraic preliminaries in SECTION 7.3 are taken from Patodi [**61**]. The proof of the Gauss-Bonnet-Chern formula in SECTION 7.3 is taken from Hsu [**43**]. Geometric background for the Dirac operator can be found in Gilkey [**31**] and Friedrich [**28**]. The earliest probabilistic proof of the Atiyah-Singer index theorem is due to Bismut [**5**]. Since then many other probabilistic proofs have emerged. The proof presented here is contained in an unpublished manuscript by the author in 1985.

CHAPTER 8. The monography Malliavin [**57**] contains a wealth of information on general stochastic analysis, its CHAPTER XI being especially relevant here. It also contains an extensive bibliography of recent literature on stochastic analysis. Another approach to path space analysis is presented in Stroock [**69**]. The proof of THEOREM 8.1.5 is taken from Janson [**49**], pp.

226–227. The quasi-invariance of the Wiener measure on the path space of a Riemannian manifold was first proved by Driver [**20**], and completed by Hsu [**41**] (see also Enchev and Stroock [**26**]). The various gradient formulas presented in SECTION 8.3 all originated in Bismut [**6**]; the proofs presented here are from Hsu [**45**]. This is also the reference from which most of this chapter is derived. The integration by parts formula in the path space over a general compact Riemannian manifold was proved in Driver [**20**]. THEOREM 8.5.1 on the martingale representation for Brownian motion on a Riemannian manifold originated in Fang [**27**], where the THEOREM 8.7.3 on the existence of a spectral gap for the Ornstein-Uhlenbeck operator was first proved. The equivalence of logarithmic Sobolev inequality and hypercontractivity was proved by Gross [**34**]; see also a general discussion on the subject by the same author [**35**]. The proof of the logarithmic Sobolev inequality for the Gaussian measure in THEOREM 8.6.1 is due to Bakry. The presentation of general properties of logarithmic Sobolev inequalities and hypercontractivity in SECTION 8.7 draws heavily from Deuschel and Stroock [**15**]. A logarithmic Sobolev inequality for the path space of a Riemannian manifold with bounded Ricci curvature was first proved by Hsu [**42**], [**44**] (see also an extrinsic proof by Aida and Elworthy [**1**]). The method used here is due to Capitaine, Hsu and Ledoux [**7**].

General Notations

Notation	Definition
$\mathscr{B}(X)$	Borel σ-field of a metric space X
$\mathscr{B}_*$	Borel filtration of $W(M) = \{\mathscr{B}_t\}$
$B(x;R)$	(geodesic) ball of radius R centered at x
$\square_M$	Hodge-de Rham Laplacian on M
$\langle X, Y\rangle$	co-variation of semimartingales X and Y
$\langle X\rangle$	quadratic variation of $X = \langle X, X\rangle$
$C^\infty(M)$	smooth functions on M
C_x	cutlocus of x
$\circ$	Stratonovich stochastic integral, $X \circ dY = XdY + \frac{1}{2}d\langle X, Y\rangle$
$c(X)$	Clifford multiplication by X (in CHAPTER 7)
d	exterior differentiation
$d_M(x,y)$	distance between x and y on M
D	gradient operator on $P_o(M)$
D	Dirac operator (in CHAPTER 7)
D_h	Cameron-Martin vector field on $P_o(M)$
δ	dual of exterior differentiation
Δ_M	Laplace-Beltrami operator on $M = -(d\delta + \delta d)$
$\Delta_{\mathscr{O}(M)}$	Bochner's horizontal Laplacian on $\mathscr{O}(M) = \sum_{i=1}^d H_i^2$
A^*	dual operator of A
$e(\omega)$	lifetime (explosion time) of a path ω
$\mathrm{End}(V)$	space of linear transforms on V
$\exp_o$	exponential map based at o
$\mathscr{F}_*$	filtration of σ-fields $= \{\mathscr{F}_t\}$

Notation	Definition
$\mathscr{F}(M)$	frame bundle of M
$\mathscr{F}_*^X$	filtration generated by process $X = \{\mathscr{F}_t^X\}$
$\Gamma(E)$	space of sections of vector bundle E
Γ_{ij}^k	Christoffel symbols
$\Gamma(f,g)$	Γ of f and $g = L(fg) - fLg - gLf$
$GL(d,\mathbb{R})$	set of real nonsingular $(d \times d)$ matrices
H_i	fundamental horizontal vector field on $\mathscr{F}(M)$
$\mathscr{H}$	Cameron-Martin space
$i(X)$	interior product with X
i_x	injectivity radius at x
i_K	injectivity radius on $K = \min\{i_x : x \in K\}$
$I(J,J)$	index form of a vector field J
$K_M(x)$	set of sectional curvatures at a point $x \in M$
L	Orstein-Uhlenbeck operator (in CHAPTER 8)
$\mathscr{M}(d,l)$	space of $(d \times l)$ matrices
$M^\dagger$	transpose of a matrix M
$\widehat{M}$	one-point compactification of a manifold $M = M \cup \{\partial_M\}$
$\lvert \cdot \rvert_{\mathscr{H}}$	Cameron-Martin norm
∇	connection and covariant differentiation
$\nabla^2 f$	Hessian of f
$\nabla^H G$	horizontal gradient of $G = \{H_1 G, \ldots, H_d G\}$
$O(d)$	$(d \times d)$ orthogonal group
$\mathfrak{o}(d)$	$(d \times d)$ anti-symmetric matrices
$\mathscr{O}(M)$	orthonormal frame bundle of M
Ω	curvature form
$\mathrm{Pf}(A)$	Pfaffian of an anti-symmetric matrix A
$P_o(M)$	space of paths on M starting from o with time length 1
$\mathbb{P}_x$	law of Brownian motion starting from x
$\mathbb{P}_{x,y;t}$	law of Brownian bridge from x to y with time length t
$p_M(t,x,y)$	heat kernel on a Riemannian manifold M
$\{P_t\}$	heat semigroup$= e^{t\Delta_M/2}$
$\{\mathscr{P}_t\}$	Ornstein-Uhlenbeck semigroup$= e^{tL/2}$
$\mathit{\Pi}$	second fundamental form
$\mathbb{R}^N$	euclidean space of dimension N
$\mathbb{R}_+$	set of nonnegative real numbers $= [0,\infty)$,
$R(X,Y)Z$	curvature tensor evaluated at X, Y, Z
$\mathrm{Ric}_M(x)$	set of Ricci curvatures at a point $x \in M$

Notation	Definition
Ric_u	Ricci transform at a frame $u \in \mathscr{O}(M)$
$\mathscr{S}_+(d)$	$(d \times d)$ symmetric positive definite matrices
$\mathbb{S}^n$	n-sphere
$\mathrm{Spin}(d)$	Spin group
$\mathscr{S}(M)$	spin bundle over a spin manifold M
$\mathscr{S}\mathscr{P}(M)$	$\mathrm{Spin}(d)$-principal bundle over M
X^*	horizontal lift of $X \in TM$ to the frame bundle $\mathscr{F}(M)$
τ_D	first exit time of $D = \inf\{t : X_t \notin D\}$
θ_t	shift operator in a path space: $(\theta_t\omega)_s = \omega_{t+s}$
T_K	first hitting time of $K = \inf\{t : X_t \in K\}$
$\widetilde{\theta}$	scalarization of a tensor θ
Trace	supertrace (in CHAPTER 7)
TM	tangent bundle of a manifold M
T_xM	tangent space of M at x
T_x^*M	cotangent space of M at x
$\wedge_x^p M$	space of p-forms at a point $x \in M$
$W(M)$	space of paths on M with lifetimes

Bibliography

1. Aida, S. and Elworthy, K. D., Differential calculus on path and loop spaces, I. Logarthmic Sobolev inequalities on path spaces, *C. R. Acad. Sci.* (Paris), **321**, Serié I (1995), 97-102.
2. Azencott, R. et al., *Géodésiques et diffusions en temps petit*, Astérisque, **84–85** (1981).
3. Berger, M., Gauduchon, P. and Mazet, E., *Le Spectre d'une Variété Riemannienne*, *Lect. Notes in Math.*, **194**, Springer-Verlag (1971).
4. Bishop, R. L. and Crittenden, R. J., *Geometry of Manifolds*, Academic Press (1964).
5. Bismut, J.M., The Atiyah-Singer Theorems: A Probabilistic Approach, *J. of Func. Anal.*, Part I: **57** (1984), 56–99; Part II: **57** (1984), 329–348.
6. Bismut, J-M., *Large Deviations and the Malliavin Calculus*, Birkhäuser (1984).
7. Capitaine, M., Hsu, E. P., and Ledoux, M., Martingale representation and a simple proof of logarithmic Sobolev inequalities on path spaces, *Electronic Communications in Probability*, **2** Paper no. 7 (1997), 71-81.
8. Chavel, I. *Eigenvalues in Riemannian Geometry*, Academic Press (1984).
9. Cheeger, J. and Ebin, D. G., *Comparison Theorems in Differential Geometry*, North-Holland/Kodansha (1975).
10. Chen, M. F. and Wang, F. Y., Application of coupling method to the first eigenvalue on manifold, *Science in China (A)*, **37**, no. 1 (1994), 1–14.
11. Chung, K. L., *Lectures from Markov Processes to Brownian Motion*, Grund. Math. Wiss. **249**, Springer-Verlag (1982).
12. Copson, E. T., *Asymptotic Expansions*, Cambridge University Press (1965).
13. Courant, R. and Hilbert, D., *Methods of Mathematical Physics*, V.1, Interscience, New York (1962).
14. Cranston, M., Gradient estimates on manifolds using coupling, *J. of Func. Anal.*, **99** (1991), 110-124.
15. Deuschel, D. J.-D. and Stroock, D. W., *Large Deviations*, Academic Press (1989).
16. Debiard, A., Gaveau, B., and Mazet, E., Théorèmes de comparaison en géométrie riemannienne, *Publ. RIMS, Kyoto Univ.*, **12** (1976), 391–425.

17. Do Carmo, M. P., *Riemannian Geometry*, 2nd edition, Birkhäuser (1993).
18. Donnelly, H. and Li, P., Lower bounds for the eigenvalues of Riemannian manifolds, *Michigan Math. J.*, **29** (1982), 149–161.
19. Dodziuk, J., Maximum principle for parabolic inequalities and the heat flow on open manifolds, *Indiana Math. J.*, **32**, No. 5 (1983), 703– 716.
20. Driver, B., A Cameron-Martin type of quasiinvariance for the Brownian motion on a compact maniofld, *J. of Func. Anal.*, **110** (1992), 237–376.
21. Elworthy, K. D., *Stochastic Differential Equations on Manifolds*, Cambridge University Press (1982).
22. Elworthy, K. D. and Truman, A., The diffusion equation and classical mechanics: an elementary formula, in *Stochastic Processes in Quantum Physics*, ed. S. Albeverio et al., *Lect. Notes in Physics* **173**, Springer-Verlag (1982), 136–146.
23. Elworthy, K. D., Geometric aspects of diffusions on manifolds, in *Ecole d'Été de Probabilités de Saint-Flour XV–XVII, 1985–1987*, ed. by P. Diaconis et al., *Lect. Notes in Math.*, **1382**, Springer-Verlag (1988), 227–425.
24. Emery, Michel, *Stochastic Calculus in Manifolds*, Springer-Verlag (1989).
25. Emery, Michel, Martingales continues dans les variétés differentiables, Lectures on Probability Theory and Statistics, Ecole d'Eté de Probabilités de Saint-Flour XXVIII, 1998, *Lect. Notes in Math.*, **1738** (2000), 1–84.
26. Enchev, O. and Stroock, D. W., Towards a Riemannian geometry on the path space over a Riemannian manifold, *J. of Func. Anal.*, **134** (1995), 329–416.
27. Fang, S., Un inéqualité du type Poincaré sur un espace de chemins, *C. R. Acad. Sci.* (Paris), **318**, Serié I (1994), 257-260.
28. Friedrich, T., *Dirac Operators in Riemannian Geometry*, Graduate Studies in Mathematics, Vol. 25, Amer. Math. Soc. (2000).
29. Fukushima, M., *Dirichlet Forms and Markov Processes*, North-Holland/Kodansha (1975).
30. Gallot, S., Hulin, D., and Lafontaine, J., *Riemannian Geometry*, Springer-Verlag (1990).
31. Gilkey, P., *Invariance Theory, the Heat Equation, and the Atiyah-Singer Index Theorem*, Publish or Perish, Inc., Wilmington, DE (1984).
32. Grigor'yan, A. A., On stochastic completeness of manifolds, *Soviet Math. Dokl.*, **34** (1987), 310–313.
33. Grigor'yan, A. A., Analytic and geometric background of recurrence and non-explosion of the Brownian motion on Riemannian manifolds, *Bull. Amer. Math. Soc.* (New Series), **36**, No. 2 (1999), 135–249.
34. Gross, L., Logarithmic Sobolev inequalities, *Amer. J. of Math.*, **97** (1975), 1061–1083.
35. Gross, L., Logarithmic Sobolev inequalities and contractivity properties of semigroups, in *Dirichlet Forms*, ed. by E. Fabes et al. *Lect. Notes in Math.*, **1563**, Springer-Verlag (1993), 54–82.
36. Hackenbroch, W. and Thalmaier, A., *Stochastische Analysis*, B. G. Teubner, Stuttgart (1994).
37. Hsu, P. and March, P., The limiting angle of certain Riemannian Brownian motions, *Comm. Pure Appl. Math.*, **38**, 755–768 (1985).
38. Hsu, P., Heat semigroup on a complete Riemannain manifold, *Ann. of Probab.*, **17**, no. 3 (1989), 1248–1254.

39. Hsu, P., Brownian bridges on Riemannian manifolds, *Probab. Theory and Rel. Fields*, **84** (1990), 103–118.

40. Hsu, P., Heat kernel on noncomplete manifolds, *Indiana Math. J.*, **39**, no.2 (1990), 431–442.

41. Hsu, E. P., Quasiinvariance of the Wiener measure and integration by parts in the path space over a compact Riemannian manifold, *J. of Func. Anal.*, **134** (1995) 417–450.

42. Hsu, E. P., Une inegalité logarithmique de Sobolev sur l'espace de chemins d'une variété riemannienne, *C. R. Acad. Sci.* (Paris), **320**, Serié I (1995), 1009–1012.

43. Hsu, E. P., Stochastic local Gauss-Bonnet-Chern theorem, *Journal of Theo. Probab.*, **10**, no. 4 (1997), 819–834.

44. Hsu, E. P., Logarithmic Sobolev inequalities on the path space of a Riemannian manifold, *Comm. Math. Phy.*, **189** (1997), 9–16.

45. Hsu, E. P., Analysis on path and loop spaces, in *Probability Theory and Applications*, IAS/Park City Mathematics Series, Vol. 6, edited by E. P. Hsu and S. R. S. Varadhan, AMS (1999), 277–347.

46. Hsu, E. P., Estimates of the derivatives of the heat kernel, *Proceedings of AMS*, **127** (1999), 3739-3744.

47. Ichihara, K., Curvature, geodesics and the Brownian otion on a Riemannian manifold I, *Nagoya Math. J.*, **87** (1982), 101–114.

48. Ikeda, N. and Watanabe, S., *Stochastic Differential Equations and Diffusion Processes*, 2nd edition, North-Holland/Kodansha (1989).

49. Janson, S., *Gaussian Hilbert Spaces*, Cambridge University Press (1997).

50. Jost, J., *Riemannian Geometry and Geometric Analysis*, Springer-Verlag (1995).

51. Kendall, W. S., Nonnegative Ricci curvature and the Brownian coupling property, *Stochastics* **19** (1986), 111–129.

52. Kendall, W. S., The radial part of Brownian motion on a manifold: a semimartingale property, *Ann. Probab.*, **15**, no. 4 (1987), 1491–1500.

53. Kobayashi, S. and Nomizu, K., *Foundations of Differential Geometry*, Vol. I, Interscience Publishers, New York (1963).

54. Leclercq, É., The asymptotic Dirichlet problem with respect to an elliptic operator on a Cartan-Hadamard manifold with unbounded curvatures, *C. R. Acad. Sci.*, **325**, Série I (1997), 857–862.

55. Li, P., *Lecture Notes on Differential Geometry*, Notes of the Series of Lectures Held at the Seoul National University, Vol. 6, Seoul National University (1993).

56. McKean, H.P. and Singer, I.M., Curvature and the eigenvalues of Laplacian, *J. of Diff. Geo.*, **1** (1967), 43–69.

57. Malliavin, P., *Stochastic Analysis*, Springer-Verlag (1997).

58. March, P., Brownian motion and harmonic functions on rotationally symmetric manifolds, *Ann. Probab.*, **14**, No. 4 (1986), 793-804.

59. Molchanov, S. A., Diffusion processes and Riemannian geometry, *Russian Math. Surveys*, **30**, No. 1 (1975), 1–63.

60. Norris, J.R., Path integral formulae for heat kernels and their derivatives, *Probab. Theory Relat. Fields*, **94** (1993), 525–541.

61. Patodi, V.K., Curvature and the eigenforms of the Laplace operator, *J. of Diff. Geo.*, **5** (1971), 233–249.

62. Phillips, R. S. and Sarason, L., Elliptic parabolic equations of second order, *J. Math. and Mech.*, **17**, 891-917 (1967).

63. de Rham, G., *Riemannian Manifolds*, Springer-Verlag (1984).

64. Schoen, R. and Yau, S.-T. *Lectures on Differential Geometry*, International Press, Cambridge, MA (1994).

65. Schwartz, Laurent, *Semimartingales and Stochastic Calculus on Manifolds*, Les Presses de l'Université de Montréal, Montréal, Canada (1984).

66. Sheu, S.-Y., Some estimates of the transition density function of a nondegenerate diffusion Markov process, *Ann. Probab.*, **19**, no. 2 (1991), 538–561.

67. Shiga, T. and Watanabe, S., Bessel diffusions as a one-parameter family of diffusion processes, *Z. Wahr. verw. Geb.*, **27** (1973), 37–46.

68. Stroock, D. W. and Varadhan, S. R. S., *Multidimensional Diffusion Processes*, Springer-Verlag (1979).

69. Stroock, D. W., *An Introduction to the Analysis of Paths on a Riemannian Manifold*, Mathematical Surveys and Monographs, Vol. 74, Amer. Math. Soc. (2000).

70. Varopoulos, N., Potential theory and diffusion on Riemannian manifolds, in *Conference on Harminic Analysis in Honor of Antoni Zygmund*, Vol. II, Wadsworth Math. Ser., Wadsworth, Belmont, CA (1983), 821–837

71. Warner, F. W., *Foundations of Differential Geometry and Lie Groups*, Springer (1983).

72. Zhong, J. Q. and Yang, H. C., Estimates of the first eigenvalue of a compact Riemannian manifold, *Scientia Sinica*, **27**, no. 12 (1984), 1251–1265.

Index